AF352903

The Structural and Functional Study of Efflux Pumps Belonging to the RND Transporters Family from Gram-Negative Bacteria

The Structural and Functional Study of Efflux Pumps Belonging to the RND Transporters Family from Gram-Negative Bacteria

Editors

Isabelle Broutin
Attilio V Vargiu
Henrietta Venter
Gilles Phan

MDPI • Basel • Beijing • Wuhan • Barcelona • Belgrade • Manchester • Tokyo • Cluj • Tianjin

Editors

Isabelle Broutin
Faculty of Pharmacy
University Paris Cité
Paris
France

Attilio V Vargiu
Physics Department
University of Cagliari
Monserrato
Italy

Henrietta Venter
Clinical and Health Science,
Health and Biomedical
Innovation
University of South Australia
Adelaide
Australia

Gilles Phan
Faculty of Pharmacy
University Paris Cité
Paris
France

Editorial Office
MDPI
St. Alban-Anlage 66
4052 Basel, Switzerland

This is a reprint of articles from the Special Issue published online in the open access journal *Antibiotics* (ISSN 2079-6382) (available at: www.mdpi.com/journal/antibiotics/special_issues/Efflux_Pumps).

For citation purposes, cite each article independently as indicated on the article page online and as indicated below:

LastName, A.A.; LastName, B.B.; LastName, C.C. Article Title. *Journal Name* **Year**, *Volume Number*, Page Range.

ISBN 978-3-0365-3946-1 (Hbk)
ISBN 978-3-0365-3945-4 (PDF)

Cover image courtesy of Isabelle Broutin

© 2022 by the authors. Articles in this book are Open Access and distributed under the Creative Commons Attribution (CC BY) license, which allows users to download, copy and build upon published articles, as long as the author and publisher are properly credited, which ensures maximum dissemination and a wider impact of our publications.

The book as a whole is distributed by MDPI under the terms and conditions of the Creative Commons license CC BY-NC-ND.

Contents

About the Editors

Isabelle Broutin

Isabelle Broutin is currently the team leader of the group "Membrane transport and signaling" at the university Paris Cité. She is a member of different commissions of trust, such as the direction comity of the doctoral schools MTCI (University Paris Cité), the scientific peer review comity for SOLEIL synchrotron (France), and the local scientific council (CSL) of the faculty of pharmacy.

She followed a formation of physicists at Orsay University, France, before becoming interested in structural biology. After obtaining a Ph.D. in 1993 on the resolution of the 3D structure of a serine collagenase from a cattle parasite, she performed a post-doctoral study in Uppsala, Sweden, on the 3D structure of CRBP. She was recruited in 1997 to the National Center of Scientific Research (CNRS), France, and was involved in the study of cellular signal transduction proteins (Grb2, Grb14, VEGFR, PRLR). She directed the group "signaling and membrane transport" in 2008, and since 2016 she has been Deputy Director of the CiTCoM laboratory, grouping structural biologists with chemists with the common goal to develop therapeutic drugs targeting diseases that are considered major public health problems. Her team works at the comprehension at the molecular level of the mechanistic of the different protein actors involved in antibiotic resistance, with recognized expertise in structural biology of membrane proteins. She has published over 80 manuscripts, book chapters, and reviews, and her recent work has provided important information for the comprehension of the assembly and opening mechanism of RND tripartite efflux pumps, as illustrated by the structure of the MexAB-OprM efflux pump from *P. aeruginosa* (Glavier, Nat Com 2020).

Attilio V Vargiu

Attilio Vargiu is currently an Assistant Professor of Applied Physics at the University of Cagliari. He obtained his Master's degree in Physics (cum laude) at the same University in 2003. Next, he received a scholarship for a Ph.D. in Statistical and Biological Physics at the International School for Advanced Studies (ISAS/SISSA) in Trieste, obtaining the title in 2008. Between 2008 and 2019, he traveled across Italy (University of Cagliari), Germany (Jacobs University, Bremen), The Netherlands (Utrecht University), and the U.S.A. (University of California, Berkeley) to perform his research. He has almost 20 years of experience in simulating biological macromolecules, peptides, small molecules, membranes, and their assemblies using a large number of advanced computational methods, from quantum-based to coarse-grained to machine-learning-based approaches. In 2010, he published the first study on the main multi-drug transporters involved in microbial resistance to antibiotics. Since then, he has published tens of papers on this subject within an ever-growing network of collaborators in Europe and the U.S. He contributed to setting up a database of conformations and dynamical physicochemical descriptors of antibiotics and other molecules related to microbial research. Since 2015, he broadened his interest to study the self-assembly of peptides for technological and biomedical applications, focusing on the role of chirality in their supramolecular features. Since 2019, he has been developing an algorithm to accurately predict the structures of challenging protein-ligand complexes, as well as exploiting machine learning techniques to identify key molecular descriptors related to antibiotic permeation and to identify native protein–protein complexes.

Henrietta Venter

Rietie Venter currently leads the Antimicrobial Resistance group in Clinical and Health Sciences at UniSA. Her research focuses on antimicrobial resistance in microbes—one of the most serious threats in healthcare today. The Venter group works on projects targeted at understanding and preventing the development and dissemination of antimicrobial resistance. Rietie is also heading an antimicrobial drug discovery program aimed at finding new therapeutics against drug-resistant pathogens.

Rietie obtained her BSc (Hons) and Master's degrees with distinction from the University of the Free State in South Africa before securing a scholarship to do a Ph.D. in the UK. After completing her Ph.D. at the University of Leeds in the beautiful Yorkshire Dales, she moved to Cambridge, where she spent twelve years conducting research on multidrug transporters, first as a post-doc and later running her own research group as a Royal Society Dorothy Hodgkin Fellow in the Department of Pharmacology. Not content with moving continents once in a lifetime, she left the ancient buildings and immaculate college lawns of Cambridge for the sun and sea in Australia after sixteen years in the UK.

Gilles Phan

Gilles Phan was trained in chemical biology at University of Orsay and Ecole Normale Supérieure Cachan. He gained experience during his Ph.D. at the University of Paris in X-ray crystallography by solving the structure of the membrane channel OprM of the multidrug efflux pump MexAB-OprM from *P. aeruginosa*. In 2008, he extended his expertise in structural biology during his post-doc at Birkbeck College and University College of London, in the group of Pr. Gabriel Waksman, where he worked on the bacterial pili biogenesis of the chaperone-usher pathway. Since 2013, he has been a member of the team of Dr. Isabelle Broutin at CiTCoM lab (Cible Thérapeutique et Conception de Médicaments, Paris) as a professor assistant in structural biology and biochemistry, at the School of Pharmacy of Paris. He is now working on the assembly and inhibition strategies of RND efflux pumps and the two-component systems involved in antibiotic resistance.

Preface to "The Structural and Functional Study of Efflux Pumps Belonging to the RND Transporters Family from Gram-Negative Bacteria"

Antimicrobial-resistant bacterial infections are a major and costly public health concern. Several pathogens are already pan-resistant, representing a major cause of mortality in patients suffering from nosocomial infections. Drug efflux pumps, which remove compounds from the bacterial cell, thereby lowering the antimicrobial concentration to sub-toxic levels, play a major role in multidrug resistance.

In this Special Issue, we present up-to-date knowledge of the mechanism of RND efflux pumps, the identification and characterization of efflux pumps from emerging pathogens and their role in antimicrobial resistance, and progress made on the development of specific inhibitors. This collection of data could serve as a basis for antimicrobial drug discovery aimed at inhibiting drug efflux pumps to reverse resistance in some of the most resistant pathogens.

Isabelle Broutin, Attilio V Vargiu, Henrietta Venter, and Gilles Phan
Editors

Editorial

The Structural and Functional Study of Efflux Pumps Belonging to the RND Transporters Family from Gram-Negative Bacteria

Attilio Vittorio Vargiu [1], Gilles Phan [2], Henrietta Venter [3] and Isabelle Broutin [2,*]

1   Department of Physics, University of Cagliari, S.P. 8 km 0.700, 09042 Monserrato, Italy; vargiu@dsf.unica.it
2   Laboratoire CiTCoM, Faculté de Santé, Université Paris Cité, CNRS, 75006 Paris, France; gilles.phan@parisdescartes.fr
3   Health and Biomedical Innovation, Clinical and Health Sciences, University of South Australia, Adelaide, SA 5000, Australia; rietie.venter@unisa.edu.au
*   Correspondence: isabelle.broutin@u-paris.fr

Citation: Vargiu, A.V.; Phan, G.; Venter, H.; Broutin, I. The Structural and Functional Study of Efflux Pumps Belonging to the RND Transporters Family from Gram-Negative Bacteria. *Antibiotics* **2022**, *11*, 429. https://doi.org/10.3390/antibiotics11040429

Received: 3 March 2022
Accepted: 21 March 2022
Published: 23 March 2022

Publisher's Note: MDPI stays neutral with regard to jurisdictional claims in published maps and institutional affiliations.

Copyright: © 2022 by the authors. Licensee MDPI, Basel, Switzerland. This article is an open access article distributed under the terms and conditions of the Creative Commons Attribution (CC BY) license (https://creativecommons.org/licenses/by/4.0/).

Antimicrobial-resistant bacterial infections are a major and costly public health concern. Several pathogens are already pan-resistant, representing a major cause of mortality in patients suffering from nosocomial infections. Drug efflux pumps, which remove compounds from the bacterial cell, thereby lowering the antimicrobial concentration to sub-toxic levels, play a major role in multidrug resistance (MDR).

Gram-negative bacteria are particularly resistant, and some are identified by the World Health Organization as the pathogens most urgently in need of new antimicrobial drug discovery. The most clinically relevant efflux pumps in Gram-negative bacteria belong to the Resistance Nodulation cell Division (RND) family, which form a tripartite macromolecular assembly spanning both membranes and the periplasmic space of Gram-negative organisms.

Along with functional studies and in silico approaches, many structures of the individual components, as well as the fully assembled pumps from several pathogens, have been solved. Nevertheless, many questions concerning the assembly and the mechanism of efflux remain, while there are still no efflux pump inhibitors available in clinical treatment.

This Special Issue is a representation of the current knowledge regarding the mechanism, regulation, dissemination, etc., of these efflux pumps.

Bernal et al. [1] explored the putative pathway used by charged substrates in the AcrAD-TolC RND efflux pump of *Salmonella typhimurium* (St) using site-directed mutational analysis, combined with complementation and minimum inhibition concentration (MIC) measurements, in an *E. coli* Δ*acrB*Δ*acrD* strain. They highlighted several amino acids of the deep binding pocket, the access pocket, and the TM1/TM2 groove region as being involved in the specificity of AcrD efflux for aminoglycosides and dianionic β-lactams. They also confirmed that St-AcrD is able to function with *E. coli* AcrA and TolC, and reported temocillin, dicloxacillin, cefazolin and fusidic acid as substrates of AcrD, contrary to piperacillin, which is not a substrate. Their results underline the need to explore single mutation variants *in cellulo*, instead of a simple extrapolation of the effect based on a model.

As the efflux of drugs in Gram-negative bacteria requires the formation of a long tripartite assembly to pass through the two membranes, several groups have attempted to understand the assembly process.

Boyer et al [2]. investigated the dynamic and selectivity of the *Pseudomonas aeruginosa* MexAB-OprM efflux pump assembly. They studied the effect of the gain of function mutation Q93R, identified in the periplasmic adaptor protein (PAP) MexA, and its capacity to form a functional assembly with different outer membrane factors (OMF) (OprM, OprN, TolC, and OprM-ΔCter truncated of its 13 last amino acids of unresolved structure). Using several biophysical approaches (size-exclusion chromatography, biolayer interferometry, negative staining electron microscopy) combined with minimum inhibition concentration

(MIC) measurements, they highlighted a modulation effect in the assembly process efficacy, involving molecular determinants other than the tip-to-tip OMF–PAP interface.

Rajapaksha et al. [3] studied the assembly process of the *E. coli* AcrAB-TolC efflux pump in cellulo. They studied the effect of the overexpression of the proteins forming the AcrAB-TolC efflux pump bearing "loss of function mutations", using MIC measurements for different antibiotics in the context of gene-deleted and wild-type *E. coli* strains. They observed no significant drop in the efflux activity, indicating that the RND pump assembly process in Gram-negative bacteria is a precisely controlled process that prevents the formation of functionless complexes. In addition, they performed competition experiments, which highlighted that the dissociation kinetics of the AcrAB complex are very slow. Together, these results indicate that the assembly of the AcrAB-TolC complex has a proof-read mechanism that effectively eliminates the formation of a futile pump complex.

Webber et al. [4] investigated the AcrAB-TolC efflux pump opening mechanism. They tried to explain how the drug transport induces assembly and opening of the AcrA and TolC partners. They compared the resting and transport states of the AcrAB-TolC pump, for which several structures exist in the presence and absence of a substrate. They indicate that the "assembled opened" conformation state has lower energy than the resting state, suggesting the need for energy input to perform the conformation changes that are required to go back to the resting state after the efflux of a substrate. From this observation, and by using distance matrices supplemented with evolutionary coupling data and buried surface area measurements, they propose a new allosteric model for the function of the pump.

AcrAB-TolC and MexAB-OprM are the two most-studied efflux pumps, but important information can also be acquired from the characterization of RND efflux pumps by looking at emerging pathogens and their role in antimicrobial resistance.

Mateus et al. [5] characterized three efflux systems found in an emergent enteropathogen, *Aliarcobacter butzleri*, looking at their involvement in resistance and virulence. They identified the substrates of AreABC, AreDEF and AreGHI by performing MIC measurements and ethidium bromide accumulation experiments, and analyzed their implication in the resistance to oxidative and bile stress, in bacterial fitness, their impact on motility and biofilm formation ability, and their ability to survive in human serum and adhere to intestinal cells. They show that these three RND pumps could be considered putative targets for new therapeutic strategies to fight infections with this emerging pathogen.

In contrast with most pathogens, only AcrB has been studied for its impact on resistance in *E. coli*. Schuster et al. [6] studied the contribution of the MdtF pump, functioning with MdtE and TolC, to drug resistance in an MDR *E. coli* isolated from a patient. By comparing the MIC values of different drugs and the accumulation and the efflux of different dyes in the original strain and in the deleted strain of the genes coding for AcrB, TolC, or both AcrB and MdtF, they showed a limited contribution of MdtF to the antibiotic resistance profile of this MDR *E. coli* isolate, but a remarkable capacity to export dyes.

Finally, as MDR bacteria are spreading wordlwide, it is most important to evaluate their diversity and predominance in clinics.

Scoffone et al. [7] listed the RND efflux pumps of the different Gram-negative bacterial species found in the lung of 70,000 patients worldwide suffering from cystic fibrosis (CF). They focused their review on the four species that are predominantly encountered (*Pseudomonas aeruginosa*, *Burkholderia cenocepacia*, *Achromobacter xylosoxidans*, *Stenotrophomonas maltophilia*). For each species, they formed the link between the CF lung environment modifications that can append during the patients' life and the over-expression of specific RND pumps that participate in the bacteria adaptation. They also provide an overview of the efflux pumps inhibitors that are described as efficient for these pumps.

Zwama and Nishiro [8] performed a genetic evolution analysis on 135 RND efflux pumps to highlight the conserved and variable domains in the RNDs' structure. They also made an inventory of all the mutants that have been described in RND transporters from different clinically, environmentally and laboratory-evolved Gram-negative bacterial strains (*Escherichia coli*, *Salmonella enterica*, *Neisseria gonorrhoeae*, *Pseudomonas aeruginosa*,

Acinetobacter baumannii and *Legionella pneumophila*), which displayed increased antimicrobial resistance. They show that the TM domain, bearing the motor, is largely conserved, as well as three residues located in the loops linking this domain to the binding domain. The binding domain presents more versatility. However, some strategic amino acid positions are highlighted as possible clues to the selectivity for certain antibiotic families. In conclusion, they underlined the worrying resistance adaptation of several bacterial strains by acquiring mutations directly in the RND-coding genes.

Davin-Regli et al. [9] provided an overview of the prevalence of the main efflux pumps observed in clinical practice and listed the clinical impact situation for some of the most concerning Gram-negative species. They highlighted the need to develop new efflux pump inhibitors, and provided an overview of the different methods used to measure the inhibition power of tested molecules.

This collection of data could serve as a basis for antimicrobial drug discovery that aims to inhibit drug efflux pumps to reverse resistance in some of the most resistant pathogens.

Funding: A.V.V. received support from the National Institutes of Allergy and Infectious Diseases project number AI136799; I.B. and G.P. received support from the French national research agency ANR (Mistec ANR-17CE11-0028). H.V. received support from the National Health and Medical Research Council of Australia (grant number GN 1147538).

Conflicts of Interest: The authors declare no conflict of interest.

References

1. Bernal, J.C.; El-Delik, J.; Göttig, S.; Pos, K.M. Characterization and Molecular Determinants for β-Lactam Specificity of the Multidrug Efflux Pump AcrD from *Salmonella typhimurium*. *Antibiotics* **2021**, *10*, 1494. [CrossRef] [PubMed]
2. Boyer, E.; Dessolin, J.; Lustig, M.; Decossas, M.; Phan, G.; Cece, Q.; Durand, G.; Dubois, V.; Sansen, J.; Taveau, J.-C.; et al. Molecular Determinants for OMF Selectivity in Tripartite RND Multidrug Efflux Systems. *Antibiotics* **2022**, *11*, 126. [CrossRef] [PubMed]
3. Rajapaksha, P.; Ojo, I.; Yang, L.; Pandeya, A.; Abeywansha, T.; Wei, Y. Insight into the AcrAB-TolC Complex Assembly Process Learned from Competition Studies. *Antibiotics* **2021**, *10*, 830. [CrossRef] [PubMed]
4. Webber, A.; Ratnaweera, M.; Harris, A.; Luisi, B.F.; Ntsogo Enguéné, V.Y. A model for allosteric communication in drug transport by the AcrAB-TolC tripartite efflux pump. *Antibiotics* **2022**, *11*, 52. [CrossRef] [PubMed]
5. Mateus, C.; Nunes, A.R.; Oleastro, M.; Domingues, F.; Ferreira, S. RND Efflux Systems Contribute to Resistance and Virulence of *Aliarcobacter butzleri*. *Antibiotics* **2021**, *10*, 823. [CrossRef] [PubMed]
6. Schuster, S.; Vavra, M.; Greim, L.; Kern, W. Exploring the Contribution of the AcrB Homolog MdtF to Drug Resistance and Dye Efflux in a Multidrug Resistant *E. coli* Isolate. *Antibiotics* **2021**, *10*, 503. [CrossRef] [PubMed]
7. Scoffone, V.C.; Trespidi, G.; Barbieri, G.; Irudal, S.; Perrin, E.; Buroni, S. Role of RND Efflux Pumps in Drug Resistance of Cystic Fibrosis Pathogens. *Antibiotics* **2021**, *10*, 863. [CrossRef] [PubMed]
8. Zwama, M.; Nishino, K. Ever-Adapting RND Efflux Pumps in Gram-Negative Multidrug-Resistant Pathogens: A Race against Time. *Antibiotics* **2021**, *10*, 774. [CrossRef] [PubMed]
9. Davin-Regli, A.; Pages, J.-M.; Ferrand, A. Clinical Status of Efflux Resistance Mechanisms in Gram-Negative Bacteria. *Antibiotics* **2021**, *10*, 1117. [CrossRef] [PubMed]

Article

Characterization and Molecular Determinants for β-Lactam Specificity of the Multidrug Efflux Pump AcrD from *Salmonella typhimurium*

Jenifer Cuesta Bernal [1], Jasmin El-Delik [1], Stephan Göttig [2] and Klaas M. Pos [1,*]

[1] Institute of Biochemistry, Goethe-University Frankfurt, Max-von-Laue-Str. 9, D-60438 Frankfurt am Main, Germany; jcuesta.bernal@gmail.com (J.C.B.); El-Delik@em.uni-frankfurt.de (J.E.-D.)

[2] Institute of Medical Microbiology and Infection Control, Hospital of the Goethe University, Paul-Ehrlich-Straße 40, D-60596 Frankfurt am Main, Germany; Stephan.Goettig@kgu.de

[*] Correspondence: pos@em.uni-frankfurt.de

Citation: Cuesta Bernal, J.; El-Delik, J.; Göttig, S.; Pos, K.M. Characterization and Molecular Determinants for β-Lactam Specificity of the Multidrug Efflux Pump AcrD from *Salmonella typhimurium*. *Antibiotics* **2021**, *10*, 1494. https://doi.org/10.3390/antibiotics10121494

Academic Editors: Isabelle Broutin, Attilio V Vargiu, Henrietta Venter and Gilles Phan

Received: 13 October 2021
Accepted: 1 December 2021
Published: 6 December 2021

Publisher's Note: MDPI stays neutral with regard to jurisdictional claims in published maps and institutional affiliations.

Copyright: © 2021 by the authors. Licensee MDPI, Basel, Switzerland. This article is an open access article distributed under the terms and conditions of the Creative Commons Attribution (CC BY) license (https://creativecommons.org/licenses/by/4.0/).

Abstract: Gram-negative Tripartite Resistance Nodulation and cell Division (RND) superfamily efflux pumps confer various functions, including multidrug and bile salt resistance, quorum-sensing, virulence and can influence the rate of mutations on the chromosome. Multidrug RND efflux systems are often characterized by a wide substrate specificity. Similarly to many other RND efflux pump systems, AcrAD-TolC confers resistance toward SDS, novobiocin and deoxycholate. In contrast to the other pumps, however, it in addition confers resistance against aminoglycosides and dianionic β-lactams, such as sulbenicillin, aztreonam and carbenicillin. Here, we could show that AcrD from *Salmonella typhimurium* confers resistance toward several hitherto unreported AcrD substrates such as temocillin, dicloxacillin, cefazolin and fusidic acid. In order to address the molecular determinants of the *S. typhimurium* AcrD substrate specificity, we conducted substitution analyses in the putative access and deep binding pockets and in the TM1/TM2 groove region. The variants were tested in *E. coli* Δ*acrB*Δ*acrD* against β-lactams oxacillin, carbenicillin, aztreonam and temocillin. Deep binding pocket variants N136A, D276A and Y327A; access pocket variant R625A; and variants with substitutions in the groove region between TM1 and TM2 conferred a sensitive phenotype and might, therefore, be involved in anionic β-lactam export. In contrast, lower susceptibilities were observed for *E. coli* cells harbouring deep binding pocket variants T139A, D176A, S180A, F609A, T611A and F627A and the TM1/TM2 groove variant I337A. This study provides the first insights of side chains involved in drug binding and transport for AcrD from *S. typhimurium*.

Keywords: antibiotic resistance; efflux pump; RND

1. Introduction

Antibiotic resistance has become a global public health concern due to the appearance of resistant strains, especially from pathogen Gram-negative bacteria, that acquired resistance determinants against many clinically used anti-infective agents. This phenomenon, known as Multidrug Resistance (MDR), can be caused by a simultaneous presence of multiple resistance mechanisms that are encoded on transferable plasmids or chromosomes [1]. Among these mechanisms, the increased active export of the drugs by multidrug efflux pumps can cause simultaneous resistance to several toxic compounds, representing a major challenge for new antibiotics development [2,3].

Gram-negative systems comprising inner membrane proteins from the Resistance Nodulation and cell Division (RND) superfamily play a major role in multidrug resistance because of their action of assembling into tripartite complexes that span the entire bacterial envelope and their ability to capture drugs from the periplasm and expelling these drugs towards the extracellular medium [3,4]. These RND-type tripartite systems play also other roles in biofilm formation, quorum-sensing, bile salt resistance and virulence, and their activity appears to be connected to the increase in mutations on the chromosome [5–10].

The importance of many (other) roles is evidenced by their wide distribution in all domains of life, and several RND pumps are genotypically encoded in the same organism in most cases [11]. At least five RND multidrug efflux pump genes have been identified in the *Escherichia coli* chromosome, ten have been identified in *Klebsiella pneumoniae*, two have been identified in *Campylobacter jejuni* and six have been identified in *Salmonella typhimurium* [12]. The potential for the deployment of these transporters in numerous bacterial species of clinical concern, such as the ESKAPE pathogens, has directed RND-type tripartite transporter research efforts toward these organisms in order to understand their structural and functional basis [2,4,13].

Salmonella enterica serovar Typhimurium (*S. typhimurium*) is a Gram-negative bacterium that causes disease in both humans and animals [14]. This food-borne pathogen accounts for the high incidence of *Salmonella* infections worldwide and emerging antibiotic resistance strains have been reported, representing a potentially serious public health problem [15]. *S. typhimurium* has at least nine functional drug efflux pumps belonging to different transporter (super)families. It has been reported that RND members, such as AcrB, AcrD, AcrF, MdsB and MdtBC, play a major role in the observed *Salmonella* resistance phenotype to a wide range of toxic compounds [14,16,17].

Several studies on the expression, regulation and transcriptome profiling of single and multiple deletion strains of RND efflux pump genes in *S. typhimurium* established that the main RND pump conferring antibiotic resistance is the AcrAB-TolC system, and its inactivation results in multidrug hypersusceptibility [14,16–19]. However, its loss (i.e., Δ*acrB*) is to a certain degree compensated by the increased expression of homologous pumps AcrD and AcrF [16].

AcrD from *S. typhimurium* (St_AcrD) assembles in a tripartite complex together with the Membrane Fusion Protein (MFP) AcrA (St_AcrA) and the Outer Membrane Factor (OMF) TolC (St_TolC), i.e., the same interaction partners employed by AcrB (St_AcrB), in order to export noxious compounds out of the cell, including SDS, novobiocin and deoxycholate [20], and also more hydrophilic compounds such as β-lactams (oxacillin, nafcillin, cloxacillin, carbenicillin, sulbenicillin and aztreonam) and aminoglycosides [20,21]. In contrast, these β-lactams and aminoglycosides are weak or non-substrates of the AcrB pumps in *E. coli* (Ec_AcrB) and *S. typhimurium* [20,22,23]. A molecular dynamics approach comparing Ec_AcrB and *E. coli* AcrD (Ec_AcrD) proximal (access) and distal (deep) binding pockets indicated that their volume and shape are rather similar [24]. However, Ec_AcrD comprises more polar and charged residues in the binding pockets compared to Ec_AcrB. By using chimeric AcrB/AcrD constructs and site-directed mutagenesis, it was shown that AcrB could confer increased resistance against aztreonam, sulbeniccilin and carbenicillin when three residues in the access pocket with an overall negative charge (1^-) were exchanged for residues with an overall positive charge (2^+). However, the molecular determinants for substrate binding to the RND transporter St_AcrD (or its closest homolog Ec_AcrD, with 94% identical residues) have not been experimentally characterized.

In this study, we used site-directed mutagenesis to target single side chains within the AcrD drug-binding access pocket, the deep binding pocket and the TM1/TM2 groove for substitution. Functional analysis in presence of monoanionic and dianionic β-lactams revealed several regions that are permissive towards substitution and that can cause hyperactive variants.

2. Results

2.1. Substrate Specificity of St_AcrD

AcrD from *S. typhimurium* has been functionally characterized and was shown to be dependent on AcrA and TolC for its activity [20]. Earlier studies had shown that the deletion of the *acrD* gene from the *S. typhimurium* chromosome resulted in susceptibilities towards aminoglycosides such as amikacin, gentamicin, neomycin, kanamycin and tobramycin and that Δ*acrD* cells accumulated higher levels of dihydrostreptomycin and gentamicin compared to the parental strain [23]. AcrAD-TolC has been shown to further confer resistance towards SDS, novobiocin and various β-lactams such as oxacillin, cloxacillin, nafcillin, carbenicillin, sulbenicillin and aztreonam.

Based on known RND multidrug efflux pump structures, we assume that AcrD is active as a trimer and that the protomers also might adopt different conformations. These different conformations might present substrate binding sites for drugs sequestered from the periplasm or the inner membrane. In order to obtain more detailed structural information on St_AcrD, we utilized a direct structural comparison with Ec_AcrB by using sequence alignments and homology model building, as both proteins share 66% identical residues (Table S1, Figure S1). The putative substrate binding pockets of St_AcrD were assigned according to the location and amino acid composition of its homologues Ec_AcrB and Ec_AcrD [24]. This study addresses residues from the Access Pocket (AP), Deep Binding Pocket (DBP) and fusidic acid binding site (TM1/TM2 region), and the latter specifically binds carboxylated drugs (fusidic acid and lipophilic β-lactams) in Ec_AcrB (Figures S1 and S2) [25].

In order to test St_AcrD functionality, drug susceptibility tests were performed with *E. coli* BW25113 (DE3) Δ*acrB*Δ*acrD*, since this strain does not express RND pump components AcrB and AcrD but still produces tripartite interaction partners AcrA and TolC. The expression plasmid p7XC3H-St_AcrD (pSt_AcrD) was employed to produce wildtype St_AcrD, and the cloning vector p7XC3H-Δ*ccdB* (pControl) was used as a negative control (see Supplementary Materials and Methods). The sequence identity of the tripartite complexes AcrAD-TolC of *E. coli* and *S. typhimurium* was high (Ec_AcrD vs. St_AcrD 94%, Ec_AcrA vs. St_AcrA 97% and Ec_TolC vs. St_TolC 90%). Hence, there is a fair assumption that St_AcrD assembles into a fully active tripartite efflux pump together with the components Ec_AcrA and Ec_TolC (encoded on the chromosome) and, therefore, complement the AcrB/AcrD deficient phenotype in *E. coli* [26].

Functionally active phenotypes of St_AcrD could be detected by the plate dilution assay, as cells that produced wildtype St_AcrD were able to grow at higher cell dilutions compared to the negative control (Δ*acrB*Δ*acrD*) in presence of anionic β-lactam (Figure 1). This included substrates such as oxacillin, carbenicillin, nafcillin and aztreonam, which were shown previously to be transported by the pump [20]. These observations also confirm the formation of a hybrid tripartite system St_AcrD-Ec_AcrA-Ec_TolC in *E. coli* [26]. Interestingly, we could identify several hitherto unreported substrates such as temocillin, dicloxacillin, cefazolin and fusidic acid. Furthermore, piperacillin does not appear to be a substrate for the AcrD pump (Figure 1).

The substrate specificity for St_AcrD as determined by the plate dilution method could be confirmed by Minimal Inhibitory Concentration (MIC) determinations [27] performed by the antibiotic gradient test for clinically relevant antibiotics (Table 1). In this assay, *E. coli* BW25113 (DE3) Δ*acrB*Δ*acrD* harboring pSt_AcrD conferred resistance against temocillin, oxacillin, aztreonam, cephalotin and cefazolin, exhibiting increases between 8-fold to 16-fold with respect to MIC values.

2.2. Site-Directed Mutagenesis of Substrate Binding Pocket Residues

In order to experimentally characterize substrate binding pockets in St_AcrD and to identify residues that are essential for β-lactam transport, single alanine mutants (Ala-mutants) were produced by site-directed mutagenesis. In total, 21 residues were selected and classified within three groups: deep binding pocket (DBP) residues (N136, T139, D176, Y178, S180, K274, D276, Y277, Y327, F609, S610, T611, S614 and F627), access pocket (AP) residues (R568, R625 and G672) and TM1/TM2 region residues (I27, I337, I338 and V341) (Figures S1 and S2).

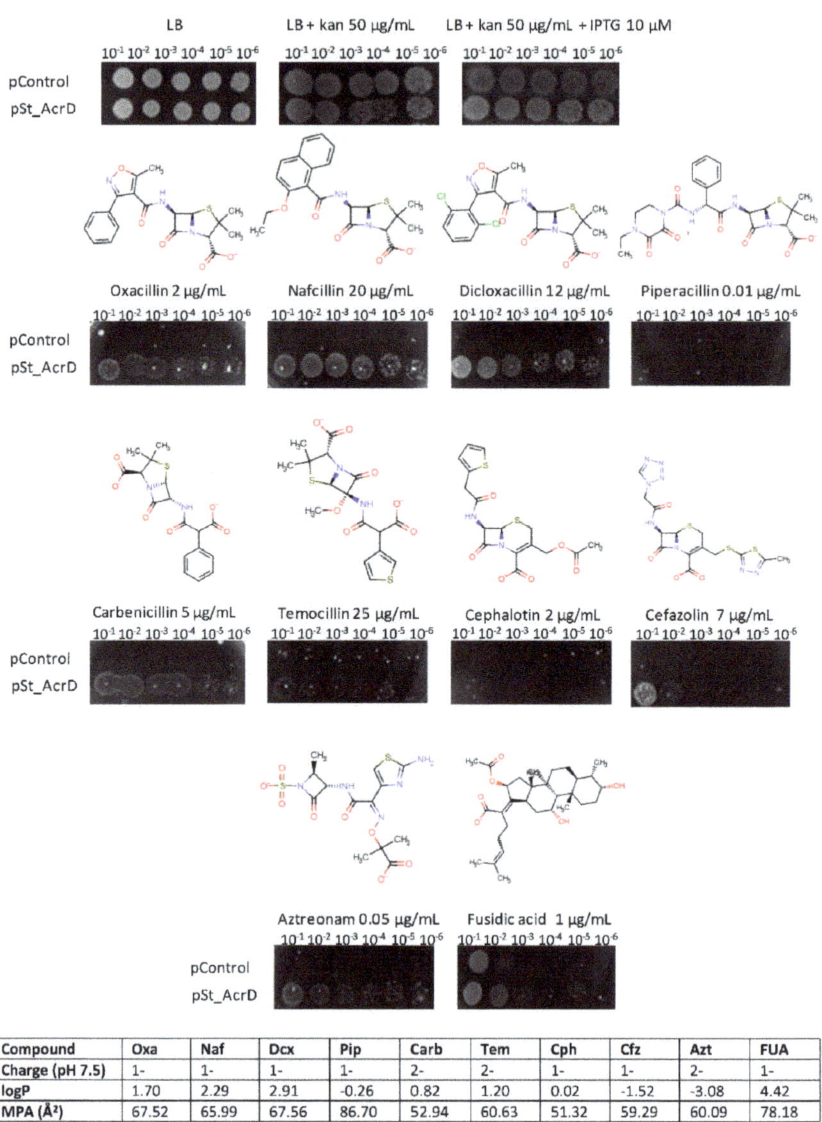

Compound	Oxa	Naf	Dcx	Pip	Carb	Tem	Cph	Cfz	Azt	FUA
Charge (pH 7.5)	1-	1-	1-	1-	2-	2-	1-	1-	2-	1-
logP	1.70	2.29	2.91	-0.26	0.82	1.20	0.02	-1.52	-3.08	4.42
MPA (Å²)	67.52	65.99	67.56	86.70	52.94	60.63	51.32	59.29	60.09	78.18

Figure 1. Determination of St_AcrD substrate specificity by agar plate dilution assay. Serial dilutions of normalized *E. coli* BW25113 (DE3) Δ*acrB*Δ*acrD* harboring the plasmids pControl (empty vector, negative control) or St_AcrD (pSt_AcrD) were spotted on LB agar plates supplemented with 50 µg/mL kanamycin, 10 µM IPTG and the indicated antibiotic and concentration. Physicochemical properties of tested substrates are indicated below (https://chemicalize.com/#/, accessed on 8 May 2018). Oxa: Oxacillin; Naf: Nafcillin; Dcx: Dicloxacillin; Pip: Piperacillin; Carb: Carbenicillin; Tem: Temocillin; Cph: Cephalotin; Cfz: Cefazolin; Azt: Aztreonam; FUA: Fusidic acid; logP: logarithm of the partition coefficient; MPA: minimal projection area.

Table 1. MIC determination of *E. coli* BW25113 (DE3) Δ*acrB*Δ*acrD* harboring pSt_AcrD or pControl. MIC determinations were performed by employing antibiotic gradient strips on Mueller–Hinton agar plates, supplemented with 50 µg/mL kanamycin and 10 µM IPTG. Experiments were performed in duplicate.

	MIC (µg/mL)	
Antibiotic	**pControl**	**pSt_AcrD**
Ampicillin	0.75	1
Temocillin	0.75	6
Piperacillin	0.047	0.064
Oxacillin	0.125	2
Penicillin	8	8
Mecillinam	0.023	0.032
Aztreonam	0.032	0.25
Cefotaxime	<0.016	<0.016
Cefoxitin	0.5	1
Cefepime	≤0.016	0.016
Ceftriaxone	0.023	0.016
Cefuroxime	0.125	0.125
Ceftaroline	≤0.016	0.016
Ceftazidime	0.047	0.047
Cefazolin	2	16
Cephalotin	2	12
Cefpodoxime	0.064	0.094
Imipenem	0.125	0.25
Doripenem	0.032	0.064
Meropenem	0.023	0.023
Tetracycline	0.25	0.25
Fusidic acid	2	4
Erythromycin	0.75	1.5
Chloramphenicol	0.5	0.38

The susceptible *E. coli* BW25113 (DE3) Δ*acrB*Δ*acrD* strain was transformed with plasmids encoding the indicated St_AcrD Ala-substitution variants and tested for activity in plate dilution assay (Figures S3–S5). Cells carrying Ala-variants grew to the same extent under non-selective conditions compared to cells harboring pSt_AcrD (WT) and pControl (Figure S3). The plate dilution assay is a preferred method for the analysis of subtle activity changes due to single-site substitution, as it is a direct visualization of the bacterial growth rate. In the experiments shown in Figure S4, the LB agar plates in addition contained kanamycin as selective antibiotic for the selection marker on the pSt_AcrD plasmid, whereas the LB agar plates in the experiments shown in Figure S5 did not contain kanamycin.

All variants were tested in the presence of the four β-lactams, oxacillin, carbenicillin, temocillin and aztreonam, in order to identify whether mutants could complement the *E. coli* susceptible phenotype, as was shown for wildtype St_AcrD. Either two similar or two opposite phenotypes were observed compared to the wildtype: Hyperactive mutants that grew on higher antibiotic concentrations or exhibited more copious growth as observed by denser and more intense spots on the LB-agar-plate. On the other hand, compromised mutants are those that were not able to grow to the same extent as cells producing wildtype St_AcrD (Figures 2, S4 and S5).

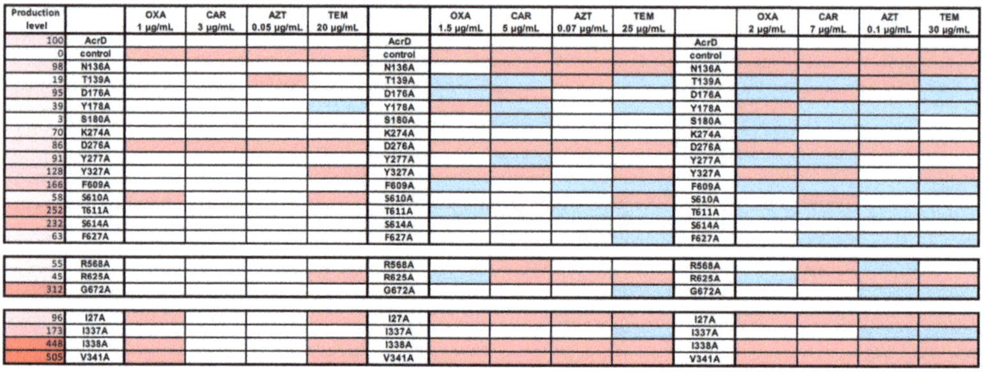

Production level		OXA 1 µg/mL	CAR 3 µg/mL	AZT 0.05 µg/mL	TEM 20 µg/mL		OXA 1.5 µg/mL	CAR 5 µg/mL	AZT 0.07 µg/mL	TEM 25 µg/mL		OXA 2 µg/mL	CAR 7 µg/mL	AZT 0.1 µg/mL	TEM 30 µg/mL
100	AcrD					AcrD					AcrD				
0	control					control					control				
98	N136A					N136A					N136A				
19	T139A					T139A					T139A				
95	D176A					D176A					D176A				
39	Y178A					Y178A					Y178A				
3	S180A					S180A					S180A				
70	K274A					K274A					K274A				
86	D276A					D276A					D276A				
91	Y277A					Y277A					Y277A				
128	Y327A					Y327A					Y327A				
166	F609A					F609A					F609A				
58	S610A					S610A					S610A				
252	T611A					T611A					T611A				
232	S614A					S614A					S614A				
63	F627A					F627A					F627A				
55	R568A					R568A					R568A				
45	R625A					R625A					R625A				
312	G672A					G672A					G672A				
96	I27A					I27A					I27A				
173	I337A					I337A					I337A				
448	I338A					I338A					I338A				
505	V341A					V341A					V341A				

Figure 2. Relative susceptibilities toward β-lactams of *E. coli* Δ*acrB*Δ*acrD* complemented by AcrD variants. Oxacillin (OXA), carbenicillin (CAR), aztreonam (AZT) and temocillin (TEM) were used at the concentrations indicated. Based on visual inspection of the number of spots on the plates for each variant shown in Figures S4 and S5, each variant was indicated as compromised (red shade), hyperactive (blue shade) or showed the same phenotype compared to cells harbouring wildtype AcrD. Relative production levels of each variant are indicated on the left based on the values determined via Western Blot analysis (Figure S7). The wildtype production level was set to 100, the control was set to 0.

2.2.1. Deep Binding Pocket Variants

Depending on the antibiotic stress applied, approximately half of the DBP variants exhibited similar growth compared to cells with wildtype St_AcrD; however, hyperactive (T139A, D176A, Y178A, S180A, F609A, T611A and F627A) and sensitive mutants (N136A, D276A and Y327A) were identified (Figures 2, S4 and S5). Most of these observations were consistent for the substrates tested (either sensitive or hyperactive), which could indicate that monoanionic and dianionic β-lactams might interact with the same residues located in the DBP. In contrast, T139A shows hyperactivity towards oxacillin, carbenicillin and temocillin but was not able to transport aztreonam. This suggests that the hydroxyl group of T139 is essential for aztreonam translocation along the substrate pathway. Likewise, Y178A confers a hyperactive phenotype, except toward oxacillin, where growth is impaired (Figure 2). Moreover, F609A conferred hyperactivity, whereas the homologous Ec_AcrB substitution variant F610A was previously shown to confer most considerable sensitivity towards all tested AcrB substrates. In fact, for AcrB, the F610A substitution had the strongest effect on the MIC values of all DBP substitutions tested [28]. One interpretation for the latter variant might be that the lack of the phenyl group facilitates transport of dianonic substrates in this St_AcrD variant, permitting accommodation of the more hydrated hydrophilic compounds in DBP [24]. On the other hand, a decrease in polar environment in the DBP (T139A, D176A, S180A and T611A) also resulted in hyperactive phenotypes. These residues are distributed in the core of the DBP PN2 and PC1 subdomains of St_AcrD, and their Ec_AcrB counterparts interact directly with doxorubicin and minocycline in the DBP, as shown in co-crystal structures [29] (Figure S6). Most distal in the DBP, D276 seems to be involved either in direct interaction with β-lactams or their transport through the exit tunnel, as the D276A substitution resulted in the most pronounced sensitivity towards all four β-lactams. Notably, an *E. coli* Δ*acrB* strain complemented with an Ec_AcrB D276C variant showed wildtype-like MIC values for erythromycin and novobiocin [30]. Residues in the other sensitive variants, i.e., N136A, Y327A and the aztreonam sensitive T139A variant, were all proximally located in the PN2 subdomain and close to the AP, where substrate recognition of anionic β-lactams has been suggested to take place [21].

Protein production of St_AcrD WT and Ala-variants in *E. coli* was detected by Western blot analysis (Figures 2 and S7). The activity of the tripartite efflux system will not only depend on the concentration of the RND component but also on the available periplasmic concentration of interaction partners AcrA and TolC. Sensitive mutants N136A, D276A and

Y327A were produced at similar levels compared to wildtype St_AcrD; hence, a decrease in activity is most likely directly related to the substitution. On the other hand, the increased activity observed for the mutants F609A and T611A might be partially explained by the overproduction of these variants compared to St_AcrD WT. However, there appears to be no clear correlation between activity and production levels as another variant, S614A, was overproduced to the same extent as T611A and showed mostly wildtype-like resistance phenotypes (Figures 2, S4 and S5). Furthermore, despite the low level protein production of T139A, Y178A and even more surprisingly S180A, higher wildtype resistances could be observed under most conditions (Figure 2). This indicates that the concentration of St_AcrD in cells was not limiting for exhibiting wildtype phenotypes, whereas it cannot be excluded that overproduction might still result in higher resistance phenotypes.

2.2.2. Access Pocket Variants

The AP variant R625A conferred decreased activity against dianionic β-lactams, such as carbenicillin, temocillin and aztreonam, but not the monoanionic oxacillin (Figures 2, S4 and S5). This is in line with the observation that R625 is important for the recognition of negatively charged β-lactams in Ec_AcrD [21]. Despite the fact that Western blot analysis shows reduced expression of this mutant (Figures 2 and S7), R568A conferred a clear hyperactive phenotype towards aztreonam (at 0.1 mg mL^{-1}) and moderate hyperactivity towards temocillin, whereas a somewhat decreased resistance towards carbenicillin was observed (Figures 2, S4 and S5). The G672A variant conferred a hyperactive phenotype towards temocillin and aztreonam and showed higher production levels compared to the wildtype (Figures 2 and S7).

2.2.3. Substitution Variants in the TM1/TM2 Groove

The I27A variant exhibited the most sensitive phenotype for all tested antibiotics (Figures 2, S4 and S5). On the other hand, the St_AcrD I337A variant appears to confer a hyperactive phenotype (for aztreonam and temocillin). In contrast, the counterpart I337A substitution in Ec_AcrB resulted in a marked increased sensitivity towards β-lactam antibiotics, including oxacillin [25,31]. The overproduction of St_AcrD I337A (1.7-fold compared to St_AcrD WT, Figure S7) may possibly obscure its potential reduced ability to transport β-lactams. On the other hand, variants I338A and V341A, both conferring substantially reduced resistance towards all tested β-lactams, showed far more extensive overproduction in *E. coli* (4.5-fold and 5-fold, respectively). Thus, a clear correlation between overproduction and mutant phenotype was not warranted. Another consideration might be that substitutions can cause improper protein folding and insertion in the membrane, causing a large signal via Western blot analysis, which does not necessarily indicate proper folding. For wildtype St_AcrD and five other RND efflux pumps, correct folding was analyzed by using RND-GFP fusions (Supplementary Materials and Methods), where proper folding is indicated by the GFP fluorescent signal [32] (Figure S8). Subsequent in-gel fluorescence in combination with Western Blot analysis indicated that the ratio of well-folded, fluorescent St_AcrD and misfolded proteins is approximately 1:1 (Figure S9). Moreover, AcrD could be purified from the AcrD-overproducing cells and results in a clear monodisperse peak indicating no aggregation in a solubilized state even after storage at 4 °C or 17 °C for one week (Figure S10). Comparison with purified AcrB indicates that AcrD is present as a monomer in detergent solution (Figure S11). Future studies are necessary in order to confirm the ratio between well-folded and misfolded proteins inside the cell membrane of each of the St_AcrD variants.

In summary, St_AcrD with Ala-substituted residues N136, D276 and Y327 in the deep binding pocket confers an overall sensitive phenotype towards the tested antibiotics, whereas T139A and Y178A confer a selective and higher resistance phenotype, as growth in aztreonam (T139A) or oxacillin (Y178A) was impaired. F609A conferred an overall higher resistance phenotype towards all tested substrates. This observation might possibly be in line with the reduction in hydrophobic binding pocket environment resulting in the enhancement of more hydrophilic substrate transport. Ala-substitutions made in the TM1/TM2 groove, proposed to be the initial binding site for carboxylated drugs in AcrB [25,31], conferred higher susceptibility to all four β-lactams tested (Figures 2 and S5). As an exception, I337A was shown to confer either wildtype-like or reduced susceptibility towards these substrates.

3. Discussion

Recently, a major emphasis was placed on structure elucidation of different RND proteins [4,13]. Nevertheless, structure–function relationships are far from clear and are in need of analysis via mutational analysis, biochemical/biophysical substrate binding, transport studies and molecular dynamic simulations. In particular, the latter technique not only requires high-resolution structures but is also reliant on additional experimental data for a supply of calculation setups with restraints derived from experiments. The results obtained from the simulation data then can be fed back into the experimental design [33].

Here, we addressed the substrate specificity for AcrD from *Salmonella typhimurium* and conducted functional analysis of AcrD variants with Ala-substituted binding pocket residues. For wildtype AcrD, we could identify temocillin, dicloxacillin, cefazolin and fusidic acid as substrates for the pump, whereas piperacillin is not a substrate (Figure 1, Table 1). The results for wildtype AcrD between plate dilution and MIC (antibiotic gradient) assays were congruent, except for cephalotin where resistance was not observed in the plate dilution assay, whereas the antibiotic gradient test showed a clear increase in MIC in AcrAD-TolC producing cells (Figure 1, Table 1).

Amongst the antimicrobials for which AcrD conferred less susceptibility, lipophilicity showed a wide distribution on the basis of the logarithm of the partition coefficient, logP [34]. St_AcrD can not only transport very hydrophilic β-lactams such as aztreonam but also the very hydrophobic antimicrobial fusidic acid. Both substrates have the presence of a carboxylic acid moiety as a common feature [25]. As an exception, piperacillin was not transported by St_AcrD despite its low logP. Piperacillin exhibits the largest minimal projection area (MPA) among the tested substrates such that it can be assumed that this characteristic might hamper its transport [35]. However, the latter property is not an impediment for the substrate to be exported by Ec_AcrB (through TM1/TM2 region). Furthermore, the volume of the substrate binding pockets of both proteins is expected to be similar and large enough to accommodate the largest reported substrates [24]. It is, hence, probable that additional non-conserved residues in the substrate translocation pathway of St_AcrD might play a substantial role. An additional consideration is the protonation state of the transported compounds at neutral pH. In contrast to AcrB, which preferentially transports monoanionic species of β-lactams, St_AcrD is able to transport monoanionic and dianionic β-lactams. Thus, it might be possible that the substrate pathway of both charged species towards the binding pockets might differ. This has been recently shown for AcrB comprising at least four channels toward the drug binding sites. These channels displayed preferences for substrates on the basis of their physicochemical properties. [21,31].

In order to characterize the role of residues inside substrate binding pockets, we conducted substitution analysis of St_AcrD. Residues that are part of the putative translocation pathway and for the putative substrate binding sites were identified (Figure S2) and substituted by Ala for a systematic characterization. Targeted residues were principally polar or charged, as it was expected that they are involved in electrostatic interactions with the negatively charged substrates. Additionally, several more hydrophobic amino acids were also investigated, including residues in the transmembrane domain, as β-lactams are partially immersed in the outer leaflet of inner membrane of *E. coli* [34].

We observed that substitutions did not only reduce resistance (conferred higher susceptibility) but also indicated hyperactivity for some of the substituted variants (Figure 2). Substrate specificity determinants appear not only to be limited to residues located in the AP as previously reported [21,31] but also include residues from the DBP and the TM1/TM2 region (Figure 3). One of the most surprising observations was related to the variants F609A and I337A, which exhibited a hyperactive phenotype, in strong contrast with its *E. coli* AcrB variant counterparts (F610A and I337A), which were shown to confer increased susceptibility [25,28]. For F609A, the hyperactivity might be in line with a reduction in the hydrophobic binding pocket environment resulting in the enhancement of the transport of hydrophilic substrates. For I337A, which shows a moderate overproduction compared to wildtype (Figure 2), the difference in phenotype between AcrB and AcrD is less straightforward to interpret. These results also emphasize the importance of studying single substitution variants of other RND pumps and directly comparing their effects (rather than extrapolate from one homolog RND pump to another), as AcrB and AcrD same-site substitution variants display different effects on susceptibility. This study emphasizes that substitutions of homologue residues can even result in opposite observations, as is shown for the St_AcrD F609A and I337A variants.

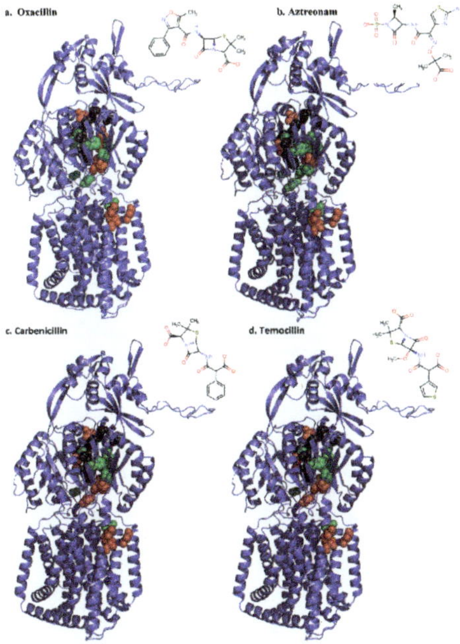

Figure 3. Putative substrate translocation pathway of the RND transporter St_AcrD. Homology model of St_AcrD localizing selected residues in DBP, AP and TM1 and TM2 regions involved in β-lactam transport. Ala-variants of selected residues exhibited hyperactive (green spheres), defective (red spheres) or similar phenotype (grey spheres) compared with wildtype St_AcrD in the presence of (**a**) oxacillin, (**b**) aztreonam, (**c**) carbenicillin and (**d**) temocillin. Figures were prepared with Pymol (https://pymol.org/, accessed on 9 May 2018).

As expected, some substitution variants of originally charged polar residues such as N136A and D276A in DBP as well as R625A in the AP exhibited a sensitive phenotype and might, therefore, be involved in anionic β-lactam export. Moreover, the I27A mutants located in TM1/TM2 and Y327A in the DBP (except for aztreonam) were unable to complement the *E. coli*-susceptible phenotype in the presence of tested compounds (Figures 2 and 3). This is opposed to previous suggestions that aromatic residues in DBP would not influence specificity towards anionic β-lactams [21]. Nevertheless, the effect of the substitution can also be due to secondary structural effects since the geometry of adjacent residues might be influenced by altered physicochemical properties. We also observed a high number of single-site variants displaying a hyperactive phenotype, which in some cases could be associated with increased protein production in the cell (Figure 2).

The effect of overproduction, i.e., higher number of RND molecules might be directly correlated to the observed activity. Nevertheless, in cases such as T139A, S180A and Y178A, production levels were much lower compared to WT expression, yet these variants confer a considerable reduced susceptibility towards some of the antibiotics tested. Future studies should establish the cause of variable production of the variant proteins in the cell and whether the observed results are correlated with correct insertion in the inner membrane (i.e., as a well-folded and active transporter).

Although this study focused on β-lactam specificity, the presented system and constructed Ala-variants are potentially available for investigating additional substrates and inhibitors. The identification of key residues and interactions involved in substrate recognition along the translocation pathway, including the switch loop [29,36] and the most proximal residues in the AP, would help to understand the determinants of polyspecificity of RND multidrug transporters and the mechanisms by which substrate uptake occurs.

4. Materials and Methods

4.1. Site-Directed Mutagenesis

Site-directed mutagenesis of the *St_acrD* gene to produce Ala-variants constructs was performed by inverse PCR with 2 ng of p7XC3H-St_AcrD and forward/reverse primers (0.5 µM each) (Table S2) Phusion Flash High Fidelity Mix (containing modified Phusion Hot Start II Polymerase, Thermo Fisher, Dreieich, Germany) using an initial denaturation step for 2 min at 98 °C, followed by 30 cycles of denaturation/annealing/extension at 98 °C/65 °C→50 °C (touchdown)/72 °C for 20/30/130 s and a final extension step for 10 min at 72 °C. The resulting PCR product was digested with 10 U *DpnI* in 1× Fast Digest buffer for 2 h at 37 °C, followed by enzyme inactivation at 80 °C for 20 min. The PCR product was purified with the DNA Clean and Concentrator Kit (Zymogen, Freiburg, Germany) and employed in a simultaneous phosphorylation and ligation reaction: 10 µL of PCR product, 1 U T4 ligase, 1× T4 ligase buffer and 10 U T4 polynucleotide kinase. The reaction was incubated o/n at room temperature and later employed to transform *E. coli* MC1061 chemically competent cells. Single colonies were selected for o/n cultures and plasmid isolation with QIAprep Spin Miniprep Kit (QIAGEN, Hilden, Germany).

4.2. Antimicrobial Susceptibility Testing: Plate Dilution Assay and MIC Determination

Freshly transformed *E. coli* BW25113 (DE3) Δ*acrB*Δ*acrD* cells with expression constructs p7XC3H to produce St_AcrD or its Ala-variants were used to inoculate 2 mL of LB supplemented with 50 µg/mL kanamycin. Cultures were incubated o/n at 37 °C and 180 rpm. On the next morning, cultures were normalized to OD_{600} = 1 and serially diluted from 10^{-1} to 10^{-6} in 96-well plates and kept at 4 °C until use. LB-agar plates containing the tested antibiotics were prepared by dissolving the antibiotics (1, 1.5 and 2 µg/mL oxacillin; 3, 5 and 7 µg/mL carbenicillin; 20, 25 and 30 µg/mL temocillin; and 0.05, 0.07 and 0.1 µg/mL aztreonam) and 10 µM IPTG solutions into hand-warmed LB-agar supplemented with 50 µg/mL kanamycin. Medium was poured in single well Omnitray Plate (Thermo Fisher, Dreieich, Germany). A volume of 4 µL of the cell suspension dilutions previously prepared was spotted on LB-agar-antibiotic plates. Once the drops were dry, plates were incubated 16 h at 37 °C. On the next day, the results were evaluated visually, and images were taken with Image Quant LAS4000 Imager (GE Healthcare, Solingen, Germany). The determination of the minimum inhibitory concentration (MIC) was performed by employing antibiotic gradient strips (Liofilchem, Roseto degli Abruzzi, Italy). Mueller–Hinton plates supplemented with 50 µg/mL kanamycin and 10 µM IPTG were inoculated with a fresh bacterial suspension equivalent to a 0.5 McFarland standard. After application of antibiotic gradient strips, plates were incubated for 20 h under aerobic conditions at 36 ± 1 °C, and MIC was determined according to the manufacturer's recommendation (at crossing point of bacterial lawn and strip).

Supplementary Materials: The following are available online at https://www.mdpi.com/article/10.3390/antibiotics10121494/s1, Supplementary Materials and Methods, Figure S1. Homology model of St_AcrD, Figure S2. Putative substrate translocation pathway in St_AcrD and localization of the selected side chains for substitution, Figure S3. Plate dilution assay with *E. coli* BW25113 (DE3) Δ*acrB*Δ*acrD* harboring St_AcrD WT or the indicated Ala-substitution variants, Figure S4. Resistance profiles against β-lactams for *E. coli* BW25113 (DE3) Δ*acrB*Δ*acrD* harboring St_AcrD Ala-substitutions in the putative DBP, AP and in the TM1/TM2 region, Figure S5. Resistance profiles against β-lactams for *E. coli* BW25113 (DE3) Δ*acrB*Δ*acrD* harboring St_AcrD Ala-substitutions in the putative DBP (a. and b.) and AP and in TM1/TM2 region (c.), Figure S6. Comparison of DBP of Ec_AcrB (PDB: 4DX7) and St_AcrD (homology model), Figure S7. Production of St_AcrD WT and Ala-variants in *E. coli* BW25113 (DE3) Δ*acrB*Δ*acrD*, Figure S8. *E. coli* strains and inducer concentration screening to produce RND-GFP fusion proteins, Figure S9. *In gel fluorescence* and anti-His Western blot detection of RND-GFP fusion proteins produced in *E. coli*, Figure S10. Purification of St_AcrD, Figure S11. Oligomeric state of St_AcrD; Table S1: Comparison of identities and similarities of RND multidrug transporters of *E. coli*, *C. jejuni* and *S. typhimurium*, Table S2: Primers used for FX-cloning and site-directed mutagenesis, Table S3. Multidrug RND transporters from *Campylobacter jejuni* and *Salmonella typhimurium*, Table S4. Cloning and expression constructs of RND transporters produced by FX-cloning, Table S5. Summary of optimized conditions to produce RND-GFP fusion proteins in *E. coli*.

Author Contributions: Conceptualization, J.C.B. and K.M.P.; data curation, J.C.B. and S.G.; formal analysis, J.C.B., S.G. and K.M.P.; funding acquisition, K.M.P.; investigation, J.C.B., J.E.-D. and S.G.; methodology, J.C.B. and J.E.-D.; project administration, K.M.P.; resources, K.M.P.; supervision, K.M.P.; validation, J.C.B. and S.G.; visualization, J.C.B.; writing—original draft, J.C.B. and K.M.P.; writing—review and editing, J.C.B., S.G. and K.M.P. All authors have read and agreed to the published version of the manuscript.

Funding: This work was funded by the DFG-SFB807 "Transport and Communication across Membranes". J.C.B. was a Marie Skłodowska-Curie fellow within the "Translocation" Network, project no. 607694.

Data Availability Statement: The data presented in this study are available upon request from the corresponding author.

Conflicts of Interest: The authors declare no conflict of interest.

References

1. Li, X.Z.; Nikaido, H. Efflux-mediated drug resistance in bacteria: An update. *Drugs* **2009**, *69*, 1555–1623. [CrossRef]
2. Li, X.Z.; Plesiat, P.; Nikaido, H. The challenge of efflux-mediated antibiotic resistance in Gram-negative bacteria. *Clin. Microbiol. Rev.* **2015**, *28*, 337–418. [CrossRef]
3. Colclough, A.L.; Alav, I.; Whittle, E.E.; Pugh, H.L.; Darby, E.M.; Legood, S.W.; McNeil, H.E.; Blair, J.M. RND efflux pumps in Gram-negative bacteria; regulation, structure and role in antibiotic resistance. *Future Microbiol.* **2020**, *15*, 143–157. [CrossRef] [PubMed]
4. Alav, I.; Kobylka, J.; Kuth, M.S.; Pos, K.M.; Picard, M.; Blair, J.M.A.; Bavro, V.N. Structure, assembly, and function of tripartite efflux and type 1 secretion systems in gram-negative bacteria. *Chem. Rev.* **2021**, *121*, 5479–5596. [CrossRef] [PubMed]
5. Alav, I.; Sutton, J.M.; Rahman, K.M. Role of bacterial efflux pumps in biofilm formation. *J. Antimicrob. Chemother.* **2018**, *73*, 2003–2020. [CrossRef]
6. Kvist, M.; Hancock, V.; Klemm, P. Inactivation of efflux pumps abolishes bacterial biofilm formation. *Appl. Environ. Microbiol.* **2008**, *74*, 7376–7382. [CrossRef] [PubMed]
7. Yang, S.; Lopez, C.R.; Zechiedrich, E.L. Quorum sensing and multidrug transporters in *Escherichia coli*. *Proc. Natl. Acad. Sci. USA* **2006**, *103*, 2386–2391. [CrossRef]
8. Thanassi, D.G.; Cheng, L.W.; Nikaido, H. Active efflux of bile salts by *Escherichia coli*. *J. Bacteriol.* **1997**, *179*, 2512–2518. [CrossRef]
9. Wang-Kan, X.; Blair, J.M.A.; Chirullo, B.; Betts, J.; La Ragione, R.M.; Ivens, A.; Ricci, V.; Opperman, T.J.; Piddock, L.J.V. Lack of AcrB efflux function confers loss of virulence on *Salmonella enterica* Serovar Typhimurium. *mBio* **2017**, *8*, e00968-17. [CrossRef]
10. El Meouche, I.; Dunlop, M.J. Heterogeneity in efflux pump expression predisposes antibiotic-resistant cells to mutation. *Science* **2018**, *362*, 686–690. [CrossRef]
11. Nikaido, H. RND transporters in the living world. *Res. Microbiol.* **2018**, *169*, 363–371. [CrossRef] [PubMed]
12. Elbourne, L.D.; Tetu, S.G.; Hassan, K.A.; Paulsen, I.T. TransportDB 2.0: A database for exploring membrane transporters in sequenced genomes from all domains of life. *Nucleic Acids Res.* **2017**, *45*, D320–D324. [CrossRef]

13. Klenotic, P.A.; Moseng, M.A.; Morgan, C.E.; Yu, E.W. Structural and functional diversity of resistance-nodulation-cell division transporters. *Chem. Rev.* **2021**, *121*, 5378–5416. [CrossRef] [PubMed]
14. Nishino, K.; Latifi, T.; Groisman, E.A. Virulence and drug resistance roles of multidrug efflux systems of *Salmonella enterica* serovar Typhimurium. *Mol. Microbiol.* **2006**, *59*, 126–141. [CrossRef] [PubMed]
15. McClelland, M.; Sanderson, K.E.; Spieth, J.; Clifton, S.W.; Latreille, P.; Courtney, L.; Porwollik, S.; Ali, J.; Dante, M.; Du, F.; et al. Complete genome sequence of *Salmonella enterica* serovar Typhimurium LT2. *Nature* **2001**, *413*, 852–856. [CrossRef] [PubMed]
16. Eaves, D.J.; Ricci, V.; Piddock, L.J. Expression of acrB, acrF, acrD, marA, and soxS in *Salmonella enterica* serovar Typhimurium: Role in multiple antibiotic resistance. *Antimicrob. Agents Chemother.* **2004**, *48*, 1145–1150. [CrossRef] [PubMed]
17. Buckner, M.M.; Blair, J.M.; La Ragione, R.M.; Newcombe, J.; Dwyer, D.J.; Ivens, A.; Piddock, L.J. Beyond antimicrobial resistance: Evidence for a distinct role of the AcrD Efflux pump in *Salmonella* biology. *mBio* **2016**, *7*, e01916-16. [CrossRef] [PubMed]
18. Piddock, L.J. Clinically relevant chromosomally encoded multidrug resistance efflux pumps in bacteria. *Clin. Microbiol. Rev.* **2006**, *19*, 382–402. [CrossRef]
19. Blair, J.M.; Smith, H.E.; Ricci, V.; Lawler, A.J.; Thompson, L.J.; Piddock, L.J. Expression of homologous RND efflux pump genes is dependent upon AcrB expression: Implications for efflux and virulence inhibitor design. *J. Antimicrob. Chemother.* **2015**, *70*, 424–431. [CrossRef]
20. Yamasaki, S.; Nagasawa, S.; Hayashi-Nishino, M.; Yamaguchi, A.; Nishino, K. AcrA dependency of the AcrD efflux pump in *Salmonella enterica* serovar Typhimurium. *J. Antibiot.* **2011**, *64*, 433–437. [CrossRef]
21. Kobayashi, N.; Tamura, N.; van Veen, H.W.; Yamaguchi, A.; Murakami, S. beta-Lactam selectivity of multidrug transporters AcrB and AcrD resides in the proximal binding pocket. *J. Biol. Chem.* **2014**, *289*, 10680–10690. [CrossRef] [PubMed]
22. Aires, J.R.; Nikaido, H. Aminoglycosides are captured from both periplasm and cytoplasm by the AcrD multidrug efflux transporter of *Escherichia coli*. *J. Bacteriol.* **2005**, *187*, 1923–1929. [CrossRef]
23. Rosenberg, E.Y.; Ma, D.; Nikaido, H. AcrD of *Escherichia coli* is an aminoglycoside efflux pump. *J. Bacteriol.* **2000**, *182*, 1754–1756. [CrossRef] [PubMed]
24. Ramaswamy, V.K.; Vargiu, A.V.; Malloci, G.; Dreier, J.; Ruggerone, P. Molecular rationale behind the differential substrate specificity of bacterial RND multi-drug transporters. *Sci. Rep.* **2017**, *7*, 8075. [CrossRef] [PubMed]
25. Oswald, C.; Tam, H.K.; Pos, K.M. Transport of lipophilic carboxylates is mediated by transmembrane helix 2 in multidrug transporter AcrB. *Nat. Commun.* **2016**, *7*, 13819. [CrossRef]
26. Rahman, M.M.; Matsuo, T.; Ogawa, W.; Koterasawa, M.; Kuroda, T.; Tsuchiya, T. Molecular cloning and characterization of all RND-type efflux transporters in *Vibrio cholerae* non-O1. *Microbiol. Immunol.* **2007**, *51*, 1061–1070. [CrossRef] [PubMed]
27. Wiegand, I.; Hilpert, K.; Hancock, R.E. Agar and broth dilution methods to determine the minimal inhibitory concentration (MIC) of antimicrobial substances. *Nat. Protoc.* **2008**, *3*, 163–175. [CrossRef] [PubMed]
28. Bohnert, J.A.; Schuster, S.; Seeger, M.A.; Fahnrich, E.; Pos, K.M.; Kern, W.V. Site-directed mutagenesis reveals putative substrate binding residues in the *Escherichia coli* RND efflux pump AcrB. *J. Bacteriol.* **2008**, *190*, 8225–8229. [CrossRef]
29. Eicher, T.; Cha, H.J.; Seeger, M.A.; Brandstätter, L.; El-Delik, J.; Bohnert, J.A.; Kern, W.V.; Verrey, F.; Grutter, M.G.; Diederichs, K.; et al. Transport of drugs by the multidrug transporter AcrB involves an access and a deep binding pocket that are separated by a switch-loop. *Proc. Natl. Acad. Sci. USA* **2012**, *109*, 5687–5692. [CrossRef]
30. Lu, W.; Zhong, M.; Chai, Q.; Wang, Z.; Yu, L.; Wei, Y. Functional relevance of AcrB Trimerization in pump assembly and substrate binding. *PLoS ONE* **2014**, *9*, e89143. [CrossRef]
31. Tam, H.K.; Foong, W.E.; Oswald, C.; Herrmann, A.; Zeng, H.; Pos, K.M. Allosteric drug transport mechanism of multidrug transporter AcrB. *Nat. Commun.* **2021**, *12*, 3889. [CrossRef] [PubMed]
32. Geertsma, E.R.; Groeneveld, M.; Slotboom, D.J.; Poolman, B. Quality control of overexpressed membrane proteins. *Proc. Natl. Acad. Sci. USA* **2008**, *105*, 5722–5727. [CrossRef] [PubMed]
33. Ruggerone, P.; Murakami, S.; Pos, K.M.; Vargiu, A.V. RND efflux pumps: Structural information translated into function and inhibition mechanisms. *Curr. Top. Med. Chem.* **2013**, *13*, 3079–3100. [CrossRef] [PubMed]
34. Nikaido, H.; Basina, M.; Nguyen, V.; Rosenberg, E.Y. Multidrug efflux pump AcrAB of *Salmonella typhimurium* excretes only those beta-lactam antibiotics containing lipophilic side chains. *J. Bacteriol.* **1998**, *180*, 4686–4692. [CrossRef] [PubMed]
35. Cha, H.J.; Muller, R.T.; Pos, K.M. Switch-loop flexibility affects transport of large drugs by the promiscuous AcrB multidrug efflux transporter. *Antimicrob. Agents Chemother.* **2014**, *58*, 4767–4772. [CrossRef]
36. Nakashima, R.; Sakurai, K.; Yamasaki, S.; Nishino, K.; Yamaguchi, A. Structures of the multidrug exporter AcrB reveal a proximal multisite drug-binding pocket. *Nature* **2011**, *480*, 565–569. [CrossRef] [PubMed]

Article

Molecular Determinants for OMF Selectivity in Tripartite RND Multidrug Efflux Systems

Esther Boyer [1], Jean Dessolin [1], Margaux Lustig [2], Marion Decossas [1], Gilles Phan [2], Quentin Cece [2], Grégory Durand [3], Véronique Dubois [4], Joris Sansen [1], Jean-Christophe Taveau [1], Isabelle Broutin [2], Laetitia Daury [1,*] and Olivier Lambert [1,*]

[1] CBMN UMR 5248, Bordeaux INP, CNRS, Université de Bordeaux, 33600 Pessac, France; esther.boyer@u-bordeaux.fr (E.B.); jean.dessolin@u-bordeaux.fr (J.D.); m.decossas@cbmn.u-bordeaux.fr (M.D.); joris.sansen@u-bordeaux.fr (J.S.); jc.taveau@cbmn.u-bordeaux.fr (J.-C.T.)

[2] Laboratoire CiTCoM, CNRS, Université de Paris, 75006 Paris, France; m.lustig93@gmail.com (M.L.); gilles.phan@u-paris.fr (G.P.); cece.quentin@cnrs.fr (Q.C.); isabelle.broutin@u-paris.fr (I.B.)

[3] Unité Propre de Recherche et d'Innovation, Université d'Avignon, Equipe S2CB, 84916 Avignon, France; gregory.durand@univ-avignon.fr

[4] MFP, UMR 5234, CNRS, Univiversité de Bordeaux, 33000 Bordeaux, France; veronique.dubois@u-bordeaux.fr

* Correspondence: daury@enscbp.fr (L.D.); olivier.lambert@u-bordeaux.fr (O.L.)

Citation: Boyer, E.; Dessolin, J.; Lustig, M.; Decossas, M.; Phan, G.; Cece, Q.; Durand, G.; Dubois, V.; Sansen, J.; Taveau, J.-C.; et al. Molecular Determinants for OMF Selectivity in Tripartite RND Multidrug Efflux Systems. *Antibiotics* **2022**, *11*, 126. https://doi.org/10.3390/antibiotics11020126

Academic Editor: Khondaker Miraz Rahman

Received: 29 December 2021
Accepted: 15 January 2022
Published: 18 January 2022

Publisher's Note: MDPI stays neutral with regard to jurisdictional claims in published maps and institutional affiliations.

Copyright: © 2022 by the authors. Licensee MDPI, Basel, Switzerland. This article is an open access article distributed under the terms and conditions of the Creative Commons Attribution (CC BY) license (https://creativecommons.org/licenses/by/4.0/).

Abstract: Tripartite multidrug RND efflux systems made of an inner membrane transporter, an outer membrane factor (OMF) and a periplasmic adaptor protein (PAP) form a canal to expel drugs across Gram-negative cell wall. Structures of MexA–MexB–OprM and AcrA–AcrB–TolC, from *Pseudomonas aeruginosa* and *Escherichia coli*, respectively, depict a reduced interfacial contact between OMF and PAP, making unclear the comprehension of how OMF is recruited. Here, we show that a Q93R mutation of MexA located in the α-hairpin domain increases antibiotic resistance in the MexA$_{Q93R}$–MexB–OprM-expressed strain. Electron microscopy single-particle analysis reveals that this mutation promotes the formation of tripartite complexes with OprM and non-cognate components OprN and TolC. Evidence indicates that MexA$_{Q93R}$ self-assembles into a hexameric form, likely due to interprotomer interactions between paired R93 and D113 amino acids. C-terminal deletion of OprM prevents the formation of tripartite complexes when mixed with MexA and MexB components but not when replacing MexA with MexA$_{Q93R}$. This study reveals the Q93R MexA mutation and the OprM C-terminal peptide as molecular determinants modulating the assembly process efficacy with cognate and non-cognate OMFs, even though they are outside the interfacial contact. It provides insights into how OMF selectivity operates during the formation of the tripartite complex.

Keywords: antibiotic resistance; efflux pump; RND

1. Introduction

In Gram-negative bacteria, tripartite systems of the resistance nodulation cell division (RND) superfamily are multidrug efflux systems contributing to antibiotic resistance by exporting biological metabolites and antimicrobial compounds [1–3]. These systems are composed of an inner-membrane RND transporter driven by the proton motive force, an outer-membrane factor (OMF), and a periplasmic adaptor protein (PAP) which connects the RND transporter to OMF, therefore, forming a tripartite complex with a contiguous exit duct. The assembly of these exporting systems is an important step to achieve the functional efflux process. Deciphering the assembly mechanism is a prerequisite in the development of blockers of tripartite systems that would restore the efficiency of the existing therapeutic arsenal [4].

While PAP and RND transporters encoded by the same operon operate in pairs, the rules governing the interactions of PAP with the OMF appear less restrictive [5,6].

Indeed, different PAPs are able to bind a single OMF, e.g., TolC or OprM. In *Escherichia coli* (*E. coli*), TolC can function with different couples of PAP-RND transporters but also for PAP-Major facilitator superfamily (MFS) transporters and PAP-ATP-binding cassette (ABC) transporters. In *Pseudomonas aeruginosa* (*P. aeruginosa*), OprM can interact with seven of the twelve PAP-RND systems including MexA-MexB, MexC-MexD, MexE-MexF, MexX-MexY [7–10]. This versatility of interaction does not strictly apply to OMFs. One PAP can also couple more than one OMF. MexA-MexB is functional with OprM, and partially with OprJ [11,12], MexE-MexF with OprN and OprM [9], and MexX-MexY with OprM and OprA [13]. Intra- and inter-species interchangeability of components has been also observed [14–16]. However, this component exchange is not representative of all tripartite systems and for several other OMFs, a strict selectivity of assembly seems to operate, as for OprN that interacts only with MexE-MexF [9]. Because of this duality of selectivity and promiscuity, it remains unclear how PAPs achieve to recognize and assemble with OMFs and what are the structural determinants governing the selection of OMF by PAP.

Recent cryo-electron microscopy (cryo-EM) studies of *E. coli* AcrA–AcrB–TolC and MexA–MexB–OprM tripartite complexes have shown overall similar architectures of six PAPs surrounding one RND trimer and in a tip-to-tip interaction with the OMF, which is in an open state (Figure 1) [17–19]. The six periplasmic helix-turn-helix of OMF face six PAP α-hairpins, involving mainly backbone H-bond contacts. In these tripartite complexes, the OMF–PAP arrangement exhibits a reduced interfacial contact that contradicts previous biochemical and functional data [20–27], predicting a strong binding surface between the α-hairpin domain of PAP and OMF in favor of a deep-interpenetration model [28]. Interestingly gain-of-function mutants that enable non-functional chimeric efflux pumps to function have been used to identify key residues involved in the PAP–OMF assembly. Evidence of adaptative mutations far away from the tip region of the α-hairpins of AcrA, MexA (i.e., MexA$_{Q93R}$), and *Vibrio Cholerae* VceA provide a gain of function for the chimeric AcrA–MexB–TolC, MexA–MexB–OprN, and VceA–VceB–OprM pumps [15,22,29]. Likewise, to adapt TolC to MexA–MexB, mutations that are not located at the tip of the coiled-coil domain of TolC provided a gain of function [21]. The role of these mutations which are not located in the tip-to-tip OMF–PAP contact is questioning the mechanisms of OMF recruitment in the assembly process and requires further investigations.

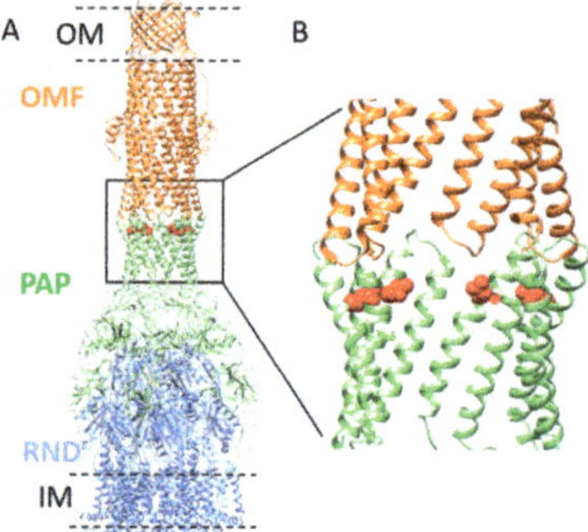

Figure 1. Model of MexA–MexB–OprM tripartite complex and position of Q93 residue in α-hairpin of MexA. (**A**) Model representation of OprM–MexA–MexB tripartite complex (PDB: 6TA5) showing OprM (OMF component) trimer (colored in orange) and MexB (RND component) trimer (colored in blue) connected by MexA (PAP component) hexamer (colored in green). The outer membrane (OM) and inner membrane (IM) are schematically drawn (black dashed lines). The position of the residue Q93 is shown in red (side chain). The position of V472 (or V455 in mature OprM sequence numbering) corresponding to the C-terminal residue solved in OprM structure is indicated on two protomers in the equatorial domain (black arrows). Residues T473–A485 are not visible in the structure. (**B**) Close-up view of the position of the Q93 residue relative to the tip-to-tip contact between OprM and MexA.

Here, we used the biolayer interferometry approach to investigate the interaction between several OMFs (OprN, TolC, OprM, and variant) and PAPs (MexA and MexA$_{Q93R}$) and electron microscopy (EM) to analyze tripartite complexes in the presence of MexB. We report the reconstitution of tripartite complexes with MexA$_{Q93R}$ and its capability to couple native OprN and TolC.

2. Results

2.1. Analysis of Mexa Binding to OMF by Biolayer Interferometry

A Q93R mutation for MexA (MexA$_{Q93R}$) conferring a gain of function with OprN [29] is located at the α-hairpin but is not described to participate in the tip-to-tip interaction with the OMF (Figure 1).

To decipher the mechanism of action of this mutant, its interaction with various OMFs, i.e., OprM, OprN, TolC, and an OprM variant (OprM$_{\Delta473-485}$) has been analyzed using the biolayer interferometry (BLI) method. Increasing concentrations of MexA$_{wt}$ and MexA$_{Q93R}$ variant were titrated to OMF immobilized by a biotinylated non-ionic amphipol (BNAPol) on a streptavidin biosensor and the association and dissociation were assessed by a shift in wavelength (Figure 2). Loading of BNAPol-OprM was performed under non-saturating concentrations (Supplementary Materials Figure S1).

BLI analysis revealed that k_{off} of MexA$_{wt}$ varied depending on the OMF ligand (Table 1). The value of k_{off}, being also indicative of residence time, suggested that the complex stability followed the order OprM > OprN > OprM$_{\Delta473-485}$ > TolC. Unlike MexA$_{wt}$, MexA$_{Q93R}$ exhibited similar k_{off} values for the four OMFs suggesting that the complex stability was not dependent on the OMF. These results revealed that the OMF binding mechanisms of MexA$_{Q93R}$ and MexA$_{wt}$ were different.

Table 1. Kinetics parameters for the OMF–PAP interaction using biolayer interferometry.

Ligand	Analyte	k_{off} (10^{-3} s^{-1})	k_{on} (10^2 M^{-1}s^{-1})	K_D (μM)
OprM$_{wt}$	MexA$_{wt}$	2.15	1.80	12.0
OprM$_{\Delta473-485}$	MexA$_{wt}$	4.58	1.03	44.0
OprN	MexA$_{wt}$	3.58	1.77	20.0
TolC	MexA$_{wt}$	5.8	1.27	45.8
OprM$_{wt}$	MexA$_{Q93R}$	2.66	0.81	32.9
OprM$_{\Delta473-485}$	MexA$_{Q93R}$	2.63	1.02	25.7
OprN	MexA$_{Q93R}$	2.38	0.88	26.9
TolC	MexA$_{Q93R}$	1.91	1.08	17.8

Data fitting using Langmuir 1:1 model.

2.2. Analysis of Oligomerization State of MexA$_{Q93R}$

Previous data have shown that MexA forms a dimer in solution and a higher oligomeric state in the crystal structure [30–32]. The substitution of a glutamine by an arginine residue in MexA$_{Q93R}$ introduced a charged amino acid that may affect protein–protein interactions. MexA$_{wt}$ and MexA$_{Q93R}$ samples were submitted to size-exclusion chromatography that showed a slight shift between elution profiles, suggesting that MexA$_{Q93R}$ retention was reduced compared with MexA$_{wt}$ (Figure 3A).

EM analysis of fractions corresponding to the MexA$_{Q93R}$ peak fraction revealed complexes regular in size (Figure 3B). The average image from single-average image analysis revealed hexagonal-shaped particles with a diameter of about 8–10 nm which is compatible with a hexameric form (Figure 3B inset). EM analysis of MexA$_{wt}$ peak fraction showed particles heterogeneous in size, reflecting the formation of aggregates when deposited on the grid (Figure 3C). This result provided evidence that MexA$_{Q93R}$ in solution formed an oligomeric form, compatible with a hexamer.

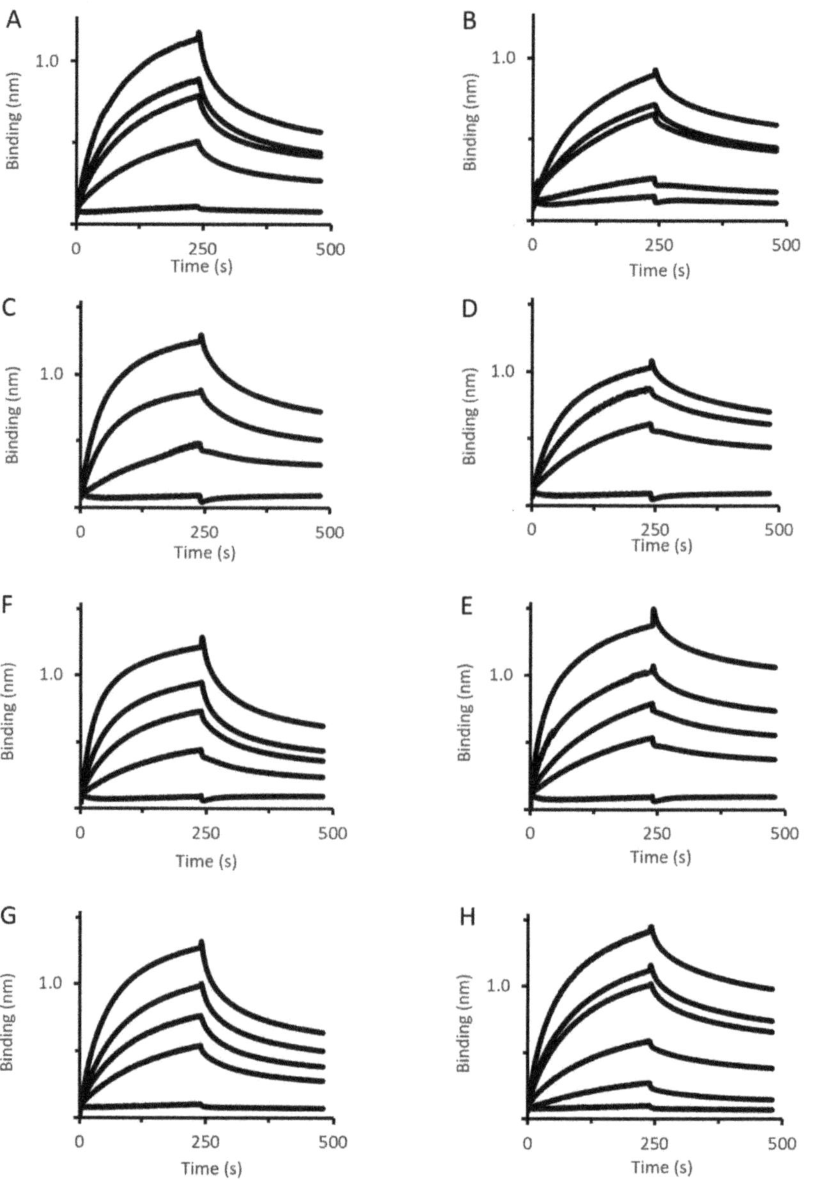

Figure 2. OMF–PAP interactions assessed by BLI. Immobilized BNAPol-OprM (**A**,**B**), BNAPol-OprN (**C**,**D**), BNAPol-TolC (**E**,**F**), BNAPol-OprM$_{\Delta473-485}$ (**G**,**H**) were exposed to different concentrations (from 0 to 100 μM) of MexA$_{wt}$ (left column) or MexA$_{Q93R}$ (right column). Interactions (association and dissociation) were assessed by a wavelength shift (nm). All reactions were performed at room temperature.

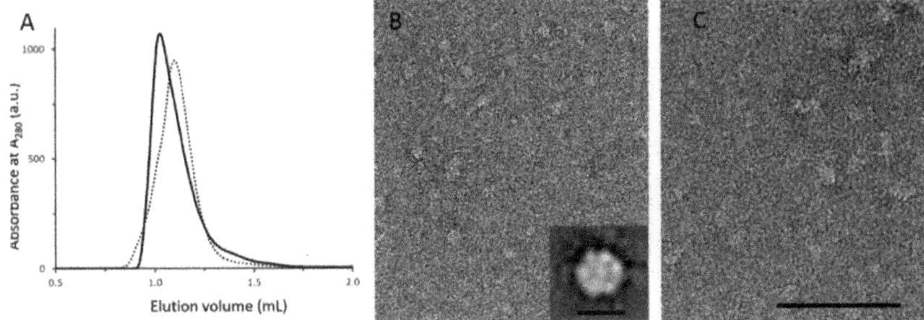

Figure 3. Analytical characterization and EM analysis of MexAQ93R and MexAwt. (**A**) Analytical size-exclusion chromatography (SEC) analysis of MexAQ93R (solid trace) and MexAwt (dotted trace) samples. (**B**) EM analysis of the SEC peak fraction of MexAQ93R exhibiting circular particles. Inset: average image showing a hexagonal-shaped particle with a diameter of about 8–10 nm. Scale bar 10 nm. (**C**) EM analysis of the SEC peak fraction of MexAwt showing heterogenous particles in size compared with (**B**). Scale bar 100 nm.

2.3. Binding Analysis of MexA Variants to MexB Using BLI

Using similar conditions as performed for MexA-OMF binding analysis, various concentrations of $MexA_{wt}$ and $MexA_{Q93R}$ variants were titrated to MexB immobilized by BNAPol on a streptavidin biosensor (Figure 4).

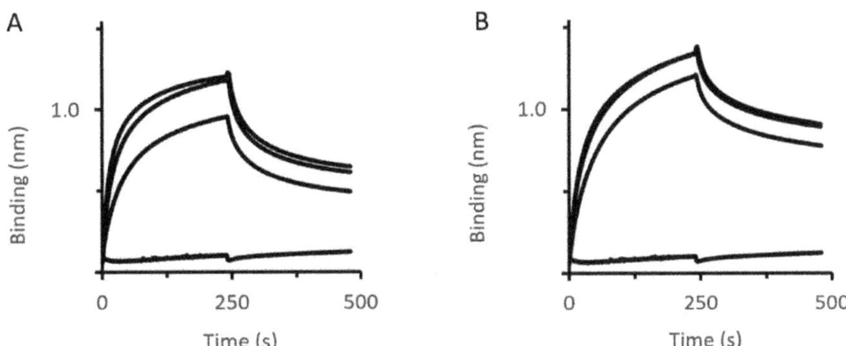

Figure 4. MexB–PAP interactions assessed by BLI. Immobilized BNAPol-MexB were exposed to different concentrations (from 0 to 200 µM) of $MexA_{wt}$ (**A**) or $MexA_{Q93R}$ (**B**). Interactions (association and dissociation) were assessed by a wavelength shift (nm). All reactions were performed at room temperature.

BLI analysis revealed that the complex stability (k_{off} value) was slightly improved with $MexA_{Q93R}$ compared with $MexA_{wt}$ (Table 2). Of note, the k_{off} values were higher than that of OprM–MexA suggesting that the MexA–MexB complex was less stable than the MexA–OprM complex.

Table 2. Kinetics parameters for the OMF–PAP interaction using biolayer interferometry.

Ligand	Analyte	k_{off} (10^{-3} s^{-1})	k_{on} (10^2 M^{-1}s^{-1})	K_D (µM)
MexB	$MexA_{wt}$	5.5	2.50	23.0
MexB	$MexA_{Q93R}$	3.0	1.73	17.4

Data fitting using Langmuir 1:1 model.

2.4. Impact of MexA$_{Q93R}$ on the Formation of Tripartite Complexes

According to the BLI experiments, the Q93R mutation for MexA dramatically changed its interaction with various OMFs, we, therefore, evaluated its impact on the formation of tripartite complexes. The four OMFs (OprM, OprN, TolC, OprM$_{\Delta473-485}$) and MexB stabilized in nanodiscs were mixed with MexA$_{wt}$ or MexA$_{Q93R}$ proteins following the method previously described [19,33]. The formation of tripartite complexes was assessed by the presence of elongated complexes observed by negative-staining EM and 2D class averaging (Figure 5 and Supplementary Materials Table S1).

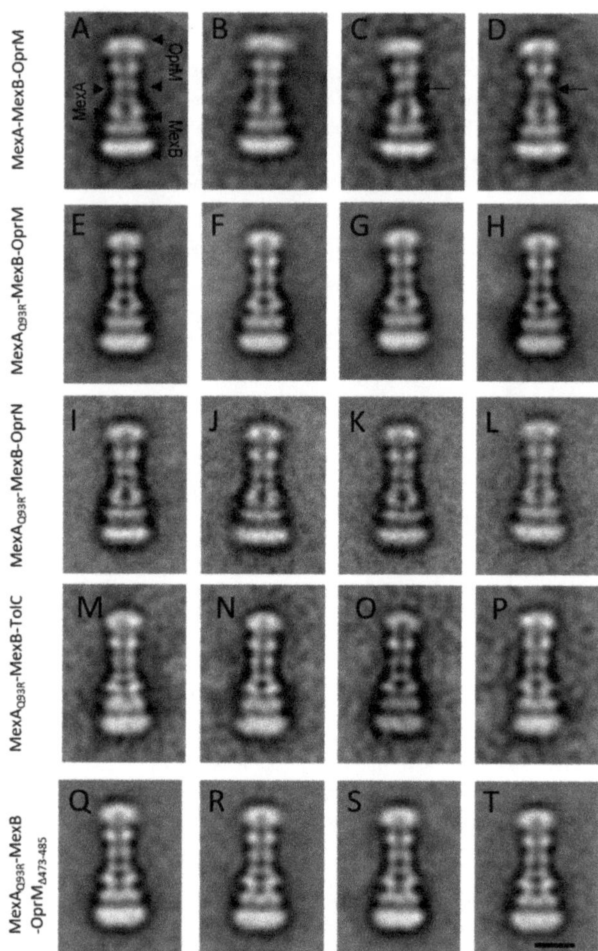

Figure 5. Single-particle analysis of tripartite complexes. Representative 2D classes of tripartite complexes MexA–MexB–OprM and derivatives observed by negative-staining EM. (**A–D**) MexA–MexB–OprM complexes. Typical classes (**A,B**) showing a continuous canal between OprM and MexA. Atypical classes (**C,D**) exhibiting a faint contact between OprM and MexA (back arrows). (**E–H**) MexA$_{Q93R}$–MexB–OprM complexes. (**I–L**) MexA$_{Q93R}$–MexB–OprN complexes. (**M–P**) MexA$_{Q93R}$–MexB–TolC complexes. (**Q–T**) MexA$_{Q93R}$–MexB–OprM$_{\Delta473-485}$ complexes. Note that when formed with MexA$_{Q93R}$, tripartite complexes exhibited an open coupled OMF whatever the considered class, unlike MexA$_{wt}$ for which several classes presented closed coupled OMF. Scale bar 10 nm.

For OprN, TolC, and OprM$_{\Delta473-485}$, tripartite complexes were formed with MexA$_{Q93R}$ while no complex was observed with MexA$_{wt}$ (Figure 5E–T). For OprM, tripartite complexes were observed with both MexA$_{wt}$ and MexA$_{Q93R}$ (Figure 5A–H). The overall architecture of these complexes was similar to that described previously [33]. The OprM facing the MexA–MexB complex with no direct contact between OprM and MexB was further resolved in a tip-to-tip interaction with MexA on the cryo-EM structure [19]. The formation of hybrid (non-cognate) OprN–MexA$_{Q93R}$–MexB complexes was in good agreement with in vivo experiments reporting a gain of function with the MexA$_{Q93R}$ variant [29]. The formation of hybrid TolC-MexA$_{Q93R}$-MexB complexes showed that the Q93R mutation for MexA extended its interaction with TolC without the need of changing any residue at the tip-to-tip interface. Note that few atypical 2D classes of tripartite MexA$_{wt}$–MexB–OprM complexes showed a faint contact between MexA and OprM (Figure 5C,D) probably as previously observed [33]. No such classes were encountered when tripartite complexes were generated with MexA$_{Q93R}$ suggesting that the complexes were more stable on EM grids.

The number of tripartite complexes has been evaluated from the micrographs and reported in Table 3. The formation of a higher number of OprM–MexA$_{Q93R}$–MexB complexes compared to OprM–MexA$_{wt}$–MexB suggested that these tripartite complexes were assembled in a more efficient manner with MexA$_{Q93R}$. This was also correlated by in vivo experiments where the minimal inhibitory concentration (MIC) values of ticarcillin and aztreonam for cells expressing OprM–MexA$_{Q93R}$–MexB were twofold and fourfold higher than for those expressing native OprM–MexA$_{wt}$–MexB (Table 4). Overall, the formation of tripartite complexes with MexA$_{Q93R}$ was significantly improved compared to MexA$_{wt}$, suggesting that MexA$_{Q93R}$ had greater capabilities than MexA$_{wt}$ to form tripartite complexes with OprM and other OMFs.

Table 3. Estimation of tripartite complexes amount from electron microscopy fields.

	PAP	
OMF	**MexAwt**	**MexA$_{Q93R}$**
OprM$_{wt}$	1146 ± 59	1981 ± 156 *a
OprM$_{\Delta473-485}$	0	589 ± 15 *b, **c
OprN	0	10 ± 0.3 **b
TolC	0	164 ± 3 **b

Complexes were counted from 3 sets of 50 micrographs. Data are the mean ± sem. Student's test significantly different (* $0.01 < p < 0.05$; ** $0.001 < p < 0.01$). [a] Compares MexA–MexB–OprM with MexA$_{Q93R}$–MexB–OprM; [b] compares MexA$_{Q93R}$–MexB–OMF with MexA$_{Q93R}$–MexB–OprM; [c] compares MexA$_{wt}$–MexB–OprM with MexA$_{Q93R}$–MexB–OprM$_{\Delta473-485}$.

Table 4. Antimicrobial susceptibility of cells expressing MexA variants.

	Minimal Inhibitory Concentration (MIC, µg/mL)	
Strain	**Ticarcillin**	**Aztreonam**
PAO1	32	4
PAO1 pUCP24-*mexAB-oprM wt*	64	8
PAO1 pUCP24-*mexA* $_{D113A}$ *mexB-oprM*	32	4
PAO1 pUCP24-*mexA* $_{Q93R}$ *mexB-oprM*	128	32
PAO1 pUCP24-*mexA* $_{D113A + Q93R}$ *mexB-oprM*	32	4

2.5. Impact of an OprM Variant on the Formation of Tripartite Complexes

In the assembly process of the tripartite complex, the OMF undergoes an important conformational change to achieve a tip-to-tip interaction with the PAP. The OMF switches from a closed state to an open state by the opening of its periplasmic helices [17–19]. Therefore, the OMF recruitment and its opening by periplasmic helices movement are two events that imply intricate interactions with PAP for which molecular details are missing.

A C-terminal-deleted mutation for OprM ($OprM_{\Delta473-485}$) was used to understand by which mechanism $MexA_{Q93R}$ promotes the assembly of tripartite complexes. $OprM_{\Delta473-485}$ did not allow the production of tripartite complexes with $MexA_{wt}$-MexB (Table 3). The inability of $MexA_{wt}$ to form a tripartite complex with $OprM_{\Delta473-485}$ correlated well with BLI experiments showing that $MexA_{wt}$ had a higher k_{off} value for $OprM_{\Delta473-485}$ than for $OprM_{wt}$, therefore exhibiting lower binding affinities (Table 1). These results showed that OprM 13 amino acids C-terminal peptide was of importance for $MexA_{wt}$ binding affinity and suggested that a reduced affinity for OprM likely impaired its recruitment, and consequently, jeopardized the formation of tripartite complexes.

By replacing $MexA_{wt}$ with $MexA_{Q93R}$, tripartite complexes were formed with $OprM_{\Delta473-485}$ meaning that $MexA_{Q93R}$ was allowed to compensate/overcome this affinity loss due to the lack of OprM C-terminal part. However, the amount of $OprM_{\Delta473-485}$–$MexA_{Q93R}$ –MexB complexes was lower than that of $OprM_{wt}$–$MexA_{Q93R}$–MexB (Table 3) suggesting that despite similar k_{off} values for $OprM_{wt}$ and $OprM_{\Delta473-485}$, $MexA_{Q93R}$ did not permit to fully compensate the lack of C-terminal part of OprM (Table 1). It seemed that $MexA_{Q93R}$ was acting more on stabilizing an OMF–PAP complex than on the recruitment step of OMF that was yet in good accordance with the BLI experiment, showing similar k_{off} values of $MexA_{Q93R}$ for the four OMFs.

2.6. The Increase in Antibiotic Resistance Is Related to a Q93R Mutation When Associated with D113 Residue

In the cryo-EM tripartite complexes, OprM interacts with six MexA molecules, the six α-hairpins of MexA forming a tight helical bundle (Figure 6A). By substituting Q93 neutral residue with R93 charged residue, the latter is closer to the adjacent D113 residue and the distance between side chains (2.96 Å) is compatible with an ionic bond (Figure 6B). Energies associated with the formation of the hexamer of MexA alone have been estimated with SymmDock. Molecular docking predicted hexameric MexA complexes with an energy score in favor of $MexA_{Q93R}$ indicating a better stabilization of the $MexA_{Q93R}$ complex (Table S2). This hexamer was assembled in a tip-to-tip interaction with OprM using PatchDock (Figure 6B). The formation of an interchain electrostatic interaction between D113 and R93 residues provided a clue on how the introduction of an arginine residue contributes to stabilizing a hexameric structure of $MexA_{Q93R}$.

To assess that the residue D113 would act in synergy with R93, an antibiotic susceptibility assay was performed. For that, the *P. aeruginosa* PAO1 strain was transformed with plasmids carrying genes encoding OprM, MexB, and MexA variants. The strain transformed with MexA–MexB–OprM is two-fold more resistant than native PAO1 which could be due to a slight increase in the level of expressed MexA–MexB–OprM system (Table 4). The introduction of the Q93R mutation in MexA resulted in a two-fold increase in the resistance of the complemented strain to ticarcillin and aztreonam. In order to evaluate the importance of the potential hydrogen bond formed between MexA-R93 and MexA-D113 (Figure 6B), the latter was mutated in alanine. Strains that expressed $MexA_{D113A}$–MexB– OprM or $MexA_{D113A + Q93R}$–MexB–OprM showed that the MICs of ticarcillin and aztreonam were two times lower than strains expressing MexA–MexB–OprM (Table 4). This result provided evidence that the pair residues D113 combined with R93 are involved in the increase in antibiotic resistance.

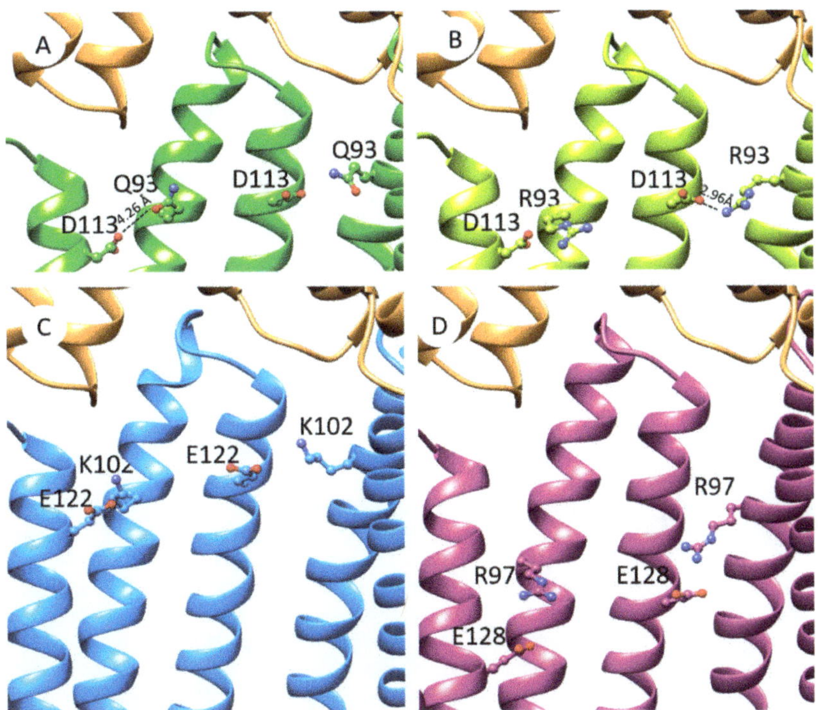

Figure 6. Hypothetic model of PAP interprotomer stabilization mediated by paired anionic-cationic residues in the tip-to-tip PAP–OprM contact. (**A**) Focus on OprM–MexA contact from cryo-EM model of OprM–MexA–MexB (PDB: 6TA5). Positions of residues Q93 and D113 are shown on MexA (colored in green). OprM is colored in orange. (**B**) Model of MexA$_{Q93R}$ (light green) interacting with OprM (orange). Distance between R93 and D113 (2.96 Å) is compatible with interprotomer electrostatic interactions. (**C**) Model of MexX (blue) interacting with OprM with putative interprotomer interactions mediated by side chains of K102–E122 residues. (**D**) Model of MexE (purple) interacting with OprM with putative interprotomer interactions mediated by R97–E128 side chains.

3. Discussion

RND efflux transporters are functional when assembled in tripartite complexes with PAP and OMF partners. Deciphering how they achieve assembly is of importance for medical treatment due to the contribution of these complexes in both multidrug resistance and virulence. With recent advances in elucidating the structure of tripartite assemblies, the OMF and PAP are coupled together via a limited protein–protein interface (so-called tip-to-tip interaction), that still does not permit untangling the tricky knots of OMF selectivity [5,6].

We show that the formation of tripartite complexes coupling OprN, TolC, and OprM$_{\Delta473-485}$ can be achieved with a MexA variant (MexA$_{Q93R}$) while it was not successful with MexA$_{wt}$. The mutated residue is located at the α-hairpin but too far for interacting directly with the OMF. Although this Q93R mutation for MexA did not originate from a pathogenic strain and presents poor clinical importance, it has been selected as a gain-of-function mutant and provides a clue for understanding the assembly process of RND tripartite systems. Indeed, it points out that putative paired anionic and cationic residues (R93, D113) between two adjacent protomers could stabilize the hexameric structure of MexA$_{Q93R}$. A comparative analysis of the amino acid sequences of other PAPs showed that similar couples of residues are present for native PAPs. MexX possesses a putative couple of residues (K102–E122) located at the same position as R93–D113 for MexA (Figure S2). In the absence of a MexX structure, a model has been predicted using the I-TASSER server [34–36] and a C6 hexamer model built with SymmDock [37]. The charged groups of K102 and E122 are at a reasonable proximity to establish electrostatic interactions suggesting that it could be used as an asset for MexX-MexY when forming a tripartite complex with OprM or/and with OprA (Figure 6C). In the MexE sequence, residues R97 and E128 are located in the α-hairpin and could form favorable electrostatic interactions between paired anionic cationic side chains (Figure 6D). Like for MexX, these interprotomer interactions mediated by these two residues may help in the formation of MexE–MexF–OprN or/and MexE–MexF–OprM complexes.

This analysis of tripartite system assembly highlights important molecular determinants for PAP–OMF interaction that are not directly involved in the tip-to-tip interaction. As OMF determinants, we have identified that the C-terminal part was of importance for forming tripartite complexes. The implication of the C-terminal part has been previously reported for functional OprM–MexA–MexB [15,38,39] and TolC–AcrA–AcrB [40,41]. Our results indicate that the deletion of 13 amino acids of the C-terminal end of OprM has a dramatic effect on the formation of tripartite OprM–MexA–MexB complexes. BLI experiments showed that MexA has a reduced affinity for $OprM_{\Delta473-485}$ suggesting that the efficacy of tripartite formation relies on the presence of this C-terminal part. This C-terminal part originates from the equatorial domain but its complete structure has not been solved in both crystal and cryo-EM structures, probably because of high flexibility. It is unlikely that its role in the assembly process occurs at the stage of the tip-to-tip interaction (too short in length) but it might participate directly or indirectly in a transient interaction with MexA, that would occur earlier than the stable tip-to-tip interaction. This transient interaction may help in OprM recruitment by MexA and altering the binding affinity of MexA for OprM decreases the efficacy of tripartite complex formation. Our results did not provide details on the protein interfaces involved in this step. However, biochemical and functional data previously suggested lateral contacts between α-hairpin of PAP and OMF helices and could well fit in an enlarged assembly sequence with transient interactions preceding the tip-to-tip contact.

As a MexA determinant, the Q93R mutation successfully produced tripartite complexes with cognate and non-cognate OMFs. Interestingly, bacteria were less susceptible to antibiotics with $MexA_{Q93R}$ than with $MexA_{wt}$, and the amount of tripartite complexes was increased. This mutation promotes the hexameric organization of MexA mediated by a putative interprotomer ionic bond (Figure 6). During the assembly process, this mutation likely promotes or stabilizes the formation of the six-helix bundle of MexA contacting OprM, which may trigger OprM opening and/or stabilize the tip-to-tip contacts. Improving the efficiency of the opening/stabilization of OMF-PAP in a tip-to-tip contact likely allows compensating for the lack of the C-terminal part for $OprM_{\Delta473-485}$ needed for the previous transient interaction described above. This hypothesis is in good accordance with the previous study on VceA–VceB–OprM complex assembly, reporting on the role of the C-terminal domain of OprM and VceA α-hairpin [15]. In addition, this Q93R mutation extends the capability of MexA to assemble with non-cognate OprN and TolC partners. According to protein–protein docking, they are predicted to interact with a lower energy binding (Table S3). The PAP–OMF interface also imposes an OMF selectivity that can be overcome by reinforcing PAP self-assembly capability.

4. Materials and Methods

4.1. Material and Reagents

1-palmitoyl-2-oleoyl-*sn*-glycero-3-phosphocholine (POPC) was purchased from Avanti Polar Lipids (Alabaster, AL, USA). Sodium cholate hydrate, octyl-β-D-glucopyranoside (OG), and *n*-Dodecyl β-D-maltoside (DDM) were purchased from Sigma-Aldrich (St Louis, MI, USA). SM2 Bio-Beads were obtained from Bio-Rad (Hercules, CA, USA). Superdex 200 PC 3.2/30 and Superose 6 Increase 3.2/300 columns were purchased from Cytivia (Freiburg, Germany). EM grids (Cu 300 mesh) were purchased from Agar Scientific (Stansted, UK).

High precision streptavidin biosensors (SAX) for BLI analysis were purchased from Sartorius (Göttingen, Germany).

4.2. Protein Preparation

Two membrane scaffold proteins, MSP1D1 and MSP1E3D1 (genetic constructs available from AddGene, Cambridge, MA, USA) expressed and purified from bacteria, were used to make OMFs and MexB nanodiscs respectively. Proteins (MexB, MexA$_{wt}$ and MexA$_{Q93R}$, OprM and OprM$_{\Delta473-485}$, OprN and TolC) were expressed and purified from bacteria as previously described [33,42]. After purification, protein buffers contained 1.5% OG for TolC, 0.03% DDM for MexB and 0.05% DDM for MexA, OprN, and OprM.

4.3. Membrane Protein Stabilization with Amphipols

BNAPols (biotinylated non-ionic amphipols) were synthesized by free radical telomerization of an amphiphilic monomer, carrying two glucose moieties and a single undecyl alkyl chain, in the presence of a thiol-based transfer agent bearing a single azido group. The biotin function was subsequently connected to the polymer through a Huisgen cycloaddition as previously described [43]. The BNAPol used in the study had an average molecular mass of ~14.9 kDa and a number-average degree of polymerization of ~20. The extent of grafting of the biotin group was estimated to be ~40% per polymer chain. The membrane protein solution was mixed with BNAPol solution at a 2:1 BNAPol:membrane protein mass ratio for 2 h at 4 °C in a 10 mM Tris/HCl, pH 7.4, 100 mM NaCl 0.01% NaN$_3$, and 0.02% DDM buffer. Detergent was removed by the addition of SM2 Bio-beads with gentle shaking for 3 h at 4 °C. After centrifugation, the mixture was subjected to size-exclusion chromatography (Superdex 200 PC 3.2/30) equilibrated with 10 mM Tris/HCl, pH 7.4, 100 mM, NaCl 0.01% NaN$_3$ buffer at 0.05 mL min^{-1}.

4.4. Binding Analysis Using BLI

Each binding assay was performed with BLItz™ device (ForteBio Inc., Fremont, CA, USA) at room temperature in 10 mM Tris/HCl, pH 7.4 100 mM NaCl 0.01% NaN$_3$, and 0.05% DDM buffer. OMFs and MexB, stabilized into BNAPols, were immobilized on SAX biosensors and exposed to a range of MexA concentrations from 0 to 200 µM. BLItz Pro™ software (version 1.2.1.5, ForteBio Inc. Fremont, CA, USA) was used to fit the curves and to process the data.

4.5. Formation of Tripartite Complexes

POPC lipids were dissolved in chloroform, then dried under vacuum using a rotatory evaporator. The lipid film was suspended in the reconstitution buffer (10 mM Tris/HCl, pH 7.4, 100 mM NaCl) and subjected to 6 rounds of 5′ sonication at 5 watts. Lipid concentration was quantified by phosphate analysis [44].

Tripartite complexes were assembled according to the protocol previously described [33] with slight modifications. Briefly, insertion of OMFs (i.e., OprM, OprN, TolC) in nanodiscs and MexB in nanodiscs using MSP1D1 and MSP1E3D1, respectively, was performed as follows. OMF and MexB solutions were mixed with POPC liposomes and MSP solution at a final lipid/MSP/protein molar ratio of 23:1:0.6 for OMFs (except for TolC, 31:1:2.4) and 32:1:0.5 for MexB in a 10 mM Tris/HCl, pH 7.4, 100 mM NaCl and 15 mM Na-cholate solution. Detergent was removed by the addition of SM2 Bio-Beads into the mixture shaken overnight at 4 °C. Tripartite complexes were assembled in vitro by mixing the OMF and MexB solution with MexA$_{wt}$ or MexA$_{Q93R}$ solution, at a MexA:MexB:OMF ratio of 10:1:1 in 10 mM Tris/HCl, pH 7.4, 100 mM NaCl 0.01% NaN$_3$ and 0.02% DDM buffer. Mixtures were incubated at 20 °C shaking for 7 days before EM grid preparation.

4.6. Analysis of MexA Oligomerisation State

MexA$_{wt}$ and MexA$_{Q93R}$ in purification buffer were subjected to size-exclusion chromatography on a Superdex 200 PC 3.2/30 column, equilibrated with 10 mM Tris/HCl, pH 7.4, 100 mM NaCl 0.01% NaN$_3$ and 0.05% DDM buffer at 0.05 mL min^{-1}.

4.7. Electron Microscopy Acquisition and Image Analysis

For EM grid preparation, a diluted mixture of the sample suspension was deposited on a glow-discharged carbon-coated copper 300 mesh grids and stained with 2% uranyl acetate (w/v) solution. Images were acquired on a Tecnai F20 electron microscope (ThermoFisher Scientific, Waltham, MA, USA)) operated at 200 kV using an Eagle 4k_4k camera (ThermoFisher Scientific). Image alignment and two-dimensional averages were performed with Eman2 [45] using a reference-free alignment procedure. For MexA$_{Q93R}$, MexA-MexB-OprM, MexA$_{Q93R}$-MexB-OprM, MexA$_{Q93R}$-MexB-TolC, and MexA-MexB-OprM$_{\Delta473-485}$, a total of 11,572, 19,260, 46,145, 1191, and 14,025 particles, respectively, were automatically selected and processed for class averaging. For MexA$_{Q93R}$-MexB-OprN, 1236 particles were manually selected and processed like the others. For assessing the occurrence of tripartite complexes, 150 micrographs were randomly collected per grid. The number of complexes was estimated by manual picking on a set of 50 micrographs. The experiment was conducted in triplicate and expressed as the mean and standard error of the mean (sem).

4.8. Model Simulation and Score Evaluation

The SymmDock server [37,46] was used to produce C6 hexamer MexA$_{wt}$ (PDB: 6TA5) and MexA$_{Q93R}$ after mutating Q93 to R93 with Discovery Studio Visualizer (BIOVIA, San Diego, CA, USA). MexA$_{Q93A}$ and MexA$_{D113A}$ hexamers were generated using the same procedure. The PatchDock server [37] was used to simulate MexA hexamer-OMF trimer assembly, with fully rigid multimers. The FireDock algorithm allowed a refinement of the obtained complexes and estimated the binding energy (Figure S3). During this refinement, the previous complex is modified in order to enhance the interface between the proteins. OprN (PDB: 5IUY) was modeled in an open state with Modeller [47], based on OprM (6TA5 chain A). OprM, modeled OprN, and TolC (PDB: 5NG5) were symmetrized with SymmDock before being submitted to PatchDock. MexX and MexE monomeric chains were obtained from the I-TASSER server and submitted to SymmDock to generate a hexameric form. Examination of the proximity between pairs of residues in adjacent chains was examined and K102 and E122 in MexX and R97 and E128 in MexE presented possible interactions.

4.9. Measurement of Antibiotic Susceptibility

The complete coding sequence corresponding to the operon *mexA-mexB-oprM* from *P. aeruginosa* PAO1 (472024–477790) (Accession No. GCF_000006765.1) was amplified by high-fidelity PCR and cloned into the pUCP24 plasmid by assembly. Then, specific mutations (D113A, Q93R, and D113A + Q93R) were inserted by site-directed mutagenesis following the recommendations of the supplier (New England Biolabs France, Evry, France). Recombinant plasmids were transferred into *E. coli*-competent cells (DH10B) by heat shock and cultured at 30 °C to avoid unspecific recombination. The sequence of the cloned and mutated *mexA-mexB-oprM* was verified by Sanger sequencing. Recombinant plasmids were then transferred into the PAO1 strain by electroporation. The recombinant strains were selected on MH medium supplemented with 10 µg/mL gentamicine. The mutated plasmid-borne efflux system was compared with the wild-type plasmid-borne one to assess the impact of the mutations. MICs to ticarcillin and aztreonam were performed following CLSI recommendations.

5. Conclusions

In conclusion, we provide evidence that the OMF selectivity does not rely only on molecular determinants of the tip-to-tip OMF–PAP interface described in the tripartite complexes, but also on additional molecular determinants on PAP and OMF that allosterically modulate the formation of tripartite complexes. Further investigations are needed to fully elucidate the molecular mechanisms underlying the formation of RND tripartite complexes.

Supplementary Materials: The following are available online at https://www.mdpi.com/article/10.3390/antibiotics11020126/s1, Figure S1: Association/dissociation curves for the loading of OprM onto the streptavidin biosensor, Figure S2: Alignment of primary sequences of several PAPs from *P. aeruginosa*, Figure S3: Model simulation of PAP hexamer and OMF–PAP complex, Table S1: Details about average images from 2D classification of tripartite complexes and MexA$_{Q93R}$, Table S2: Hexamer assembly of MexA and variant modelled by SymmDock, Table S3: Scoring of OMF–MexA interaction.

Author Contributions: Conceptualization, E.B., L.D., I.B. and O.L.; methodology, E.B., M.D., J.D., M.L., G.P. and Q.C.; software, J.-C.T. and J.S.; validation, V.D., G.D., I.B. and O.L.; formal analysis, E.B., L.D. and O.L.; investigation, E.B., L.D., I.B. and O.L.; resources, O.L. and I.B.; data curation, E.B., L.D., J.-C.T., J.D. and O.L.; writing—original draft preparation, E.B. and O.L.; writing—review and editing, E.B., L.D., G.P., I.B. and O.L.; supervision, O.L.; project administration, O.L.; funding acquisition, I.B. and O.L. All authors have read and agreed to the published version of the manuscript.

Funding: This work was funded by French national research agency ANR (Mistec ANR-17CE11-0028 and GHScReen2, ANR-17-CE18-0022-03). E.B. is a recipient fellowship from ANR Mistec.

Data Availability Statement: The data presented in this study are available on request from the corresponding authors.

Acknowledgments: This work has benefited from the facilities of UMS3033/US001-IECB. The authors thank SMALTIS company for MIC measurements and Klaas M Pos for providing TolC proteins. We thank Céline Gounou for technical assistance. We warmly thank Gilles Joucla and Christophe Quétard for assistance with BLI experiments.

Conflicts of Interest: The authors declare no conflict of interest.

References

1.   Zgurskaya, H.I.; Nikaido, H. Bypassing the Periplasm: Reconstitution of the AcrAB Multidrug Efflux Pump of Escherichia Coli. *Proc. Natl. Acad. Sci. USA* **1999**, *96*, 7190–7195. [CrossRef] [PubMed]
2.   Nikaido, H. Multidrug Resistance in Bacteria. *Annu. Rev. Biochem.* **2009**, *78*, 119–146. [CrossRef]
3.   Li, X.-Z.; Plésiat, P.; Nikaido, H. The Challenge of Efflux-Mediated Antibiotic Resistance in Gram-Negative Bacteria. *Clin. Microbiol. Rev.* **2015**, *28*, 337–418. [CrossRef] [PubMed]
4.   Venter, H. Reversing Resistance to Counter Antimicrobial Resistance in the World Health Organisation's Critical Priority of Most Dangerous Pathogens. *Biosci. Rep.* **2019**, *39*, BSR20180474. [CrossRef]
5.   Symmons, M.F.; Marshall, R.L.; Bavro, V.N. Architecture and Roles of Periplasmic Adaptor Proteins in Tripartite Efflux Assemblies. *Front. Microbiol.* **2015**, *6*, 513. [CrossRef] [PubMed]
6.   Alav, I.; Kobylka, J.; Kuth, M.S.; Pos, K.M.; Picard, M.; Blair, J.M.A.; Bavro, V.N. Structure, Assembly, and Function of Tripartite Efflux and Type 1 Secretion Systems in Gram-Negative Bacteria. *Chem. Rev.* **2021**, *121*, 5479–5596. [CrossRef]
7.   Schweizer, H.P. Efflux as a Mechanism of Resistance to Antimicrobials in Pseudomonas Aeruginosa and Related Bacteria: Unanswered Questions. *Genet. Mol. Res.* **2003**, *2*, 48–62.
8.   Masuda, N.; Sakagawa, E.; Ohya, S.; Gotoh, N.; Tsujimoto, H.; Nishino, T. Substrate Specificities of MexAB-OprM, MexCD-OprJ, and MexXY-OprM Efflux Pumps in Pseudomonas Aeruginosa. *Antimicrob. Agents Chemother.* **2000**, *44*, 3322–3327. [CrossRef]
9.   Maseda, H.; Yoneyama, H.; Nakae, T. Assignment of the Substrate-Selective Subunits of the MexEF-OprN Multidrug Efflux Pump of Pseudomonas Aeruginosa. *Antimicrob. Agents Chemother.* **2000**, *44*, 658–664. [CrossRef]
10.   Phan, G.; Picard, M.; Broutin, I. Focus on the Outer Membrane Factor OprM, the Forgotten Player from Efflux Pumps Assemblies. *Antibiotics* **2015**, *4*, 544–566. [CrossRef]
11.   Srikumar, R.; Li, X.Z.; Poole, K. Inner Membrane Efflux Components Are Responsible for Beta-Lactam Specificity of Multidrug Efflux Pumps in Pseudomonas Aeruginosa. *J. Bacteriol.* **1997**, *179*, 7875–7881. [CrossRef] [PubMed]
12.   Yoneyama, H.; Ocaktan, A.; Gotoh, N.; Nishino, T.; Nakae, T. Subunit Swapping in the Mex-Extrusion Pumps in Pseudomonas Aeruginosa. *Biochem. Biophys. Res. Commun.* **1998**, *244*, 898–902. [CrossRef] [PubMed]

13. Singh, M.; Sykes, E.M.E.; Li, Y.; Kumar, A. MexXY RND Pump of Pseudomonas Aeruginosa PA7 Effluxes Bi-Anionic β-Lactams Carbenicillin and Sulbenicillin When It Partners with the Outer Membrane Factor OprA but Not with OprM. *Microbiology* **2020**, *166*, 1095–1106. [CrossRef] [PubMed]
14. Vediyappan, G.; Borisova, T.; Fralick, J.A. Isolation and Characterization of VceC Gain-of-Function Mutants That Can Function with the AcrAB Multiple-Drug-Resistant Efflux Pump of Escherichia Coli. *J. Bacteriol.* **2006**, *188*, 3757–3762. [CrossRef]
15. Bai, J.; Mosley, L.; Fralick, J.A. Evidence That the C-Terminus of OprM Is Involved in the Assembly of the VceAB-OprM Efflux Pump. *FEBS Lett.* **2010**, *584*, 1493–1497. [CrossRef]
16. Weeks, J.W.; Nickels, L.M.; Ntreh, A.T.; Zgurskaya, H.I. Non-Equivalent Roles of Two Periplasmic Subunits in the Function and Assembly of Triclosan Pump TriABC from Pseudomonas Aeruginosa. *Mol. Microbiol.* **2015**, *98*, 343–356. [CrossRef]
17. Wang, Z.; Fan, G.; Hryc, C.F.; Blaza, J.N.; Serysheva, I.I.; Schmid, M.F.; Chiu, W.; Luisi, B.F.; Du, D. An Allosteric Transport Mechanism for the AcrAB-TolC Multidrug Efflux Pump. *elife* **2017**, *6*, e24905. [CrossRef]
18. Tsutsumi, K.; Yonehara, R.; Ishizaka-Ikeda, E.; Miyazaki, N.; Maeda, S.; Iwasaki, K.; Nakagawa, A.; Yamashita, E. Structures of the Wild-Type MexAB-OprM Tripartite Pump Reveal Its Complex Formation and Drug Efflux Mechanism. *Nat. Commun.* **2019**, *10*, 1520. [CrossRef]
19. Glavier, M.; Puvanendran, D.; Salvador, D.; Decossas, M.; Phan, G.; Garnier, C.; Frezza, E.; Cece, Q.; Schoehn, G.; Picard, M.; et al. Antibiotic Export by MexB Multidrug Efflux Transporter Is Allosterically Controlled by a MexA-OprM Chaperone-like Complex. *Nat. Commun.* **2020**, *11*, 4948. [CrossRef]
20. Lobedanz, S.; Bokma, E.; Symmons, M.F.; Koronakis, E.; Hughes, C.; Koronakis, V. A Periplasmic Coiled-Coil Interface Underlying TolC Recruitment and the Assembly of Bacterial Drug Efflux Pumps. *Proc. Natl. Acad. Sci. USA* **2007**, *104*, 4612–4617. [CrossRef]
21. Bokma, E.; Koronakis, E.; Lobedanz, S.; Hughes, C.; Koronakis, V. Directed Evolution of a Bacterial Efflux Pump: Adaptation of the E. Coli TolC Exit Duct to the Pseudomonas MexAB Translocase. *FEBS Lett.* **2006**, *580*, 5339–5343. [CrossRef]
22. Krishnamoorthy, G.; Tikhonova, E.B.; Zgurskaya, H.I. Fitting Periplasmic Membrane Fusion Proteins to Inner Membrane Transporters: Mutations That Enable Escherichia Coli AcrA to Function with Pseudomonas Aeruginosa MexB. *J. Bacteriol.* **2008**, *190*, 691–698. [CrossRef] [PubMed]
23. Marshall, R.L.; Bavro, V.N. Mutations in the TolC Periplasmic Domain Affect Substrate Specificity of the AcrAB-TolC Pump. *Front. Mol. Biosci.* **2020**, *7*, 166. [CrossRef]
24. Tikhonova, E.B.; Yamada, Y.; Zgurskaya, H.I. Sequential Mechanism of Assembly of Multidrug Efflux Pump AcrAB-TolC. *Chem. Biol.* **2011**, *18*, 454–463. [CrossRef]
25. Weeks, J.W.; Celaya-Kolb, T.; Pecora, S.; Misra, R. AcrA Suppressor Alterations Reverse the Drug Hypersensitivity Phenotype of a TolC Mutant by Inducing TolC Aperture Opening. *Mol. Microbiol.* **2010**, *75*, 1468–1483. [CrossRef] [PubMed]
26. Hayashi, K.; Nakashima, R.; Sakurai, K.; Kitagawa, K.; Yamasaki, S.; Nishino, K.; Yamaguchi, A. AcrB-AcrA Fusion Proteins That Act as Multidrug Efflux Transporters. *J. Bacteriol.* **2016**, *198*, 332–342. [CrossRef]
27. Stegmeier, J.F.; Polleichtner, G.; Brandes, N.; Hotz, C.; Andersen, C. Importance of the Adaptor (Membrane Fusion) Protein Hairpin Domain for the Functionality of Multidrug Efflux Pumps. *Biochemistry* **2006**, *45*, 10303–10312. [CrossRef] [PubMed]
28. Symmons, M.F.; Bokma, E.; Koronakis, E.; Hughes, C.; Koronakis, V. The Assembled Structure of a Complete Tripartite Bacterial Multidrug Efflux Pump. *Proc. Natl. Acad. Sci. USA* **2009**, *106*, 7173–7178. [CrossRef] [PubMed]
29. Eda, S.; Maseda, H.; Yoshihara, E.; Nakae, T. Assignment of the Outer-Membrane-Subunit-Selective Domain of the Membrane Fusion Protein in the Tripartite Xenobiotic Efflux Pump of Pseudomonas Aeruginosa. *FEMS Microbiol. Lett.* **2006**, *254*, 101–107. [CrossRef]
30. Ferrandez, Y.; Monlezun, L.; Phan, G.; Benabdelhak, H.; Benas, P.; Ulryck, N.; Falson, P.; Ducruix, A.; Picard, M.; Broutin, I. Stoichiometry of the MexA-OprM Binding, as Investigated by Blue Native Gel Electrophoresis. *Electrophoresis* **2012**, *33*, 1282–1287. [CrossRef]
31. Akama, H.; Matsuura, T.; Kashiwagi, S.; Yoneyama, H.; Narita, S.-I.; Tsukihara, T.; Nakagawa, A.; Nakae, T. Crystal Structure of the Membrane Fusion Protein, MexA, of the Multidrug Transporter in Pseudomonas Aeruginosa. *J. Biol. Chem.* **2004**, *279*, 25939–25942. [CrossRef] [PubMed]
32. Higgins, M.K.; Bokma, E.; Koronakis, E.; Hughes, C.; Koronakis, V. Structure of the Periplasmic Component of a Bacterial Drug Efflux Pump. *Proc. Natl. Acad. Sci. USA* **2004**, *101*, 9994–9999. [CrossRef]
33. Daury, L.; Orange, F.; Taveau, J.-C.; Verchère, A.; Monlezun, L.; Gounou, C.; Marreddy, R.K.R.; Picard, M.; Broutin, I.; Pos, K.M.; et al. Tripartite Assembly of RND Multidrug Efflux Pumps. *Nat. Commun.* **2016**, *7*, 10731. [CrossRef]
34. Yang, J.; Yan, R.; Roy, A.; Xu, D.; Poisson, J.; Zhang, Y. The I-TASSER Suite: Protein Structure and Function Prediction. *Nat. Methods* **2015**, *12*, 7–8. [CrossRef] [PubMed]
35. Yang, J.; Zhang, Y. I-TASSER Server: New Development for Protein Structure and Function Predictions. *Nucleic Acids Res.* **2015**, *43*, W174–W181. [CrossRef] [PubMed]
36. Roy, A.; Kucukural, A.; Zhang, Y. I-TASSER: A Unified Platform for Automated Protein Structure and Function Prediction. *Nat. Protoc.* **2010**, *5*, 725–738. [CrossRef]
37. Schneidman-Duhovny, D.; Inbar, Y.; Nussinov, R.; Wolfson, H.J. PatchDock and SymmDock: Servers for Rigid and Symmetric Docking. *Nucleic Acids Res.* **2005**, *33*, W363–W367. [CrossRef]
38. Yoshihara, E.; Eda, S.; Kaitou, S. Functional Interaction Sites of OprM with MexAB in the Pseudomonas Aeruginosa Multidrug Efflux Pump. *FEMS Microbiol. Lett.* **2009**, *299*, 200–204. [CrossRef] [PubMed]

39. Bai, J.; Bhagavathi, R.; Tran, P.; Muzzarelli, K.; Wang, D.; Fralick, J.A. Evidence That the C-Terminal Region Is Involved in the Stability and Functionality of OprM in *E. coli*. *Microbiol. Res.* **2014**, *169*, 425–431. [CrossRef]

40. Yamanaka, H.; Izawa, H.; Okamoto, K. Carboxy-Terminal Region Involved in Activity of Escherichia Coli TolC. *J. Bacteriol.* **2001**, *183*, 6961–6964. [CrossRef]

41. Yamanaka, H.; Nomura, T.; Morisada, N.; Shinoda, S.; Okamoto, K. Site-Directed Mutagenesis Studies of the Amino Acid Residue at Position 412 of Escherichia Coli TolC Which Is Required for the Activity. *Microb. Pathog.* **2002**, *33*, 81–89. [CrossRef] [PubMed]

42. Ntsogo Enguéné, Y.V.; Phan, G.; Garnier, C.; Ducruix, A.; Prangé, T.; Broutin, I. Xenon for Tunnelling Analysis of the Efflux Pump Component OprN. *PLoS ONE* **2017**, *12*, e0184045. [CrossRef] [PubMed]

43. Bosco, M.; Damian, M.; Chauhan, V.; Roche, M.; Guillet, P.; Fehrentz, J.-A.; Bonneté, F.; Polidori, A.; Banères, J.-L.; Durand, G. Biotinylated Non-Ionic Amphipols for GPCR Ligands Screening. *Methods* **2020**, *180*, 69–78. [CrossRef]

44. Rouser, G.; Fleischer, S.; Yamamoto, A. Two Dimensional Then Layer Chromatographic Separation of Polar Lipids and Determination of Phospholipids by Phosphorus Analysis of Spots. *Lipids* **1970**, *5*, 494–496. [CrossRef]

45. Tang, G.; Peng, L.; Baldwin, P.R.; Mann, D.S.; Jiang, W.; Rees, I.; Ludtke, S.J. EMAN2: An Extensible Image Processing Suite for Electron Microscopy. *J. Struct. Biol.* **2007**, *157*, 38–46. [CrossRef]

46. Schneidman-Duhovny, D.; Inbar, Y.; Nussinov, R.; Wolfson, H.J. Geometry-Based Flexible and Symmetric Protein Docking. *Proteins* **2005**, *60*, 224–231. [CrossRef] [PubMed]

47. Sali, A.; Blundell, T.L. Comparative Protein Modelling by Satisfaction of Spatial Restraints. *J. Mol. Biol.* **1993**, *234*, 779–815. [CrossRef]

Article

Insight into the AcrAB-TolC Complex Assembly Process Learned from Competition Studies

Prasangi Rajapaksha [†], Isoiza Ojo [†], Ling Yang, Ankit Pandeya, Thilini Abeywansha and Yinan Wei [*]

Department of Chemistry, University of Kentucky, Lexington, KY 40506, USA; prasangi_iro@uky.edu (P.R.); Isoiza.ojo@uky.edu (I.O.); ling.yang@uky.edu (L.Y.); pandeya.ankit@uky.edu (A.P.); thilini.abeywansha@uky.edu (T.A.)
* Correspondence: yinan.wei@uky.edu
† These authors contributed equally to this work.

Abstract: The RND family efflux pump AcrAB-TolC in *E. coli* and its homologs in other Gram-negative bacteria are major players in conferring multidrug resistance to the cells. While the structure of the pump complex has been elucidated with ever-increasing resolution through crystallography and Cryo-EM efforts, the dynamic assembly process remains poorly understood. Here, we tested the effect of overexpressing functionally defective pump components in wild type *E. coli* cells to probe the pump assembly process. Incorporation of a defective component is expected to reduce the efflux efficiency of the complex, leading to the so called "dominant negative" effect. Being one of the most intensively studied bacterial multidrug efflux pumps, many AcrA and AcrB mutations have been reported that disrupt efflux through different mechanisms. We examined five groups of AcrB and AcrA mutants, defective in different aspects of assembly and substrate efflux. We found that none of them demonstrated the expected dominant negative effect, even when expressed at concentrations many folds higher than their genomic counterpart. The assembly of the AcrAB-TolC complex appears to have a proof-read mechanism that effectively eliminated the formation of futile pump complex.

Keywords: RND pump; dominant negative effect; assembly; protein-protein interaction; mutation

Citation: Rajapaksha, P.; Ojo, I.; Yang, L.; Pandeya, A.; Abeywansha, T.; Wei, Y. Insight into the AcrAB-TolC Complex Assembly Process Learned from Competition Studies. *Antibiotics* **2021**, *10*, 830. https://doi.org/10.3390/antibiotics10070830

Academic Editors: Isabelle Broutin, Attilio V. Vargiu, Henrietta Venter and Gilles Phan

Received: 12 May 2021
Accepted: 6 July 2021
Published: 8 July 2021

Publisher's Note: MDPI stays neutral with regard to jurisdictional claims in published maps and institutional affiliations.

Copyright: © 2021 by the authors. Licensee MDPI, Basel, Switzerland. This article is an open access article distributed under the terms and conditions of the Creative Commons Attribution (CC BY) license (https://creativecommons.org/licenses/by/4.0/).

1. Introduction

Antimicrobial resistance, especially multidrug resistance in bacterial pathogens, is among the top 10 global threats to humanity [1]. Among the large array of different defense mechanisms adapted by bacteria, the overexpression of efflux pumps has a significant role in conferring multidrug resistance. AcrAB-TolC is one of the most extensively studied efflux pump systems in Gram-negative bacteria, playing a crucial role in the multidrug resistance in bacteria such as *Eshcherichia coli* [2–4]. AcrAB-TolC is a member of the resistance nodulation division (RND) superfamily. AcrAB-TolC efflux pump confers resistance to a broad spectrum of antimicrobial compounds including β-lactams, tetracycline, novobiocin, and fluroquinolones [5,6]. This tripartite efflux transporter consists of three major protein components [7,8], an outer membrane channel TolC, a periplasmic adaptor protein (PAP) AcrA, and an inner membrane proton-driven antiporter AcrB [9–12]. TolC forms a channel that spans the outer membrane and acts as the exit pathway of substrates translocated from the inner membrane and the periplasmic space. AcrA has function in stabilizing the connection between the two membrane components TolC and AcrB [13,14]. The RND transporter protein AcrB is responsible for substrate recognition and energy transduction. Upon binding of a substrate, AcrB uses the energy from the proton flow down its concentration gradient through a proton translocation pathway in the transmembrane domain to drive the conformational change necessary to move the substrate upward toward the exit tunnel [4,15]. TolC is shared by several efflux systems, hence *E. coli* strains deficient in TolC are more sensitive to a wider variety of chemicals (e. g. detergents, drugs, bile salts, and organic solvents) [16,17].

With the dedication of many research groups, the structure and mechanism of drug efflux by the RND pumps have been brought to light. The first crystal structure of the pump component was determined for TolC by Koronakis et al. [18] in 2000. TolC is a trimer with an overall length of 140 Å with 40 Å in the β-barrel domain mainly composed of β strands, and 100 Å in the periplasmic domain mainly composed of α-helices. The periplasmic end of the TolC tunnel is sealed at the resting state, which likely opens by an allosteric protein–protein interaction mechanism [19]. In 2002, Murakami et al. first reported the crystal structure of AcrB, followed by the proposal of the functional rotation mechanism [20–22]. Later in 2006, Mikolosko and coworkers determined the crystal structure of AcrA. In contrast to the trimeric TolC and AcrB, AcrA forms a hexamer in the pump assembly [23]. The assembled pump structure was first proposed as the "deep interpenetration model", which shows that AcrB and TolC have direct interactions with AcrA wrapped around on the outside to strengthen the interaction [24]. More recently, Wang et al. proposed a new model based on Cryo-EM studies, known as the "tip-to-tip model". In this model, AcrA hairpins form a barrel-like conformation, contacting TolC in a tip-to-tip arrangement [25,26]. The recent determination of the complex structure first by cryo-EM, then by X-ray crystallography, confirmed the tip-to-tip model [7,19,20,27–30]. Energy does not seem to be required to assemble the AcrAB-TolC complex and AcrAB could interact with the TolC channel to form a AcrAB-TolC complex even in the absence of known substrate [31]. The dynamic process that leads to the formation of the complex is still elusive.

The dominant negative effect describes the phenomena in which an excess of a functionless mutant of a protein in the presence of its wild type counterpart, reduces the observed activity due to competition from the mutant for interaction with functional partners of the protein of interest. In the AcrAB-TolC complex, over-expression of functionless AcrB or AcrA mutant in wild type *E. coli* strains is expected to drastically reduce the assembly of functional efflux complex, and thus reduce the efflux activity and increase the sensitivity to substrate compounds. However, we tested the overexpression of several functionally defective mutants in the wild type *E. coli* strain, but did not observe the expected level of reduction. We speculate that the assembly of the AcrAB-TolC is a precisely controlled process involving delicate proof-reading procedures.

2. Results

2.1. AcrB Mutants Defective in Proton Transport

Several key residues have been identified in the AcrB transmembrane domain, forming the proton translocation pathway [20,30,32–38]. We have created single alanine replacement mutations at each of these sites to obtain mutants AcrB-D407A, AcrB-D408A, AcrB-K940A, AcrB-T978A, and AcrB-R971A [34,37]. The plasmid encoding of these mutants was first transformed into the BW25113Δ*acrB* strain to examine their efflux activity (Table 1). As expected, the minimum inhibitory concentrations (MICs) of most examined substrates against the strains containing the mutants were the same as those against the strains without plasmids. The only mutant that displayed significant activity is T978A, which remained partially active. As a positive control, we showed that transformation with a plasmid encoding the wild type AcrB completely restored the efflux activity with MIC values similar to a wild type BW25113 strain. Next, we transformed these plasmids into the wild-type BW25113 strain and measured the MIC. We expected the AcrB mutants to compete with the genomic AcrB in binding and interaction with genomic AcrA and/or TolC, thus reduce the number of functional efflux complexes and subsequently the drug susceptibility. However, we observed a two-fold reduction of MIC value in some cases, and no reduction in others, which is consistent with an earlier study reporting the modest reduction of substrate susceptibility when AcrB-D407A was over-expressed in a wild -type *E. coli* strain (Table 1) [39].

Table 1. MIC values (μg/mL) of BW25113 or BW25113Δ*acrB* strains containing the indicated plasmid encoding AcrB mutants defective in proton translocation pathway.

Substrate [1]	NOV	ERY	TPP	EtBr	R6G	NA
BW25113Δ*acrB* containing						
/	4	4	4	8	8	1
WT	128	64	256	128	256	4
D407A	4	4	4	8	8	ND
D408A	4	4	4	8	8	ND
K940A	8	4	4	8	8	ND
R971A	8	4	4	8	16	ND
T978A	8	8	16	32	128	ND
BW25113 containing						
/	256	64	1024	512	1024	4
WT	512	128	1024	512	1024	4
D407A	128	64	1024	256	512	4
D408A	128	64	1024	512	1024	4
K940A	256	64	1024	512	1024	4
R971A	128	32	1024	512	1024	4
T978A	256	64	1024	512	1024	2

[1] NOV, novobiocin. ERY erythromycin, TPP tetraphenylphosphonium, EtBr ethidium bromide. R6G, rhodamine 6G, NA, nalidixic acid. ND, not determined.

We conducted the MIC assay under the basal expression condition without induction to avoid the potential artifact that may arise from over-expression. To examine how much of each mutant actually expressed under the basal condition, we prepared samples from BW25113 containing different plasmids, and compared the expression level from the plasmid to the level of genomic AcrB. Even without induction, under our experimental condition, the plasmid-encoded mutants were expressed at a much higher level (5–20 folds higher) compared to the level of the genomic AcrB (Figure 1a). This result indicates that the lack of impact on drug susceptibility is not due to the lack of expression. Even presented at a large excess, the mutants were not effective in disrupting the normal efflux activity.

Single mutations on the proton relay pathway do not significantly affect the overall structure of AcrB. The crystal structure of a couple of these mutants have been determined using X-ray crystallography [34,40]. All mutants form trimeric structures similar to those in the wild-type AcrB [34]. To examine if a mutant defective in proton translocation could still bind AcrA, we used two inter-subunit disulfide bonds as our yardsticks to probe the interaction between AcrA and AcrB. We first constructed a plasmid expressing both AcrA and AcrB (pBAD33-AcrAB), then introduced a pair of cysteines, one in AcrA and the other in AcrB: AcrA-P57C/AcrB-N191C, and AcrA-T217C/AcrB-S258C. These residues are predicted to be close to each other according to the "Disulfide by Design 2.0" (http://cptweb.cpt.wayne.edu/DbD2/ (accessed on 6 January 2019)) [41]. The formation of disulfide bond linked AcrA-AcrB complex was confirmed using anti-AcrA and anti-AcrB Western blot (Figure 1b). A high molecular weight complex could be detected in both blots, which disappeared upon incubation with β-mercaptoethanol (BME). The disulfide bond-linked species migrated slightly differently in the gel, likely due to differences in the conformations of the two complexes under the gel running condition. Mutation and disulfide bond formation did not significantly impair efflux activity, as revealed in the MIC measurement (Table 2). Next, we introduced the D408A mutation into both constructs to examine the effect of this additional mutation on the formation of disulfide bond linked AcrA-AcrB complex. If the D408A mutation had a significant impact on the interaction between AcrB and AcrA, we expect to see a reduction of the intensity of the high molecular weight AcrA-AcrB complex. As shown in Figure 1c, disulfide bond formation was to a similar level in both constructs, suggesting that the additional D408A mutation did not have a significant impact on AcrA-AcrB interaction.

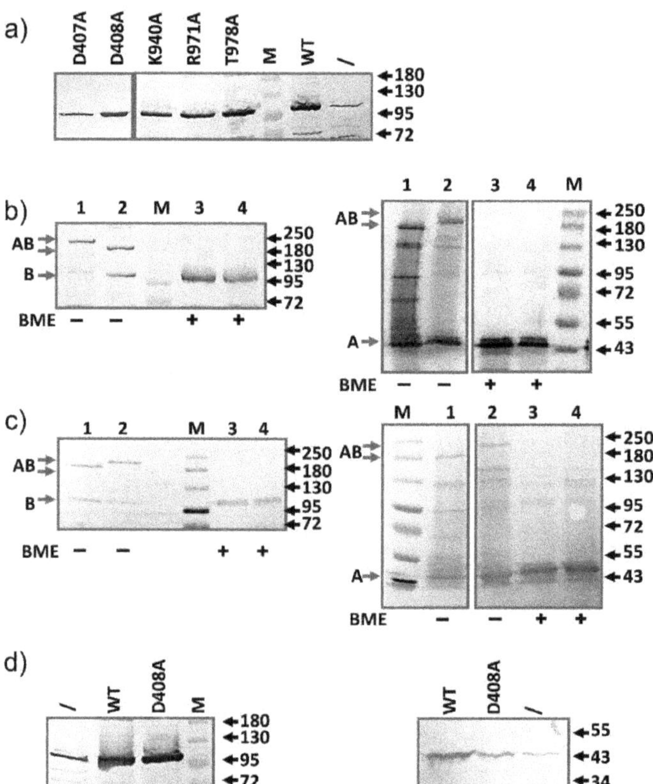

Figure 1. Characterization of mutants defective in the proton translocation pathway. (**a**) The point mutation did not affect expression level. Anti-AcrB Western blot analysis of the expression of all five mutants and the wild-type AcrB from plasmid transformed into BW25113. Sample prepared from plasmid-free BW25113 (\) was also prepared and loaded to serve as a control to highlight the difference in expression levels. (**b**) Anti-AcrB and Anti-AcrA Western blot analyses revealing the formation of disulfide bonded AcrA-AcrB complexes, which was reduced after incubation with BME. AcrA-P57C/AcrB-N191C (lane 1 and 3), AcrA-T217C/AcrB-S258C (lane 2 and 4). (**c**) Similar to b, with the additional D408A mutation introduced into the constructs. AcrA-P57C/AcrB-N191C/AcrB-D408A (lane 1 and 3), AcrA-T217C/AcrB-S258C/AcrB-D408A (lane 2 and 4). Molecular weight markers are labeled as "M" and the molecular weight of bands (kD) were indicated on the right. The expected bands for AcrA, AcrB, and disulfide bond linked AcrA-AcrB are marked on the left of the gels as A, B, and AB, respectively. (**d**) Anti-AcrB (left) and anti-AcrA (right) Western blot analysis of BW25113 expressing plasmid pBAD33-AcrAB (WT) or pBAD33-AcrAB-D408A (D408A). Samples prepared from BW25113 not containing plasmid was used as the control (/). For anti-AcrA Western blot, plasmid-containing samples were diluted 4-fold before being loaded into the gel.

Table 2. MIC values (µg/mL) of BW25113 and BW25113Δ*acrAB* strains containing the indicated plasmid encoding gene for both AcrA and AcrB.

Substrate	NOV	ERY	TPP	EtBr	R6G
BW25113Δ*acrAB* containing					
/	4	4	8	4	16
WT	32	32	128	128	32
AcrA/AcrB-D408A	4	2	8	8	8
AcrAP57C/AcrBN191C	32	16	64	64	32
AcrA-T217C/AcrB-S258C	64	16	128	128	32
BW25113 containing					
/	256	32	1024	512	512
AcrAB-WT	256	64	1024	512	1024
AcrAB-D408A	256	32	1024	512	512

To examine if the additional expression of AcrA from the same plasmid as the AcrB-D408A have any impact on the competition with the genomic AcrB, we introduced plasmid pBAD33-AcrAB-D408A into the wild type BW25113 strain and examined the MIC. Both AcrA and AcrB-D408A expressed at levels much higher than their genomic counterparts (Figure 1d), and yet, no dominant negative effect was observed (Table 2).

2.2. AcrB Mutants Defective in Substrate Binding

One feature of the AcrAB-TolC complex that has drawn much research interest is their ability to efflux a large array of substrates ranging broadly in molecular weight, charge, and hydrophobicity. Many mutations have been introduced in the substrate binding pocket in AcrB to probe their impact on the efflux of different substrates [30,42]. We chose three such mutants to include in this study, F610A, I278A, and F178A, since they were reported to have the most significant impact on efflux. First, plasmids containing single residue mutations at these sites were introduced into BW25113Δ*acrB* to examine their activities (Table 3). While the F610A mutant is largely inactive, both F178A and I278A remained partially active, which is consistent with previous reports [43]. It is clear that in general, single point mutations introduced at the substrate binding site are not as detrimental as mutations introduced in the proton translocation pathway. This is reasonable when considering several residues collectively form a substrate binding site, while the proton translocation pathway is more linear.

Table 3. MIC values (µg/mL) of BW25113 and BW25113Δ*acrB* strains containing the indicated plasmid encoding AcrB mutants defective in substrate binding.

Substrate	NOV	ERY	TPP	EtBr	R6G	NA
BW25113Δ*acrB* containing						
/	4	4	4	8	8	1
WT	128	64	256	128	256	4
F610A	8	8	64	32	128	ND
F178A	64	8	128	128	256	ND
I278A	32	32	128	128	128	ND
BW25113 containing						
/	256	64	1024	512	1024	4
WT	512	128	1024	512	1024	4
F610A	512	64	1024	512	1024	2
F178A	512	64	1024	512	1024	4
I278A	512	64	1024	512	1024	4

Next, plasmids encoding these mutations were transformed into BW25113 to determine their impact on efflux activity. Similar to proton relay pathway mutants, the MIC values of the strain were not significantly affected (Table 3). The presence of AcrB mutants defective in substrate binding does not display the dominant negative phenotype either.

Next, we examined the expression of these mutants under the basal condition (Figure 2a). Similar as described above, the plasmid-encoded mutants were expressed at a much higher level compared to the level of the genomic AcrB, indicating that the lack of impact on MIC is not due to the lack of expression.

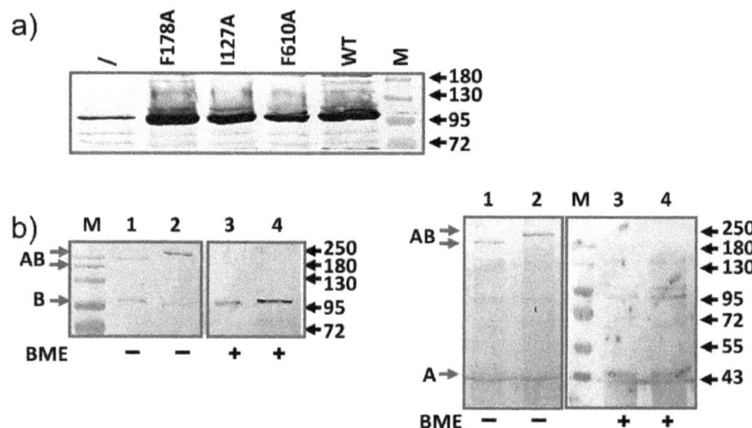

Figure 2. Characterization of AcrB mutants defective in substrate binding. (**a**) The point mutation did not affect expression level. Anti-AcrB Western blot analysis of basal expression of all three mutants and the wild type AcrB from plasmid transformed into BW25113. Sample prepared from plasmid-free BW25113 (\) was also prepared and loaded to serve as a control to highlight the difference in expression levels. (**b**) Anti-AcrB and Anti-AcrA Western blot analyses revealing the formation of disulfide bonded AcrA-AcrB complexes, which was reduced after incubation with BME. AcrA-P57C/AcrB-N191C, F610A is in lane 1 and 3, and AcrA-T217C/AcrB-S258C, F610A in lane 2 and 4. Molecular weight markers are labeled as "M" and the molecular weight of bands (kD) were indicated on the right. The expected bands for AcrA, AcrB, and disulfide bond linked AcrA-AcrB are marked on the left of the gels as A, B, and AB, respectively.

To determine if a mutation in the substrate binding pocket of AcrB (F610A) affects interaction between AcrB and AcrA, we used the disulfide bond pairs as described above and introduced an additional AcrB F610A mutation (Figure 2b). Similar as in Figure 1b, AcrA-AcrB complexes were observed, similar as in samples without the F610A mutation, suggesting that the AcrB F610A mutant still binds with AcrA.

2.3. AcrA Mutant Defective in TolC Interaction

AcrA is the periplasmic adaptor protein of the efflux system. While debate still exists concerning the conformation of AcrA in the final assembled pump, it was clear that several residues at the tip of its long α-helical hairpin loop play an important role in the interaction with TolC [44]. We created three such mutants, R128D, L132D, and S139D, and introduced plasmids encoding these mutants first into BW25113Δ*acrA* to confirm that they are not active (Table 4). As expected, the mutants were largely inactive. The plasmids were then transformed into BW25113 and substrate susceptibility of the strains were measured. Similar to the AcrB mutants, we did not observe the dominant negative effect.

Table 4. MIC values (µg/mL) of BW25113 and BW25113Δ*acrA* strains containing the indicated plasmid encoding AcrA mutants.

Substrate	NOV	ERY	TPP	R6G	NA
BW25113Δ*acrA* containing					
/	4–8	2	4	4	1
WT	64	32	256	128	4
L132D	8	2	4	4	ND
R128D	4	2	4	4	ND
S139D	4	2	8–16	4	ND
A113D	64	4	32	16	ND
A155D	8	4	8	4	ND
A113D, A155D	8	2	8	4	ND
L50C	32	8	32	32	ND
I52C	32	4	16	32	ND
E229C	32	4	32	16	ND
R225C	16	8	64	16	ND
L50C, R225C	4	8	8	4	ND
I52C, R225C	4	4	8	8	ND
I52C, E229C	4	2	4	4	ND
BW25113 containing					
/	512	64	1024	512	4
R128D	512	64	1024	512	4
L132D	512	64	1024	512	4
S139D	256	64	1024	512	4
A113D	512	64	1024	512	4
A155D	512	64	1024	512	4
A113D, A155D	512	64	1024	512	4
L50C	512	64	1024	512	4
I52C	256	64	1024	512	4
E229C	512	64	1024	256	4
R225C	512	64	1024	512	4
L50C, R225C	512	64	1024	256	4
I52C, R225C	512	128	1024	512	4
I52C, E229C	512	128	1024	512	4

Next, we examined the expression of AcrA from the plasmids. As discussed above for AcrB mutants, we did not induce expression. Both the MIC study and the expression level detection experiments were performed under the basal expression condition. Expressions of the mutants (R128D, L132D, S139D) were higher than the expression level of the genomic AcrA (Figure 3a). To make the sample intensity comparable on the Western blot analysis, samples prepared from plasmid-containing strains were diluted 4-fold before being loaded into the gel. Thus, the expression levels of the mutant AcrA constructs were 10–20 folds higher than the expression level of the wild type AcrA from the genome. While these mutants had been studied for their impact on interaction with TolC, it was not clear if they would have an impact on binding with AcrB. We used the double Cys mutants as described above to probe the interaction between AcrA-R128D and AcrB (Figure 3b). This extra mutation did not seem to disrupt the interaction between AcrA and AcrB.

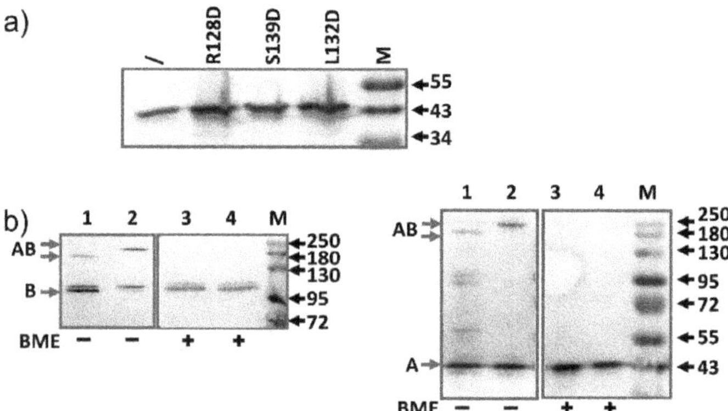

Figure 3. Characterization of AcrA mutants defective in TolC interaction. (**a**) Anti-AcrA Western blot analysis of samples prepared from BW25113 containing plasmids expressing the indicated AcrA mutant (diluted 4- folds). Sample prepared from plasmid-free BW25113 was also prepared and loaded without dilution (\) to serve as a control to highlight the difference in expression levels. (**b**) Anti-AcrB and Anti-AcrA Western blot analyses revealing the formation of disulfide bonded AcrA-AcrB complexes, which was reduced after incubation with BME. AcrA-P57C/AcrB-N191C, R128D (lane 1 and 3), AcrA-T217C/AcrB-S258C, R128D (lane 2 and 4). Molecular weight markers are labeled as "M" and the molecular weight of bands (kD) were indicated on the right. The expected bands for AcrA, AcrB, and disulfide bond linked AcrA-AcrB are marked on the left of the gels as A, B, and AB, respectively.

2.4. AcrA Mutant Defective in AcrA Assembly

According to the cryo-EM structure of the complex, AcrA forms a hexameric barrel, with each subunit contributing a long helical hairpin. We speculate that mutations introduced at neighboring sites between the hairpins would disrupt the interaction between neighboring AcrA subunits, and thus disrupt the formation of the helical barrel. We chose two sites to introduce mutation, A113 and A155. They are located at the inter-subunit interface in the middle of the long hairpin. We created both single and double mutants containing A113D and A155D. Plasmids encoding these strains were first transformed into BW25113Δ*acrA* to examine the impact of the mutation on the efflux activity (Table 4). Both mutations disrupted efflux, as revealed by a reduction of the MIC. The A155D mutation is more detrimental than A113D. Both A155D and the double mutations A113D/A155D were largely inactive. When transformed into BW25113, we still did not observe a significant reduction of MIC. The expression levels of the mutants were examined to confirm that mutants did express in excess compared to the level of the genomic AcrA under the experimental condition. As shown in Figure 4a, the expression of the mutants was similar or slightly higher than that of the genomic AcrA. We next used the inter-AcrA/AcrB disulfide bond pairs to examine the impact of A155D mutation on the interaction between AcrA and AcrB (data not shown). This additional mutation did not have a significant impact on the level of disulfide bond formation, indicating that it did not disrupt the interaction between AcrA and AcrB.

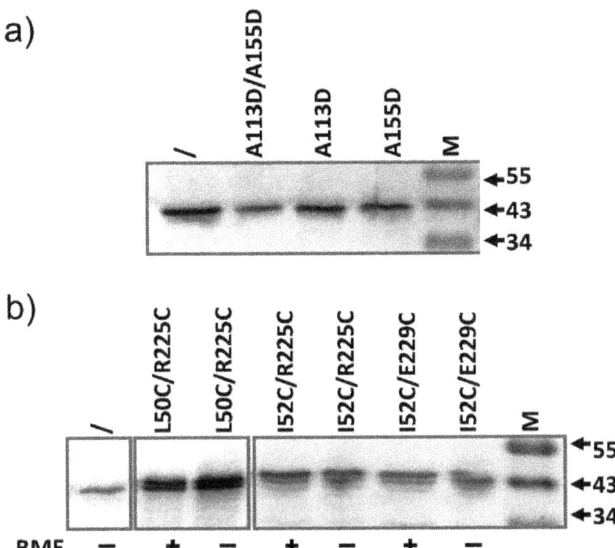

Figure 4. Anti-AcrA Western blot analysis of samples prepared from BW25113 containing plasmids expressing the indicated AcrA mutant (diluted 4-folds). Sample prepared from plasmid-free BW25113 was also prepared and loaded without dilution (\) to serve as a control to highlight the difference in expression levels. (**a**) AcrA mutants are expected to affect AcrA–AcrA interaction during pump assembly. (**b**) AcrA mutants forming intra-subunit disulfide bond to be trapped in an inactive conformation. Addition of BME did not lead to an observable change in mobility.

2.5. AcrA Mutant Defective in Conformational Change

We created three pairs of defective AcrA mutants that are trapped in a nonfunctional confirmation via a disulfide bond, L50C-R225C, I52C-R225C, and I52C-E229C [45]. Hazel et al. first created these mutants to examine their hypothesis that AcrA adopts two conformations, a *cis*-like conformation in which membrane proximal (MP) and α-helical domains point to the same direction, and a *trans*-like conformation in which they point to opposite directions. We created the corresponding single and double mutants and confirmed that the double mutants were largely inactive when transformed into BW25113Δ*acrA* (Table 4), consistent with previous report. Yet, expression of these mutants in the BW25113 strain did not have a significant impact on the efflux of AcrAB-TolC substrates. We have also examined the expression of the mutants and confirmed that they were present in the cells at a higher level than the genomic AcrA (Figure 4b).

2.6. AcrA Mutants in a Strain Containing Anchor-Free AcrA

Finally, we examined the potential contribution of the lipid anchoring of AcrA on the stability of AcrA-AcrB interaction. Toward this goal, we engineered a BW25113 strain (BW25113*spmut*) in which the signal peptide of AcrA (residue 1–24 encoded in the *acrA* gene) in the genome was replaced with the signal peptide of OmpA. The OmpA signal peptide directs the secretion of the AcrA to the periplasm after synthesis. The lipid anchoring was abolished, but the activity of AcrA was not affected [13]. We reason that removal of the lipid anchoring might favor dissociation of AcrA from AcrB in the AcrA-AcrB complex, and thus providing an opportunity for the mutant AcrA, which is lipid-anchored, to compete more effectively for binding with AcrB. We transformed above-mentioned AcrA mutant into BW25113*spmut* and measured the MIC of the strains (Table 5). We still did not observe a significant drop in efflux activity.

Table 5. MIC values (µg/mL) of BW25113*spmut* strain containing the indicated plasmid encoding AcrA mutants.

Substrate	NOV	ERY	TPP	R6G
/	512	128	2048	1024
R128D	512	128	2048	1024
L132D	512	128	2048	1024
S139D	512	128	2048	512
A113D	512	128	2048	512
A155D	512	128	2048	512
A113D, A155D	512	128	2048	512
L50C, R225C	512	128	2048	512
I52C, R225C	512	128	2048	1024
I52C, E229C	512	128	2048	1024

2.7. Slow Dissociation of the AcrAB Complex

We speculate that the AcrAB complex, once formed, dissociates very slowly. To experimentally test this speculation, we introduced a plasmid encoding AcrA bearing a histag at the C-terminus (AcrA-his) into a wild-type and the corresponding *acrA* knockout strains. We first confirmed that under our experimental condition, the genomic AcrB could not be purified using metal affinity chromatography (Figure 5a). In the absence of the plasmid, no AcrB could be detected in the eluate in anti-AcrB Western blot. In contrast, when AcrA-his was introduced into the cells, AcrB could be detected in the eluates, indicating that it was co-purified through interaction with AcrA-his. Interestingly, the AcrB band intensity was higher in the eluate prepared from the *acrA* knockout strain, indicating more AcrB were co-purified. Next, we examined the time course of co-purification. The rationale is, in the wild-type strain, genomic AcrB and AcrA form stable complexes. If the dissociation is fast, the introduced AcrA-his will quickly compete with genomic AcrA to form a complex with AcrB, and in turn enable purification of AcrB through metal affinity chromatography. Otherwise, if the dissociation is slow, then it takes much longer for the competition to happen. We monitored the formation of AcrAB complex between the genomic AcrB and plasmid-expressed AcrA-his at three-time points, right after the induction period, and 5 and 17 h after induction (Figure 5b). We performed crosslinking right before protein extraction and purification to stabilize the complexes. As a control experiment, we also examined the formation of the AcrAB complex between the genomic AcrB and plasmid-expressed AcrA-his in an *acrA* gene knock-out strain. In this case, we do not expect competition from the genomic AcrA; thus complex should form faster. We found that the formation of the complex is plateaued much faster in the *acrA* knockout strain, as intensities of the eluates prepared from 5 and 17 h into incubation were very similar. In contrast, it took much longer for AcrB to interact with AcrA-his in the wild type strain, which is likely due to the requirement of an extra step of AcrAB dissociation between the genomic AcrA and AcrB.

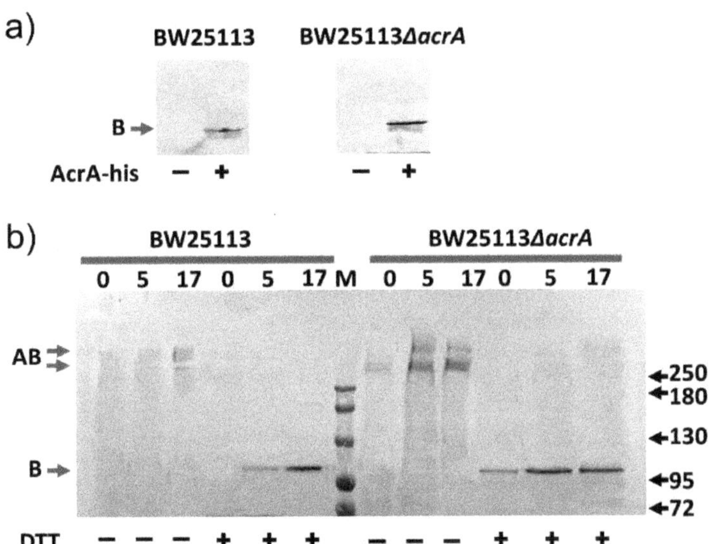

Figure 5. Co-purification of genomic AcrB with AcrA-his. (**a**) Anti-AcrB Western blot analyses of samples prepared from BW25113 or BW25113Δ*acrA* with or without plasmid-encoded AcrA-his. (**b**) Anti-AcrB Western blot analyses of samples collected after 0, 5, or 17 h of incubation following the induction of AcrA-his production in BW25113 or BW25113Δ*acrA* strains. Dithiobis-(succinimidyl proprionate) (DSP) crosslinking was performed to stabilize the AcrAB complex before protein purification. Reduction using dithiothreitol (DTT) breaks the disulfide bond in the linker of DSP and the complex into AcrA and AcrB subunits. Molecular weight markers are labeled as "M", and the molecular weight of bands (kD) were indicated on the right. The expected bands for AcrB and DSP linked AcrA-AcrB are marked on the left of the gels as B and AB, respectively.

3. Discussion

The dominant negative effect describes the situation in which the phenotype is dominated by the negative impact of the functionless mutant. The observation of the dominant negative effect has been used in many studies to investigate the mechanism of protein–protein interaction, including the identification of protein–protein interactions interface [46], determination of enzymatic activity related to oligomerization [47], and the effect of mutations in genetic disorders [48,49].

In the process of AcrAB-TolC assembly, there are many steps where the incorporation of a functionless AcrA or AcrB mutant would negatively impact the efflux activity. First, all three proteins in the system are oligomers. AcrB and TolC are obligate timers, while AcrA is believed to exist as a dimer or trimer in the free form and assembles into a hexamer in the pump complex [13,50]. While AcrA and AcrB are believed to form a complex in the absence of substrate and efflux, TolC assembles with AcrAB during active efflux. When a functionless AcrA mutant is expressed in excess in a wild type *E. coli* cell containing genomic AcrA, we expect them to compete with their genomic counterpart to engage genomic AcrB, forming non-functional interactions to reduce the overall efflux activity. Similarly, we expect functionless AcrB mutant expressed from a plasmid to compete with genomic AcrB for genomic AcrA. In addition, we expect the competition for genomic TolC will further enhance the dominant negative effect. For this competition to occur, we chose mutants that are defective due to mechanisms not directly related to the interaction between AcrA and AcrB. The structure of the AcrAB-TolC complex and location of mutants mentioned in this study are shown in Figure 6. We determined the expression level of the mutants relative to their genomic counterpart. Using serial dilution and quantitative

Western blot analysis, we found that the expression levels of the AcrB mutants were 10–20 folds of the level of the genomic AcrB, and the AcrA mutants were 2 to 20 folds of the level of the genomic AcrA. With this high level of excess, we expect to observe a strong dominant-negative effect if the mutants were actively involved in the pump assembly, competing for binding partners. We constructed five groups of AcrA and AcrB mutants, defective in different aspects. We observed that the effect of certain mutations was not always the same for different substrates. For example, T978A mutation in AcrB is detrimental to all substrates tested except for R6G, while F178A mutation in AcrB drastically reduced the MIC for ERY, but not as much for other substrates tested. This difference in mutation effects has been observed in many studies characterizing AcrA and AcrB mutants, for example, [23,42,45]. We speculate that this difference could be due to differences in the binding and interaction of specific substrates with the pump complex. The substrates vary drastically in their size, shape, structure, and charged state. As a result, the subgroup of residues that they interact with on their way to be transported are not likely to completely overlap. Therefore, point mutations introduced in AcrA and AcrB could have different impacts on specific substrates.

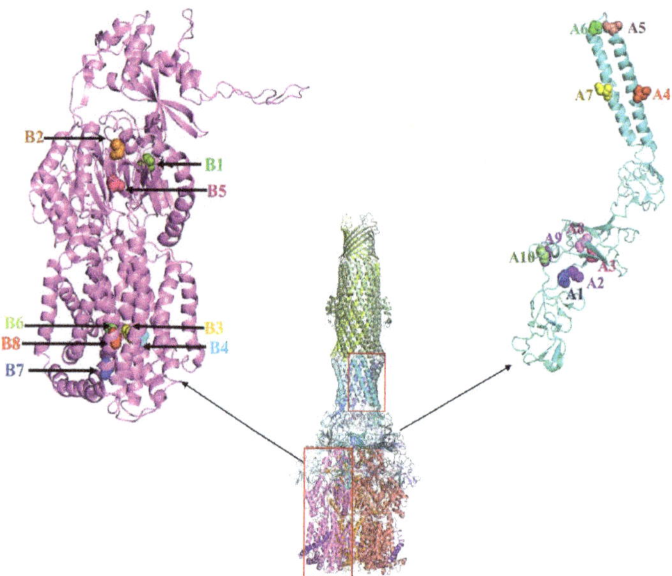

Figure 6. Structure of the AcrAB-TolC complex with the residues mutated in this study highlighted in AcrA and AcrB structures. AcrA mutations labeled as; A1-L50C, A2-I52C, A3-P57C, A4-A113D, A5-L132D, A6-S139D, A7-A155D, A8-T217C, A9-R225C, A10-E229C. AcrB mutations are labelled as; B1-F178A, B2-I278A, B3-D407A, B4-D408A, B5-F610A, B6-K940A, B7-R971A, B8-T978A. AcrAB crystal structure is created using pymol from 5N5G.pdb (https://pymol.org/2/ (accessed on 15 Marth 2021)).

Mutants defective in the proton translocation pathway still form trimers [34] and interact properly with AcrA (Figure 1c). Then, was why no dominant negative effect observed? One possibility is that the genomic AcrA and AcrB are transcribed together, sharing the same mRNA. Hence, the newly produced AcrA and AcrB could be clustered as well. As a result, the genomic AcrA and AcrB form a AcrAB complex as soon as they are translated and inserted into the membrane (AcrB) or secreted into the periplasm with lipid anchoring (AcrA). Since the local concentration of the genomic proteins are high, they associate with each other with a much higher chance than associate with a plasmid-encoded partner. Another requirement for the observed activity is that the AcrAB complex, once formed, should be resistant to dissociation. Otherwise, the high concentration of AcrB mutant in the cell membrane would be effective in competing with genomic AcrB to form a nonfunctional AcrAB complex.

We further examined the impact of over-expressing both AcrA and AcrB, in which the entire sequence coding AcrA and AcrB were inserted in a plasmid. As a result, we expect them to form AcrA-AcrBD408A complex to compete with plasmid encoded AcrAB for genomic TolC. Yet, while the expression level of the complex was ~20 fold higher than that of the genomic AcrAB, no significant reduction of MIC was observed. It appears that TolC could differentiate the two complexes, even though the only difference between them is the single residue mutation down in the transmembrane domain of AcrB.

The second group of AcrB mutants examined are defective in substrate binding. These mutations are not expected to affect AcrB structure [42,51], nor do they impact the interaction between AcrA and AcrB (Figure 2b). However, we did not observe a significant reduction of MIC.

We also examined the effect of expressing AcrA mutants on the efflux activity. The first group of residues we tested are ones that impact AcrA interaction with TolC, as they are involved in the tip-to-tip interaction with TolC [44]. We confirmed that a representative mutant in this group, R128D, still binds to AcrB (Figure 3b). Similarly, we expect the expression of these mutants will lead to competition with genomic AcrA for genomic AcrB, and the formation of a nonfunctional complex. However, no reduction of MIC was observed. To further probe the interaction of AcrA and AcrB, we created a strain of *E. coli* that is genetically modified to replace the signal peptide of AcrA with the signal peptide of OmpA (BW25113*spmut*). The resultant AcrA can still be secreted into the periplasm and is fully functional, but the lipid anchoring is lacking [13]. With this free-floating AcrA, we expect the interaction between AcrA and AcrB to be weakened, which may increase the competitiveness of the plasmid encoded AcrA, which is lipid anchored. Yet, we did not observe a reduction of MIC when the functionless mutants were expressed in strain BW25113*spmut*.

The next group of AcrA mutants contains changes at the inter-subunit interface of AcrA to disrupt formation of the functional hexametric ring. While the A155D single mutation was enough to completely abolish activity, expression of the protein in BW25113 did not lead to a reduction of MIC. Finally, we examine a group of AcrA that forms disulfide bond locked conformation that is functionally incompatible. Again, similar as other groups, over-expression of these mutants did not have a significant impact on MIC.

In conclusion, we examined the effect of plasmid-encoded AcrA and AcrB mutants in wild type *E. coli* cells, to probe the potential disruption of normal AcrA-AcrB-TolC assembly in the presence of excess mutants of AcrA or AcrB. To our surprise, none of the five groups of mutants showed the so-called "dominant negative" effect. This observation indicates that the RND pump assembly process in Gram-negative bacteria is a precisely controlled process that prevents the formation of functionless complex. An alternate explanation is the possibility that efflux depends on only a very small population of AcrAB-TolC pumps active at a given moment, as the population of active AcrAB-TolC pumps was not detectable in situ in *E. coli* [19]. If the majority of AcrAB and TolC in the cells are idle, then the effect of over-expressing functionless mutant would be greatly limited. In addition, our results suggest that dissociation kinetics of the AcrAB complex is very slow. Once formed, the complex remains bound and does not dissociate easily.

4. Materials and Methods

4.1. Bacterial Strains, Plasmids, and Growth Conditions

Escherichia coli BW25113 and BW25113Δ*acrB* were obtained from Yale *E. coli* genetic resources. BW25113Δ*acrAB*, BW25113Δ*acrA* were constructed using the *E. coli* gene deletion kit (Cat #K006, Gene Bridge) following the manufacturer's protocol. Plasmids pBAD33-AcrB and pBAD18-AcrA were created in our previous study [52]. To create the plasmid pBAD33-AcrAB, *acrAB* gene was amplified from genomic DNA of BW25113 and cloned into pBAD33 by following the fast cloning method described previously by Li et al. [53]. Plasmids pBAD33-AcrB, pBAD33-AcrAB, and pBAD18-AcrA were used as the templates to create respective mutations discussed below. Mutations were introduced using the Quikchange site-directed mutagenesis kit following manufacturer's instructions (Agilent, Santa Clara, CA, USA). BW25113spmut, in which the signal peptide of AcrA was exchanged with the signal peptide of outer membrane protein OmpA, was created using the CRISPR-Cas9 system by following the published protocol [54]. Plasmids pTargetF and pCas were a gift from Shen Yang (Addgene plasmid #62226 and #62225) [54]. Bacteria were cultured at 37 °C with shaking at 250 rpm in Luria broth (LB) media unless otherwise noted.

4.2. Drug Susceptibility Assay

The minimum inhibitory concentration (MIC) was determined for erythromycin, novobiocin, ethidium bromide (EtBr), rhodamine 6G (R6G), nalidixic acid, and tetraphenylphosphonium chloride (TPP) following the CLSI guidelines [55]. Briefly, overnight cultures of the indicated strain were diluted to a final concentration of 10^5 CFU/mL in fresh Muller Hinton Broth 2 (cation adjusted) media (Millipore Sigma, St. Louis, MO, USA) in a 48 well microtiter plate containing the indicated compounds at two-fold serial dilutions. Plates were incubated at 37 °C with shaking at 160 rpm for 17 h; the absorbance at 600 nm (OD600) were measured to identify the lowest concentrations with no observable cell growth. All experiments were repeated at least three times.

4.3. Protein Expression, SDS-PAGE and Western Blot Analysis

For expression test, 5 mL of cells were cultured overnight at 37 °C with shaking at 250 rpm. The next morning, the cell was inoculated with a 100-fold dilution into a 5 mL fresh LB media supplemented with antibiotics and grow until ~OD600 1.0. Cells were pelleted and resuspended in 1 mL phosphate buffer containing phenylmethylsulphonyl fluoride (PMSF) (1:1000 dilution of a saturated ethanol solution) and sonicated for 1 min followed by centrifugation for 10 min at 15,000 rpm. The supernatant was removed, and cell pellets were resuspended in 0.1 mL PBS containing 2% Triton-X100. The samples were incubated at room temperature with shaking for 45 min and centrifuged again for 10 min. The supernatant was used for SDS-PAGE and Western blot analysis. For studies of disulfide bond formation, iodoacetamide (IAM) was added to a final concentration of 20 mM in all buffers. To reduce disulfide bond, β-mercaptoethanol (BME) was added to a final concentration of 2% followed by incubation at room temperature for 30 min.

For AcrAB dissociation experiment, BW25113 or BW25113Δ*acrA* containing plasmid pBAD18-AcrA was cultured to the log phage (OD600 0.8) and arabinose was added to a final concentration of 0.2% (*w*/*v*) to induce the expression of AcrA for 50 min. The cells were then pelleted, washed, and resuspended with fresh LB. An aliquot of cell culture was collected, pelleted, and stored at −20 °C. The rest of the cell culture was returned to the shaker and cultured for 5 h, and another aliquot of sample was collected, pelleted, and stored at −20 °C. The last sample was collected at 17 h. OD600 of the samples were measured and used to adjust the sample volume collected to ensure that the same number of cells were used for each time point. All pellets were resuspended and sonicated to lyse the cells. After centrifugation, the pellet was extracted using PBS + 2% Triton for 2 h. The mixtures were centrifuged again and the supernatants were incubated with Ni beads for 40 min, followed by washing with the same buffer supplemented with 50 mM imidazole, and finally eluted with the same buffer supplemented with 500 mM imidazole. For DSP crosslinking experiments, the cell pellet was washed and then resuspended in PBS buffer. DSP was added to a final concentration of 1 mM, and the mixture was incubated at room temperature for 30 min. To stop the reaction, a Tris-Cl buffer (pH 8.0) was added to a final concentration of 20 mM. Cells were then pelleted, and proteins were purified similarly as described above. To break the disulfide bond in DSP, DTT was added to the sample to a final concentration of 10 mM.

For Western blot analysis, after transferring to the PVDF membrane, protein bands were detected using an Anti-AcrB polyclonal (Rabbit) antibody raised to recognize a C-terminal peptide corresponding to residues number 1036–1045 [39] or Anti-AcrA antibody, respectively. The membrane was then washed and incubated with an alkaline phosphatase conjugated goat anti-rabbit secondary antibody. The BCIP/NBT (5-bromo-4-chloro-3'-indolyphosphate and nitro-blue tetrazolium) solution was used to stain the membranes. All experiments were repeated at least three times.

Author Contributions: Conceptualization, P.R., I.O. and Y.W.; methodology, P.R., I.O., L.Y., A.P., T.A. and Y.W.; resources, Y.W.; data curation, P.R., I.O., L.Y. and Y.W.; writing—original draft preparation, P.R., I.O. and Y.W.; writing—review and editing, L.Y., A.P., T.A. and Y.W.; supervision, Y.W.; project administration, Y.W.; funding acquisition, Y.W. All authors have read and agreed to the published version of the manuscript.

Funding: This research was funded by NIH grant number 1R56AI137020, 1R21AI142063-01, NIH/NHLBI HL142640, and NIH/NIGMS GM132443, and NSF grant number CHE-1709381.

Data Availability Statement: The data presented in this study are available in the article.

Conflicts of Interest: The authors declare no conflict of interest.

References

1. WHO. Antimicrobial resistance. *Wkly. Epidemiol. Rec. Relev. Épidémiol. Hebd.* **2000**, *75*, 336.
2. Krishnamoorthy, G.; Tikhonova, E.B.; Dhamdhere, G.; Zgurskaya, H.I. On the role of TolC in multidrug efflux: The function and assembly of AcrAB-TolC tolerate significant depletion of intracellular TolC protein. *Mol. Microbiol.* **2013**, *87*, 982–997. [CrossRef]
3. Nikaido, H. Structure and mechanism of RND-type multidrug efflux pumps. *Adv. Enzym. Relat. Areas Mol. Biol.* **2011**, *77*, 1–60. [CrossRef]
4. Yamaguchi, A.; Nakashima, R.; Sakurai, K. Structural basis of RND-type multidrug exporters. *Front. Microbiol.* **2015**, *6*, 327. [CrossRef]
5. Du, D.; van Veen, H.W.; Murakami, S.; Pos, K.M.; Luisi, B.F. Structure, mechanism and cooperation of bacterial multidrug transporters. *Curr. Opin. Struct. Biol.* **2015**, *33*, 76–91. [CrossRef]
6. Nishino, K.; Nikaido, E.; Yamaguchi, A. Regulation and physiological function of multidrug efflux pumps in Escherichia coli and Salmonella. *Biochim. Et Biophys. Acta Proteins Proteom.* **2009**, *1794*, 834–843. [CrossRef]
7. Nikaido, H. Multidrug efflux pumps of gram-negative bacteria. *J. Bacteriol.* **1996**, *178*, 5853. [CrossRef]
8. Eswaran, J.; Koronakis, E.; Higgins, M.K.; Hughes, C.K.; Oronakis, V. Three's company: Component structures bring a closer view of tripartite drug efflux pumps. *Curr. Opin. Struct. Biol.* **2004**, *14*, 741–747. [CrossRef]
9. Eicher, T.; Seeger, M.A.; Anselmi, C.; Zhou, W.; Brandstätter, L.; Verrey, F.; Diederichs, K.; Faraldo-Gómez, J.D.; Pos, K.M. Coupling of remote alternating-access transport mechanisms for protons and substrates in the multidrug efflux pump AcrB. *Elife* **2014**, *3*, e03145. [CrossRef]
10. Daury, L.; Orange, F.; Taveau, J.-C.; Verchère, A.; Monlezun, L.; Gounou, C.; Marreddy, R.K.; Picard, M.; Broutin, I.; Pos, K.M.; et al. Tripartite assembly of RND multidrug efflux pumps. *Nat. Commun.* **2016**, *7*, 10731. [CrossRef]
11. Du, D.; Wang, Z.; James, N.R.; Voss, J.E.; Klimont, E.; Ohene-Agyei, T.; Venter, H.; Chiu, W.; Luisi, B.F. Structure of the AcrAB–TolC multidrug efflux pump. *Nature* **2014**, *509*, 512–515. [CrossRef]
12. Kobylka, J.; Kuth, M.S.; Müller, R.T.; Geertsma, E.R.; Pos, K.M. AcrB: A mean, keen, drug efflux machine. *Ann. N. Y. Acad. Sci.* **2020**, *1459*, 38–68. [CrossRef]
13. Zgurskaya, H.I.; Nikaido, H. AcrA is a highly asymmetric protein capable of spanning the periplasm. *J. Mol. Biol.* **1999**, *285*, 409–420. [CrossRef]
14. Zgurskaya, H.I.; Weeks, J.W.; Ntreh, A.T.; Nickels, L.M.; Wolloscheck, D. Mechanism of coupling drug transport reactions located in two different membranes. *Front. Microbiol.* **2015**, *6*, 100. [CrossRef]
15. Higgins, C.F. Multiple molecular mechanisms for multidrug resistance transporters. *Nature* **2007**, *446*, 749–757. [CrossRef]
16. Yen, M.R.; Peabody, C.R.; Partovi, S.M.; Zhai, Y.; Tseng, Y.H.; Saier, M.H. Protein-translocating outer membrane porins of Gram-negative bacteria. *Biochim. Biophys. Acta* **2002**, *1562*, 6–31. [CrossRef]
17. Lobedanz, S.; Bokma, E.; Symmons, M.F.; Koronakis, E.; Hughes, C.; Koronakis, V. A periplasmic coiled-coil interface underlying TolC recruitment and the assembly of bacterial drug efflux pumps. *Proc. Natl. Acad. Sci. USA* **2007**, *104*, 4612–4617. [CrossRef]
18. Koronakis, V.; Sharff, A.; Koronakis, E.; Luisi, B.; Hughes, C. Crystal structure of the bacterial membrane protein TolC central to multidrug efflux and protein export. *Nature* **2000**, *405*, 914–919. [CrossRef] [PubMed]
19. Shi, X.; Chen, M.; Yu, Z.; Bell, J.M.; Wang, H.; Forrester, I.; Villarreal, H.; Jakana, J.; Du, D.; Luisi, B.F.; et al. In situ structure and assembly of the multidrug efflux pump AcrAB-TolC. *Nat. Commun.* **2019**, *10*, 2635. [CrossRef] [PubMed]
20. Murakami, S.; Nakashima, R.; Yamashita, E.; Yamaguchi, A. Crystal structure of bacterial multidrug efflux transporter AcrB. *Nature* **2002**, *419*, 587. [CrossRef] [PubMed]

21. Schulz, R.; Vargiu, A.V.; Collu, F.; Kleinekathöfer, U.; Ruggerone, P. Functional rotation of the transporter AcrB: Insights into drug extrusion from simulations. *PLoS Comput. Biol.* **2010**, *6*, e1000806. [CrossRef] [PubMed]
22. Murakami, S.; Nakashima, R.; Yamashita, E.; Matsumoto, T.; Yamaguchi, A. Crystal structures of a multidrug transporter reveal a functionally rotating mechanism. *Nature* **2006**, *443*, 173–179. [CrossRef]
23. Mikolosko, J.; Bobyk, K.; Zgurskaya, H.I.; Ghosh, P. Conformational flexibility in the multidrug efflux system protein AcrA. *Structure* **2006**, *14*, 577–587. [CrossRef]
24. Symmons, M.F.; Bokma, E.; Koronakis, E.; Hughes, C.; Koronakis, V. The assembled structure of a complete tripartite bacterial multidrug efflux pump. *Proc. Natl. Acad. Sci. USA* **2009**, *106*, 7173–7178. [CrossRef]
25. Wang, Z.; Fan, G.; Hryc, C.F.; Blaza, J.N.; Serysheva, I.I.; Schmid, M.F.; Chiu, W.; Luisi, B.F.; Du, D. An allosteric transport mechanism for the AcrAB-TolC multidrug efflux pump. *Elife* **2017**, *6*, e24905. [CrossRef]
26. Marshall, R.L.; Bavro, V.N. Mutations in the TolC periplasmic domain affect substrate specificity of the AcrAB-TolC pump. *Front. Mol. Biosci.* **2020**, *7*, 166. [CrossRef] [PubMed]
27. Murakami, S. Multidrug efflux transporter, AcrB—the pumping mechanism. *Curr. Opin. Struct. Biol.* **2008**, *18*, 459–465. [CrossRef] [PubMed]
28. Nakashima, R.; Sakurai, K.; Yamasaki, S.; Nishino, K.; Yamaguchi, A. Structures of the multidrug exporter AcrB reveal a proximal multisite drug-binding pocket. *Nature* **2011**, *480*, 565–569. [CrossRef] [PubMed]
29. Pos, K.M.; Schiefner, A.; Seeger, M.A.; Diederichs, K. Crystallographic analysis of AcrB. *FEBS Lett.* **2004**, *564*, 333–339. [CrossRef]
30. Seeger, M.A.; Schiefner, A.; Eicher, T.; Verrey, F.; Diederichs, K.; Pos, K.M. Structural asymmetry of AcrB trimer suggests a peristaltic pump mechanism. *Science* **2006**, *313*, 1295–1298. [CrossRef]
31. Tikhonova, E.B.; Zgurskaya, H.I. AcrA, AcrB, and TolC of Escherichia coli form a stable intermembrane multidrug efflux complex. *J. Biol. Chem.* **2004**, *279*, 32116–32124. [CrossRef] [PubMed]
32. Zhang, X.C.; Liu, M.; Han, L. Energy coupling mechanisms of AcrB-like RND transporters. *Biophys. Rep.* **2017**, *3*, 73–84. [CrossRef] [PubMed]
33. Eicher, T.; Cha, H.-J.; Seeger, M.A.; Brandstätter, L.; El-Delik, J.; Bohnert, J.A.; Kern, W.V.; Verrey, F.; Grütter, M.G.; Diederichs, K.; et al. Transport of drugs by the multidrug transporter AcrB involves an access and a deep binding pocket that are separated by a switch-loop. *Proc. Natl. Acad. Sci. USA* **2012**, *109*, 5687–5692. [CrossRef]
34. Su, C.-C.; Li, M.; Gu, R.; Takatsuka, Y.; McDermott, G.; Nikaido, H.; Edward, W.Y. Conformation of the AcrB multidrug efflux pump in mutants of the putative proton relay pathway. *J. Bacteriol.* **2006**, *188*, 7290–7296. [CrossRef]
35. Li, X.-Z.; Plésiat, P.; Nikaido, H. The challenge of efflux-mediated antibiotic resistance in Gram-negative bacteria. *Clin. Microbiol. Rev.* **2015**, *28*, 337–418. [CrossRef] [PubMed]
36. Murakami, S.; Yamaguchi, A. Multidrug-exporting secondary transporters. *Curr. Opin. Struct. Biol.* **2003**, *13*, 443–452. [CrossRef]
37. Guan, L.; Nakae, T. Identification of Essential Charged Residues in Transmembrane Segments of the Multidrug Transporter MexB ofPseudomonas aeruginosa. *J. Bacteriol.* **2001**, *183*, 1734–1739. [CrossRef]
38. Liu, M.; Zhang, X.C. Energy-coupling mechanism of the multidrug resistance transporter AcrB: Evidence for membrane potential-driving hypothesis through mutagenic analysis. *Protein Cell* **2017**, *8*, 623–627. [CrossRef]
39. Lu, W.; Zhong, M.; Wei, Y. Folding of AcrB subunit precedes trimerization. *J. Mol. Biol.* **2011**, *411*, 264–274. [CrossRef]
40. Matsunaga, Y.; Yamane, T.; Terada, T.; Moritsugu, K.; Fujisaki, H.; Murakami, S.; Ikeguchi, M.; Kidera, A. Energetics and conformational pathways of functional rotation in the multidrug transporter AcrB. *Elife* **2018**, *7*, e31715. [CrossRef]
41. Craig, D.B.; Dombkowski, A.A. Disulfide by Design 2.0: A web-based tool for disulfide engineering in proteins. *BMC Bioinform.* **2013**, *14*, 346. [CrossRef] [PubMed]
42. Bohnert, J.A.; Schuster, S.; Seeger, M.A.; Fähnrich, E.; Pos, K.M.; Kern, W.V. Site-directed mutagenesis reveals putative substrate binding residues in the Escherichia coli RND efflux pump AcrB. *J. Bacteriol.* **2008**, *190*, 8225–8229. [CrossRef] [PubMed]
43. Kinana, A.D.; Vargiu, A.V.; Nikaido, H. Effect of site-directed mutations in multidrug efflux pump AcrB examined by quantitative efflux assays. *Biochem. Biophys. Res. Commun.* **2016**, *480*, 552–557. [CrossRef] [PubMed]
44. Kim, H.-M.; Xu, Y.; Lee, M.; Piao, S.; Sim, S.-H.; Ha, N.-C.; Lee, K. Functional relationships between the AcrA hairpin tip region and the TolC aperture tip region for the formation of the bacterial tripartite efflux pump AcrAB-TolC. *J. Bacteriol.* **2010**, *192*, 4498–4503. [CrossRef]
45. Hazel, A.J.; Abdali, N.; Leus, I.V.; Parks, J.M.; Smith, J.C.; Zgurskaya, H.I.; Gumbart, J.C. Conformational dynamics of AcrA govern multidrug efflux pump assembly. *ACS Infect. Dis.* **2019**, *5*, 1926–1935. [CrossRef]
46. Torres, V.J.; McClain, M.S.; Cover, T.L. Mapping of a domain required for protein-protein interactions and inhibitory activity of a Helicobacter pylori dominant-negative VacA mutant protein. *Infect. Immun.* **2006**, *74*, 2093–2101. [CrossRef]
47. Valente, L.; Nishikura, K. RNA binding-independent dimerization of adenosine deaminases acting on RNA and dominant negative effects of nonfunctional subunits on dimer functions. *J. Biol. Chem.* **2007**, *282*, 16054–16061. [CrossRef]
48. Chevrier, L.; De Brevern, A.; Hernandez, E.; Leprince, J.; Vaudry, H.; Guedj, A.M.; De Roux, N. PRR repeats in the intracellular domain of KISS1R are important for its export to cell membrane. *Mol. Endocrinol.* **2013**, *27*, 1004–1014. [CrossRef]
49. Mantovani, R.; Li, X.-Y.; Pessara, U.; Van Huisjduijnen, R.H.; Benoist, C.; Mathis, D. Dominant negative analogs of NF-YA. *J. Biol. Chem.* **1994**, *269*, 20340–20346. [CrossRef]
50. Zgurskaya, H.I.; Nikaido, H. Cross-linked complex between oligomeric periplasmic lipoprotein AcrA and the inner-membrane-associated multidrug efflux pump AcrB from Escherichia coli. *J. Bacteriol.* **2000**, *182*, 4264–4267. [CrossRef]

51. Soparkar, K.; Kinana, A.D.; Weeks, J.W.; Morrison, K.D.; Nikaido, H.; Misra, R. Reversal of the drug binding pocket defects of the AcrB multidrug efflux pump protein of Escherichia coli. *J. Bacteriol.* **2015**, *197*, 3255–3264. [CrossRef]
52. Lu, W.; Chai, Q.; Zhong, M.; Yu, L.; Fang, J.; Wang, T.; Li, H.; Zhu, H.; Wei, Y. Assembling of AcrB trimer in cell membrane. *J. Mol. Biol.* **2012**, *423*, 123–134. [CrossRef] [PubMed]
53. Li, C.; Wen, A.; Shen, B.; Lu, J.; Huang, Y.; Chang, Y. FastCloning: A highly simplified, purification-free, sequence-and ligation-independent PCR cloning method. *BMC Biotechnol.* **2011**, *11*, 92. [CrossRef]
54. Jiang, Y.; Chen, B.; Duan, C.; Sun, B.; Yang, J.; Yang, S. Multigene editing in the Escherichia coli genome via the CRISPR-Cas9 system. *Appl. Environ. Microbiol.* **2015**, *81*, 2506–2514. [CrossRef]
55. Humphries, R.M.; Ambler, J.; Mitchell, S.L.; Castanheira, M.; Dingle, T.; Hindler, J.A.; Koeth, L.; Sei, K. CLSI methods development and standardization working group best practices for evaluation of antimicrobial susceptibility tests. *J. Clin. Microbiol.* **2018**, *56*, e01934-17. [CrossRef] [PubMed]

Article

A Model for Allosteric Communication in Drug Transport by the AcrAB-TolC Tripartite Efflux Pump

Anya Webber [1], Malitha Ratnaweera [2], Andrzej Harris [1], Ben F. Luisi [1,*] and Véronique Yvette Ntsogo Enguéné [1,*]

[1] Department of Biochemistry, University of Cambridge, Tennis Court Road, Cambridge CB2 1GA, UK; anya.webber@costellomedical.com (A.W.); ams243@cam.ac.uk (A.H.)

[2] Department of Oncology, MRC Weatherall Institute of Molecular Medicine, University of Oxford, Oxford OX3 9DS, UK; malitha.ratnaweera@oncology.ox.ac.uk

[*] Correspondence: bfl20@cam.ac.uk (B.F.L.); yn286@cam.ac.uk (V.Y.N.E.)

Abstract: RND family efflux pumps are complex macromolecular machines involved in multidrug resistance by extruding antibiotics from the cell. While structural studies and molecular dynamics simulations have provided insights into the architecture and conformational states of the pumps, the path followed by conformational changes from the inner membrane protein (IMP) to the periplasmic membrane fusion protein (MFP) and to the outer membrane protein (OMP) in tripartite efflux assemblies is not fully understood. Here, we investigated AcrAB-TolC efflux pump's allostery by comparing resting and transport states using difference distance matrices supplemented with evolutionary couplings data and buried surface area measurements. Our analysis indicated that substrate binding by the IMP triggers quaternary level conformational changes in the MFP, which induce OMP to switch from the closed state to the open state, accompanied by a considerable increase in the interface area between the MFP subunits and between the OMPs and MFPs. This suggests that the pump's transport-ready state is at a more favourable energy level than the resting state, but raises the puzzle of how the pump does not become stably trapped in a transport-intermediate state. We propose a model for pump allostery that includes a downhill energetic transition process from a proposed 'activated' transport state back to the resting pump.

Keywords: allostery; antimicrobial resistance; conformational changes; efflux pump; energetic transition; gram-negative bacteria; pump activation

Citation: Webber, A.; Ratnaweera, M.; Harris, A.; Luisi, B.F.; Ntsogo Enguéné, V.Y. A Model for Allosteric Communication in Drug Transport by the AcrAB-TolC Tripartite Efflux Pump. *Antibiotics* **2022**, *11*, 52. https://doi.org/10.3390/antibiotics11010052

Academic Editor: Seok Hoon Jeong

Received: 29 November 2021
Accepted: 28 December 2021
Published: 1 January 2022

Publisher's Note: MDPI stays neutral with regard to jurisdictional claims in published maps and institutional affiliations.

Copyright: © 2022 by the authors. Licensee MDPI, Basel, Switzerland. This article is an open access article distributed under the terms and conditions of the Creative Commons Attribution (CC BY) license (https:// creativecommons.org/licenses/by/ 4.0/).

1. Introduction

Antimicrobial resistance rates are rising: it is predicted that by 2050 there could be ten million deaths per year due to drug-resistant infections [1]. Compounding the problem of high resistance rates, the development of new antimicrobials has stalled due to research and funding challenges [2,3]. Understanding the underlying mechanisms through which pathogens develop resistance is key to meeting this growing health challenge [4]. In gram-negative bacteria, an important multidrug resistance mechanism is the overexpression of drug efflux pumps [5]. These pumps can export diverse antibiotics, preventing them from reaching their cellular targets [6]. They also contribute to pathogenicity through extrusion of molecules involved in bacterial toxicity, quorum sensing, and biofilm formation [7]. Despite the wide range of existing pumps among bacteria showing diverse structure and activities, those systems share broad similarities including a potential requirement for allosteric switching. Their functional importance make them attractive targets for inhibitors design that either occlude or lock substrate binding sites or impede allosteric transitions [8].

Energy-dependent antibiotic efflux in bacteria was first identified more than forty years ago [9], and, to date, six families of bacterial transporters have been found to be involved in efflux [10]. These transporters are all localised in the inner membrane and—with the exception of the ATP-binding cassette (ABC) transporter family (which hydrolyses

ATP)—all act as secondary transporters, using electrochemical gradients to drive substrate transport. Of the six identified families of bacterial transporters, members of the Resistance Nodulation Division (RND), Major Facilitator Superfamily (MFS), and ABC transporter families have all been found to form tripartite pump assemblies. A tripartite efflux pump, beside the inner membrane component, contains an adaptor membrane fusion protein (MFP), present in the periplasm, and an outer membrane protein (OMP). Together they make up a transport channel spanning from the cytoplasm to the extracellular medium. X-ray crystallography has helped to resolve individual components of efflux pumps, and the structures of whole pump assemblies in vitro have been determined by cryogenic electron microscopy (cryo-EM) [11–14]. Cryogenic electron tomography (cryo-ET) of whole bacterial cells has also provided images of these macromolecular machines in situ [15,16].

Inhibition of the tripartite efflux pumps presents a potential approach to increase the efficacy of existing antibiotics, but, so far, no clinically effective inhibitors have been developed [17]. Increased understanding of the structure, assembly, and mechanism of the pumps will help efforts to produce improved inhibitors, providing a valuable tool to tackling antimicrobial resistance.

The AcrAB-TolC pump from *Escherichia coli* (Figure 1a) is part of the RND family and is one of the most well-characterised efflux pumps [18]. The assembly is formed by Acriflavine resistance protein B (AcrB) and A (AcrA) as well as tolerance to Colicins protein (TolC); these are the RND transporter, MFP, and OMP components, respectively. AcrB assembles as a trimer in the pump and substrates bind to the periplasmic poly-specific pocket in each of the three subunits [19]. Drug transport occurs via sequential conformational changes driven by a proton motive force (PMF) [20,21]. AcrB also associates with the small-factor AcrZ, which is thought to enhance extrusion of certain antibiotic classes through modulating the conformation of AcrB [22,23]. AcrA assembles as a hexameric ring in the tripartite pump to bridge AcrB with TolC [11]. AcrA protomers are found in two distinct conformations arranged in an alternating fashion around the ring. The AcrA protomer has four structural domains, each of which is generated from pairs of distinct segments of the amino acid chain (Figure 1b). The membrane proximal (MP) and beta barrel (BB) domains contact AcrB, with adjacent AcrA protomers forming distinct interactions with a single AcrB protomer, whilst the helix-loop-helix (HLH) domains form tip-to-tip interactions with the periplasmic helices of TolC. In the quaternary structure of the pump, the BB and lipoyl (Lip) domains assemble into two rings and HLH domains form a channel [11].

TolC assembles as a trimer, enclosing a channel that spans from the HLH domain of AcrA to the outer membrane. On its periplasm-facing surface, TolC contains the so-called equatorial domain, which interacts with the peptidoglycan layer in the cell wall [15]. The TolC protomer contains a structural repeat (Figure 1b), allowing the three sets of four periplasmic helices (H3, H4, H7, H8) to form quasi-equivalent interactions with the six HLH domains of AcrA [11]. The TolC helices situated between the outer membrane and the equatorial domain assemble into a unique alpha-barrel structure in the trimer which, along with the BB domain, forms a pore which is inserted into the outer membrane (Figure 1c).

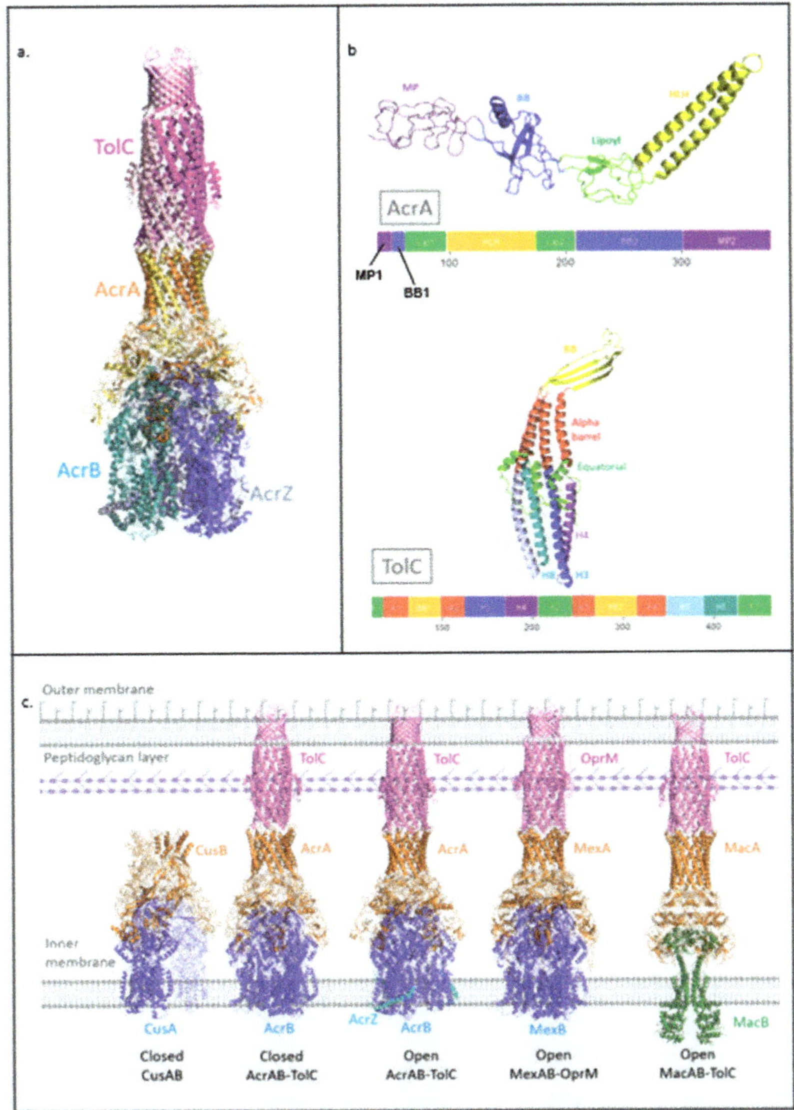

Figure 1. Structures of AcrAB-TolC and other drug efflux pump assemblies. (**a**) Transport state structure of AcrAB-TolC tripartite assembly (PDB: 5NG5). (**b**) Domain structure of AcrA and TolC protomers. For each protein, the top image shows the tertiary structure. The bottom image shows the primary sequence contribution to tertiary domain structure with equivalent colour scheme (5V5S). (**c**) High-resolution structures of RND family pump assemblies and of MacAB-TolC from the ABC transporter family (right to left: 3NE5, 5V5S, 5NG5, 6TA6, 5NIK). The CusAB trimer structure was generated from the PDB file of the monomeric unit through crystallographic symmetry using Coot [24,25]. The cell envelope includes a peptidoglycan layer and lipopolysaccharide in the outer leaflet of the outer membrane, both of which are represented schematically.

Structures of assemblies of other RND family members have also been obtained in recent years, including the metal-ion transporter CusAB [26] and the drug efflux pump MexAB-OprM from *Pseudomonas aeruginosa* [27,28], (Figure 1c). Overall, these assemblies show similar organisation to AcrAB-TolC. The OMP TolC also interacts with other partner systems from the MFS and ABC superfamilies, such as ABC transporter MacB that energises the MacAB-TolC system, where the periplasmic protein MacA serves as the MFP (Figure 1c). MacA is engaged in tip-to-tip interactions with TolC, similarly to AcrA [29]. Evidence for the importance of MFPs, in particular the HLH domain for the formation of specific and functional pump assemblies via tip-to-tip interactions with the OMP, has also been provided by experiments showing that exchanging HLH domains between different chimeric MFPs to complement their cognate OMP can restore their functionality [30].

Currently, AcrAB-TolC is the only drug efflux pump for which there are high-resolution structures of the whole pump assembly in a resting and transport state, and these offer structural insight into the pump's mechanism of action [13]. Molecular dynamics (MD) simulations have also been used to study AcrAB-TolC's conformational change, particularly the functionally rotating mechanism of AcrB [31,32]. This was complemented by a recent study of numerous X-ray structures of AcrB with different substrates that revealed several intermediate states of the transport cycle [33]. Computational modelling of dynamics of the full tripartite assembly is difficult, partly due to the large size of the complex, which limits MD simulations to short timescales [34] that cannot capture larger-scale changes. Consequently, no simulations of the whole pump assembly have been reported to date [35]. On the other hand, biochemical studies have provided some insights into pump conformational changes. For example, mutational analysis identified key residues located in the helical regions of AcrA and TolC forming tip-to-tip interactions, which are critical for pore dilation on pump activation [36].

The current model for substrate extrusion by AcrAB-TolC describes the changes to AcrB on substrate binding, but lacks detail on how these changes are communicated to AcrA and TolC, as well as the role of the proton motive force (PMF) [18]. The model proposes a functionally rotating mechanism, where an AcrB subunit shifts from a Loose (L) to Tight (T) state on substrate binding, then transitions to an Open (O) state to release the drug into the MFP-OMP pore, and finally returns to the L state. This cycle is coupled to an alternating-access mechanism for proton transport that utilises PMF to drive drug efflux [35,37]. The proton relay network, made of several ionisable residues (D407, D408, K940, and R971) is located in the transmembrane domain of AcrB. The protonation/deprotonation of these residues triggers a collective motion of transmembrane helices through the LTO conformational cycle [37]. On substrate binding to AcrB, AcrA and TolC shift from a closed (resting) to an open (transport) conformation to allow drug extrusion, and then maintain the same open state throughout AcrB conformational cycling [13].

Evaluating the communication of conformational changes in a complex multi-component system such as a tripartite assembly requires comparison of the different conformational states that is independent of reference-frame, and one approach suitable to this requirement is application of the difference distance matrix (DDM) [38]. DDM algorithms compute changes in distances between all residue pairs in two different protein states (Figure 2a). As an example, a block DDM analysis comparing the average movement of helices between deoxy- and oxy-states of the α-subunit of hemoglobin A is shown in Figure 2b. The plot clearly shows the movement of the F-helix, particularly in relation to the C-helix, which is known to be important for the protein's mechanism of action [39]; whilst, the same motion is less discernible in the aligned structure models (Figure 2c).

In addition to structural data, sequence information can also provide clues into the allosteric mechanism. One approach is to compare sequences from evolutionarily related species, identifying evolutionary couplings (ECs) through sequence covariance in multiple sequence alignment (Figure 2d) [40]. The analysis assumes that residue pairs that are physically close or allosterically coupled are likely to be conserved to maintain functional interactions. Thus, potential allosteric residues can be identified using ECs data.

Furthermore, analysis of the allosteric mechanism can also be supported by estimating energies involved in conformational transitions, and one metric proportional to energy is change in buried surface area (Figure 2e). An increase in interface size between subunits indicates more interactions between residues, and so can be used as a proxy to suggest a more favourable energy state [41]. Comparing interface areas between conformers can provide information on the relative energy needed to transition between the observed states.

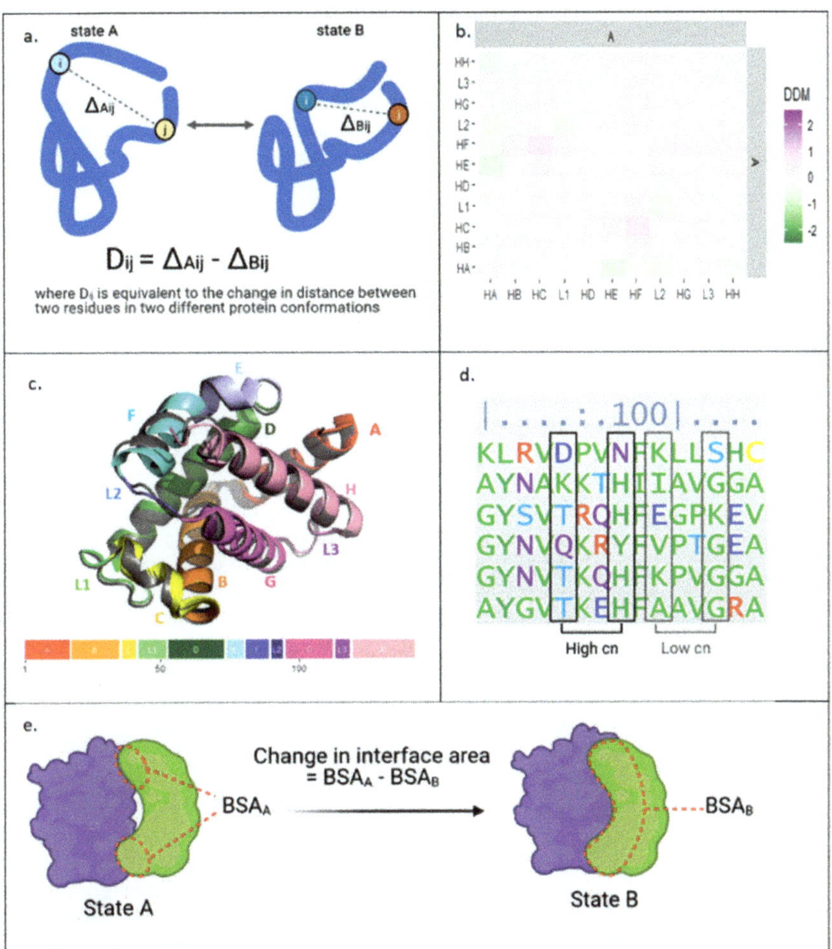

Figure 2. Bioinformatic approaches applied herein to study protein allostery. (**a**) Calculation of difference distance matrices (DDM) data from structures of a protein in two different conformations. (**b**) Block DDM for deoxy- and oxyhemoglobin-α. Squares represent average DDM changes for residues within domains specified. (**c**) Deoxyhemoglobin-α (grey, 1A3N) and oxyhemoglobin-α (multicoloured, 2DN1) aligned with C helices. Oxyhemoglobin-α's colours correlate with domain structure shown below. (**d**) Calculation of evolutionary couplings: a section of a multiple sequence alignment for haemoglobin-α. (**e**) Calculation of the change in interface area at dimer interface. HA = helix A, L1 = loop 1, BSA = buried surface area, cn = coupling number, with higher values indicating stronger coupling.

Here, we analysed available structural data to provide molecular insights into the long-distance subunit communication upon conformational changes in the efflux pump AcrAB-TolC. An analysis of the ECs in AcrA and TolC was performed alongside a DDM analysis to characterise the conformational changes of the pump and their evolutionary importance. To assess the energy required to transition between the resting and transport states of the pump, changes in interface area at subunit interfaces were measured. The results from these bioinformatic analyses were combined to propose an allosteric model for drug efflux in AcrAB-TolC.

2. Results

2.1. TolC Subunit Interface Contains Strongly Coupled Residue Pairs

To investigate allostery through an evolutionary lens, ECs data for TolC and AcrA were generated using multiple sequence alignments with ~60,000 and ~80,000 effective sequences for TolC and AcrA, respectively. EC analysis generates coupling numbers (cn) for each pair of residues, with higher cn values indicating stronger coupling. To reduce the risk of false positives, only residue pairs with coupling numbers (cn) above 1 were considered in further analysis. The remaining strongly coupled pairs were mapped onto molecular structure models, and residue pairs on the same helix were removed from the analysis, as these are unlikely to reflect allosteric interactions. Overall, we hypothesise that strongly coupled residues that are physically distant in the protein's tertiary structure could be involved in allosteric communication [42,43].

As structural data show little movement of TolC above the equatorial domain [13], EC analysis focused on the periplasmic helices. Most coupled pairs within a TolC protomer were physically close to each other (less than 10 Å apart, data not shown) and so are likely involved in maintaining the tertiary fold of the protein rather than in transferring allosteric communication. However, some strongly coupled (cn $\geq$ 2) residues were found at the interface between TolC subunits (Figure 3a). These pairs likely play a role in the quaternary assembly of the TolC trimer: D184-Q368 and Q177-S375 could form polar interactions, and V191-S361, I173-S375 and L170-T388 could be part of hydrophobic interactions. Interestingly, we found two strongly coupled (cn > 2) residues, Q164 and A382, which were rather distant; they were separated by around 20 Å, both within a protomer and between neighbouring subunits, both in the resting and the transport state.

A similar filtering rationale was applied to the ECs data for AcrA. Due to the hairpin arrangement of the AcrA domains (Figure 1b), only residue pairs that were physically close within an AcrA protomer were identified (data not shown). For both the AcrA and TolC, the ECs data were combined with DDM analysis, yielding a more systematic mapping of the ECs onto the protein structures.

2.2. ECs Correlate to Intrasubunit Movements in TolC and AcrA

To investigate the conformational changes of AcrAB-TolC in relation to sequence covariance in the evolutionary couplings, DDM analysis of single TolC and AcrA protomers was performed and compared with the ECs data. For this analysis, structures of the proteins in resting and transport states were used. The average resolution of the structures was ~6 Å; therefore, any DDM movements of less than 3 Å were ignored in subsequent analysis. As in the previous step, ECs data were filtered to only use residue pairs with coupling numbers greater or equal to 1. The combined results were plotted, showing the relative movements and evolutionary couplings between different parts of the protein (Figure 3).

The DDM for TolC (Figure 3b) showed that, as previously reported [13], there was little TolC intrasubunit movement, with changes in the distance between residue pairs often being less than 2 Å. As expected, many of the strong ECs correlated to physically close residues, visible as lines of coupled residue pairs along the diagonal and as regions of strong ECs that correlate with the alpha barrel and equatorial domains of TolC (Figures 1b and 3b). These did not show large movements in the DDM analysis. However, there was one region where the ECs and DDM appeared correlated: between residues 126–206 and 348–427, where the periplasmic helices of TolC moved closer to each other within a protomer. On closer inspection, these helix movements were slightly more complex, with simultaneous contraction and expansion (inset of Figure 3b). The largest region of contraction was between the H3-H4 loop and the H7-H8 loop; strong ECs were observed in this area, suggesting this movement might be conserved (Figure 3b, inset). Outward movement of helix H3 away from the loop of H7-H8 was also observed, but it did not correlate with any strong ECs (Figure 3b, inset).

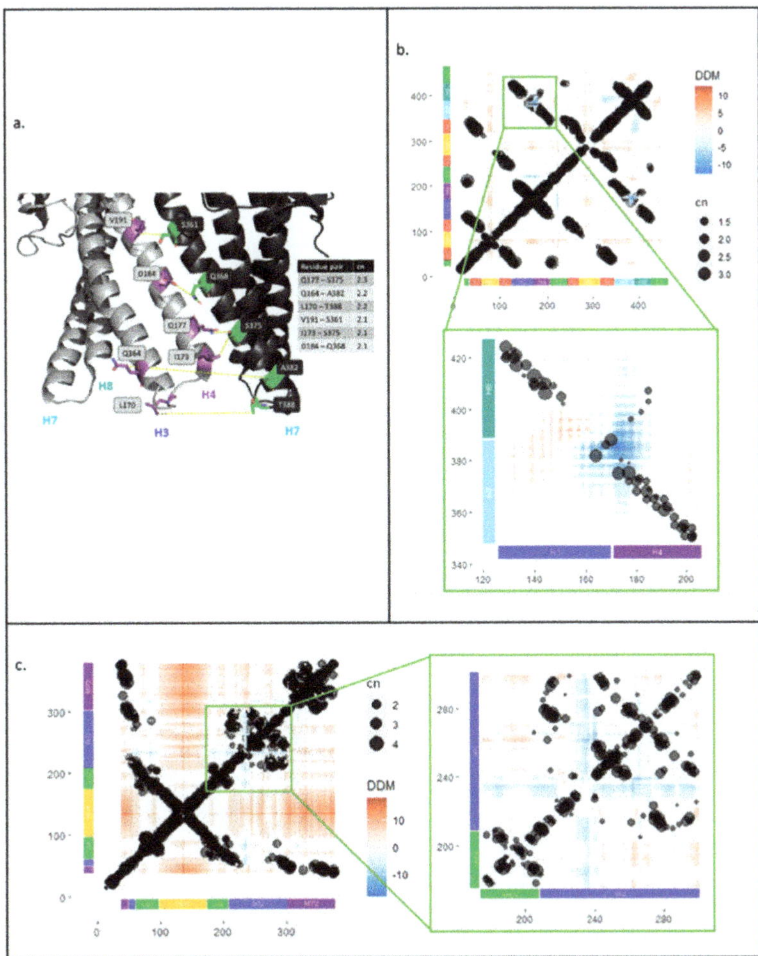

Figure 3. Strongly coupled residues at the interface between TolC subunits. Residue pairs in TolC identified from screen of evolutionary couplings data were filtered for couplings with cn $\geq$ 2 between residues in H3 and H4 (126–206) and residues in H7 and H8 (348–427) helices. (**a**) The cartoon shows helices of two TolC subunits in the transport-active open state with residue pairs between H3 and H4 helices of one protomer (light grey) and H7 helix of the second protomer (dark grey) (5NG5). Overlay of evolutionary couplings onto difference distance matrix analysis for TolC and AcrA. (**b**) Main: Difference distance matrix (DDM) and evolutionary couplings (EC) overlay for TolC protomer with domain structure on both axes. Inset: DDM and EC overlay for TolC periplasmic helices H3, H4, H7, H8. (**c**) Main: DDM and EC overlay for one AcrA subunit with domain structure on both axes. Inset: DDM and EC overlay for lipoyl and BB domains of AcrA.

The same approach was used to relate the DDM and ECs data for AcrA. A similar pattern of ECs that act to maintain the tertiary fold was observed with a line of coupled residues along the diagonal and strong couplings between the two sequence segments that fold back to make the halves of the AcrA domains (starting with MP1, BB1, Lipoyl 1, the turn at the hairpin of HLH domain, followed by Lipoyl 2, BB2, and MP2) (Figures 1b and 3c).

In our DDM analysis, AcrA appeared to be more dynamic than TolC with more extensive regions of movement (Figure 3c). The HLH domain, particularly the loop, had the largest outward movement, which is likely involved in widening of the AcrA pore for substrate extrusion, and enables it to form stronger tip-to-tip interactions with TolC. However, this movement does not correlate with any ECs. Some regions of contraction were observed, mainly in residues 230–240 in the BB domain, relative to the rest of the lipoyl and BB domains (Figure 1b). This region of contracted movement roughly colocalises with a region of strong ECs. However, upon closer inspection, the residues that move most were not strongly coupled (inset of Figure 3c).

2.3. TolC Periplasmic Helices Show Largest Outward Movement at the Quaternary Level

In the pump assembly, TolC protomers assemble into a trimer and do not just act individually. Therefore, the DDM analysis was expanded to probe quaternary changes by investigating the relative movements of protomers. To simplify the analysis, the average changes in each domain were calculated as a block (Figure 4).

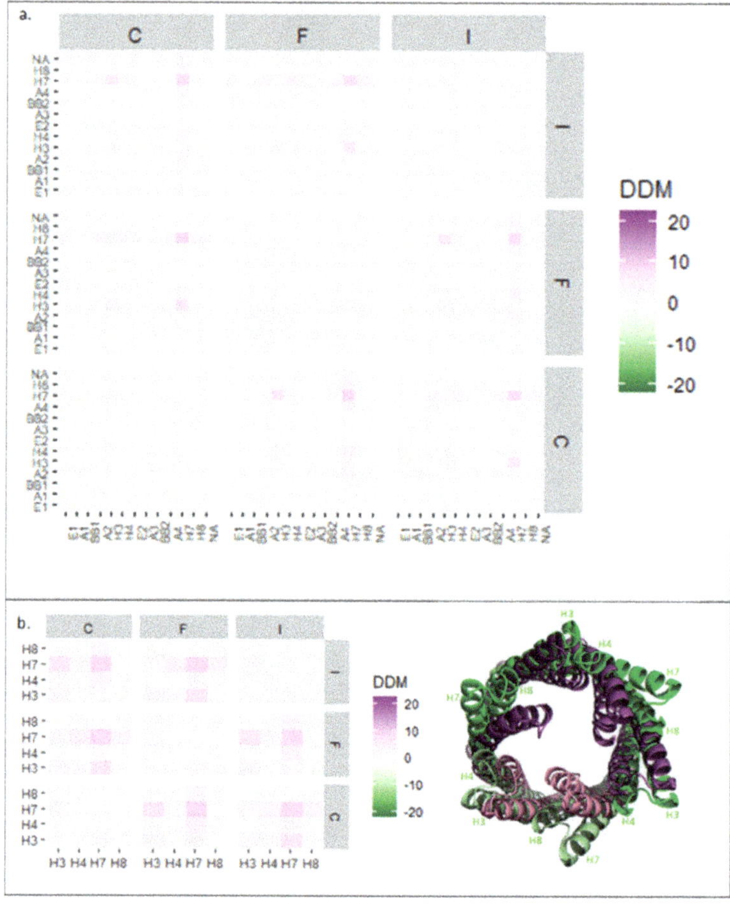

Figure 4. Block difference distance matrix (DDM) analysis of TolC domains in trimeric assembly. (**a**) Average DDM changes for all TolC domains in all three TolC protomers. (**b**) Left: Average DDM changes for TolC periplasmic helices H3, H4, H7, and H8 for all three TolC protomers. Right: TolC helices aligned in resting (purple, 5V5S) and transport (green, 5NG5) states, as viewed from above. Subunit numbering from PDB: 5NG5.

The block analysis for all the TolC domains of all three protomers (Figure 4a) showed symmetrical conformational changes on pump activation. Quaternary changes were more pronounced than those at the tertiary level (stronger colouring on the squares off the diagonal) and an overall slight contraction of the trimer was observed (pale green colouring in the distribution) (Figure 4a). However, this contraction was relatively minor, illustrating that TolC was generally static, apart from the periplasmic helices, which moved outwards at the quaternary level (Figure 4b). Helix H7 moved the most—by more than 10 Å away from its neighbouring protomer's equivalent helix (Figure 4b, right). As shown in Figure 3a, H7 formed an interface with H4 of the neighbouring subunit, so the movement of H7 is likely important for trimer assembly.

2.4. AcrA Domain Movements Corresponded to a Symmetrical Ring Contraction

As in the case of TolC, AcrA protomers assemble into a multimeric state in the pump, meriting an analysis of movement in the quaternary structure. A block DDM analysis of the AcrA hexameric ring showed symmetrical changes (Figure 5a,d), including larger changes at the quaternary level than the tertiary. Conformational compaction of the ring (pale green colouring in the distribution) was also observed throughout the plot.

To characterise the changes further, isolated AcrA domains were analysed. The two helices of the HLH domain, within a protomer, showed similar symmetrical movements (Figure 5b,d), indicating that HLH domains moved as a unit. In contrast to the expansion of the HLH domain within a single AcrA subunit (Figure 1b), at a quaternary level, the HLH domains showed a different pattern of movements. HLHs of protomers A, B, and H contracted, whereas HLHs of protomers D, E, and G expanded relative to each other. As illustrated in Figure 5d, the HLH domains moved slightly upon transition from the resting to the transport state, producing an apparent anticlockwise twisting motion.

The lipoyl and BB domains of AcrA were also analysed using the same approach (Figure 5c). The block DDM showed symmetrical, quaternary contraction. The largest contraction was observed in protomers positioned opposite each other in the ring. Super-imposed structures (Figure 5e), aligned with one subunit, showed a 45° inward rolling motion.

2.5. An Increase in the Interface Surface Area Is Observed in the Transport State

To further characterise conformational changes of AcrA and TolC on pump opening, the interface surface areas of the close (resting) and open (transport) states were analysed as a proxy for binding energy [44]. As the DDM analysis showed that quaternary changes were more pronounced than tertiary changes, differences in interface areas between subunits were measured.

The total interface area between subunits of the AcrA hexamer increased by almost 5000 Å^2 on transition from the resting to the transport state (Figure 6a). BB, lipoyl, and HLH domain contributions to interface area were calculated separately (Figure 6b). The interface area between domains of adjacent subunits increased on transition from the resting to the transport state for all three domains. We also observed an additional pattern of change: one subunit interface increasing much more than the adjacent interface, i.e., the H-G interface increased dramatically, whilst the D-H interface showed little change (Figure 6b).

For TolC, the total interface between subunits was very similar in both states, with a slightly smaller area in the transport state. As the DDM showed smaller changes for TolC than AcrA (Figure 4a), this was not surprising. However, the tip-to-tip interactions between AcrA and TolC increased by around 2000 Å^2 on transition from the resting to the transport state. This could be classified as a change from a weak to a strong affinity interaction, with the threshold typically defined as a surface area of more than 2000 Å^2 [45].

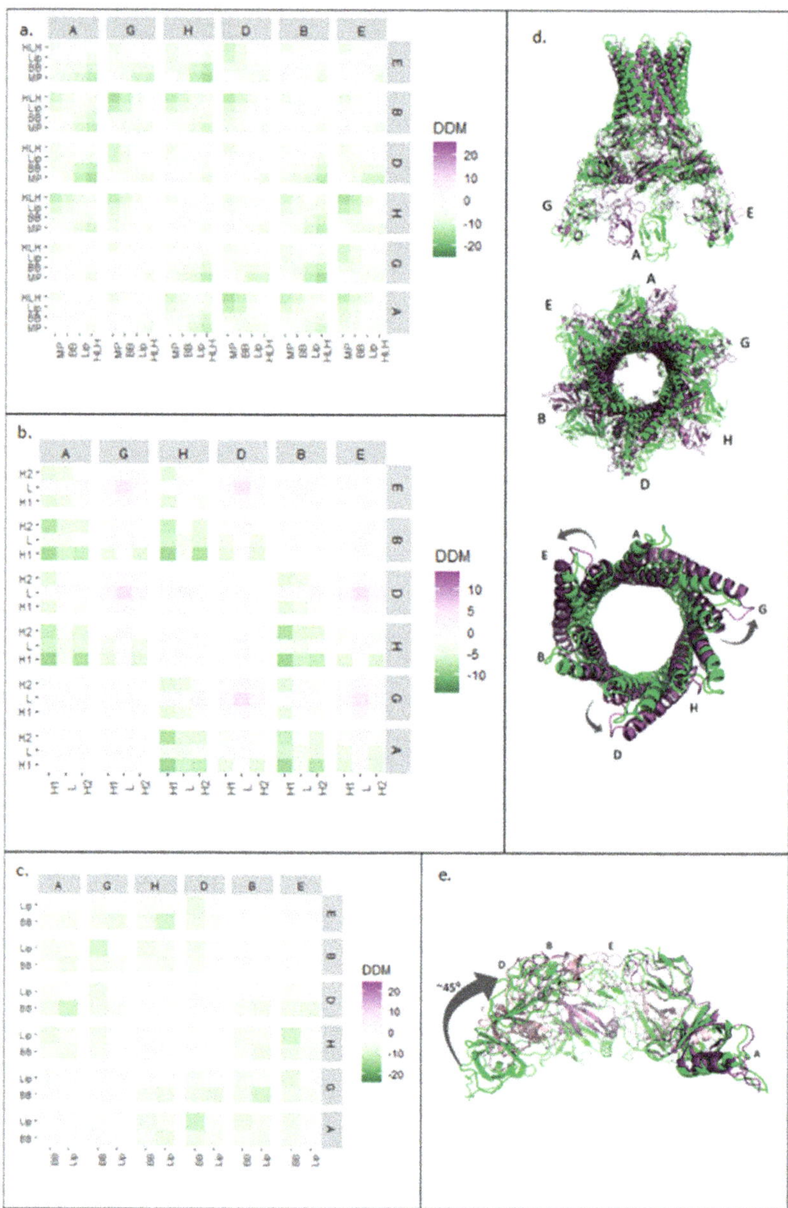

Figure 5. Block difference distance matrix (DDM) analysis of AcrA and HLH, BB, and lipoyl domains in hexameric assembly. (**a**) Average (DDM) changes for all AcrA domains in all six AcrA subunits. (**b**) Average DDM changes for HLH domain in all six AcrA protomers. (**c**) Average DDM changes for lipoyl and BB domains in all six AcrA subunits. (**d**) Resting (green, 5V5S) and transport (purple, 5NG5) structures of AcrA hexamer from the side (top) and above (middle) and of AcrA HLH domains from above (bottom). (**e**) BB and lipoyl domains from four of the six AcrA subunits aligned with subunit A in resting (green, 5V5S) and transport (purple, 5NG5) states. Subunit numbering from PDB: 5NG5. HLH: helix loop helix, BB: beta barrel.

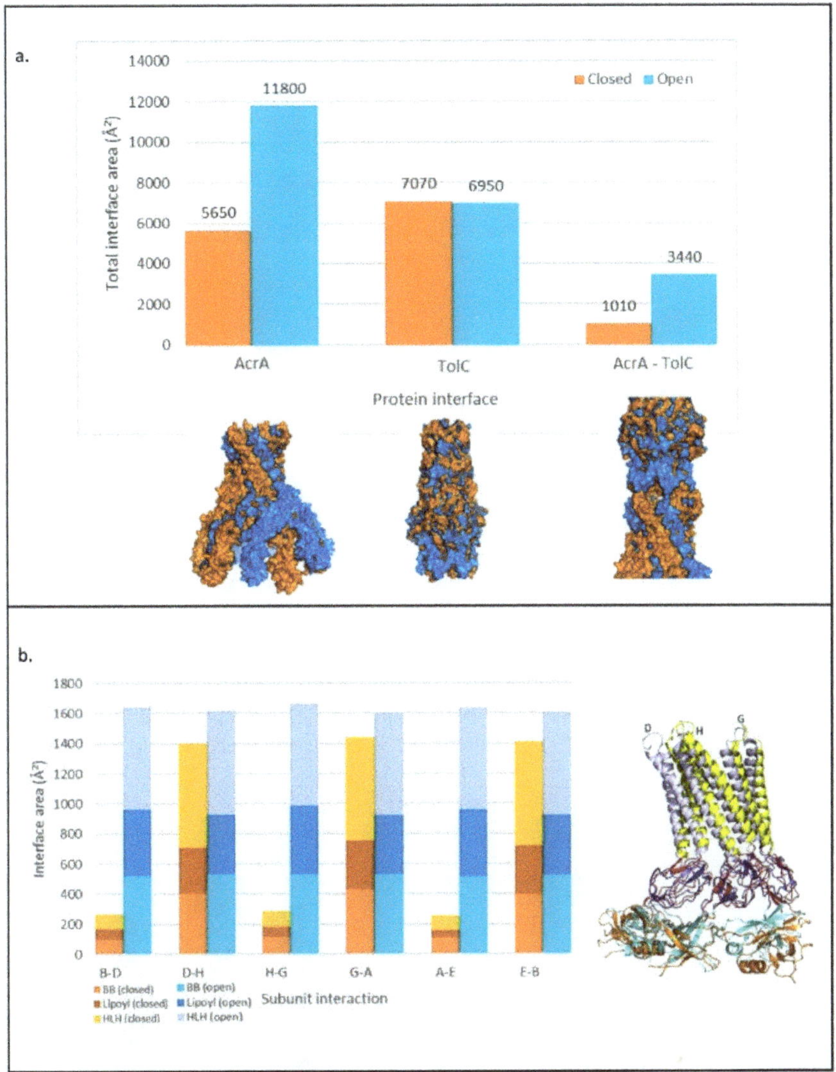

Figure 6. Changes in the interface area within and between TolC and AcrA in transport and resting pump structures. (**a**) Above: Total interface area between subunits and proteins for TolC and AcrA calculated with PDBePISA to three significant figures. Below: Surface of aligned structures of proteins in closed (resting) (orange) and open (transport) (blue) states. (**b**) Left: Change in interface area between AcrA subunits in transport and resting states, showing the contribution of BB, lipoyl, and HLH domains. Right: Three AcrA subunits aligned in resting and transport states. Colours relate to domains on the bar chart. Subunit numbering from PDB: 5NG5. HLH: helix loop helix, BB: beta barrel.

2.6. Pattern of Interface Changes Is Conserved in Homologous Pump Assemblies

To investigate whether other pump assemblies also exhibit large interface areas in the transport state, we investigated three other MFPs (CusB, MexA, and MacA) (Figures 1c and 7a). CusB structure has only been captured in a resting conformation; in the interface analysis, the interface area between subunits showed a similar pattern to resting AcrA. At all six interfaces, the interface area was relatively small, but the size oscillated between adjacent interfaces. However, this pattern was less pronounced than in resting AcrA and an analysis of a transport structure of CusB would be needed to determine if the interface areas showed a large increase in the transport state. MexA and MacA have only been captured in transport conformations; the interface analysis showed a large interface area that was of similar size to that in the transport state of AcrA. As in AcrA, the interface sizes in MexA and MacA did not oscillate between adjacent interfaces.

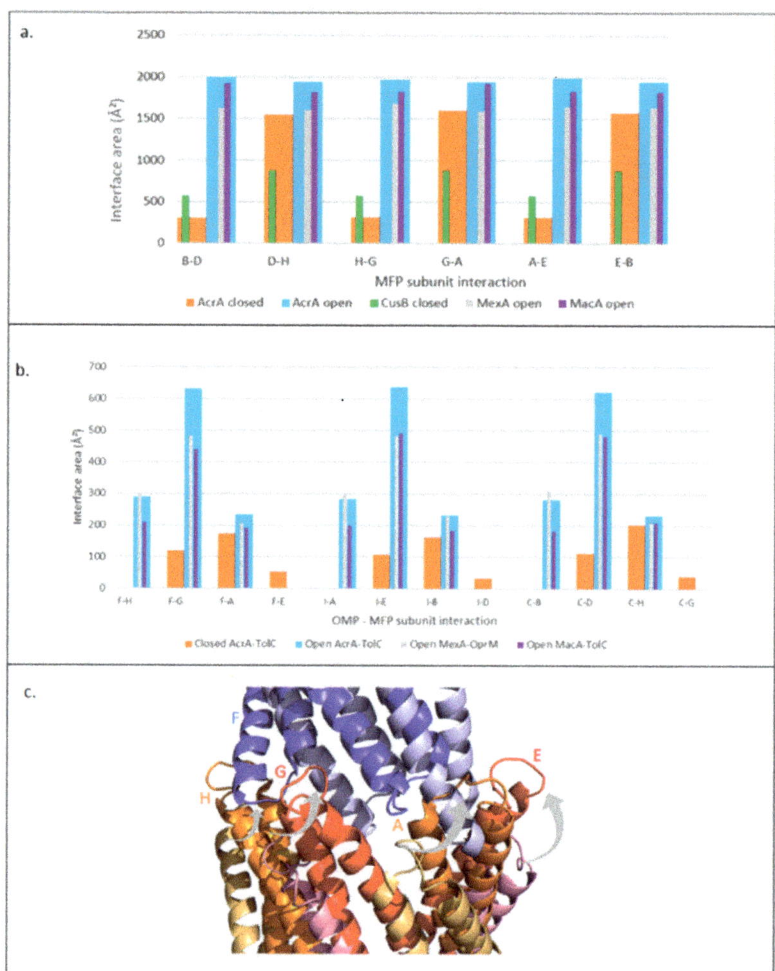

Figure 7. Trends in interface areas in other MFPs. (**a**) Interface area between subunits of different MFPs. (**b**) Tip-to-tip interface areas between OMP periplasmic helices and MFP HLH domains. OMP subunits: F, I, E. MFP subunits: H, G, A, E, B, D. (**c**) Aligned structures of AcrA-TolC tip-to-tip interaction. An individual TolC subunit is shown in the resting (light blue) and the transport (dark blue) state. Four AcrA subunits are shown in the resting (alternate pale yellow and pink) and the transport (alternate orange and red) state. Grey arrows show the direction of AcrA HLH twisting movement on pump activation. Results shown for Open MexAB-OprM (transport state) are from analysis of the 6TA6 PDB file; similar results were obtained for 6IOK and 6IOL as well (data not shown). Subunit numbering is based on AcrA 5NG5. MFP: Membrane Fusion Protein, OMP: Outer membrane protein.

We also investigated other pump assemblies to further characterise the MFP-OMP tip-to-tip interactions, assessing the trends and the contributions of individual subunits. The AcrA subunits that contribute to the interaction interface with TolC in the resting state are different to the subunits which interact in the transport state (Figure 7b,c). This is in accordance with the DDM results, where an iris-like opening motion of the HLH domain of AcrA (Figure 5b,d) and the outward movement of the TolC periplasmic helices (Figure 4b) was observed on pump activation. MexA and MacB seemed to form a similar pattern of interactions with their respective OMPs (Figure 7b), suggesting that different MFPs form assemblies through similar interactions.

3. Discussion

3.1. ECs Maintain Tertiary Fold but Role at Subunit Interfaces Is Less Clear

Previous investigations of sequence covariance in tripartite efflux pumps have provided support for the tip-to-tip interaction between the MFP and the OMP in the pump assembly [34,46], but the relationship between covariance and efflux pump dynamics has not yet been studied. In this work, we contextualised strong ECs with AcrA and TolC protein structures, finding that many strongly coupled residues act to maintain the tertiary fold, as has been extensively reported for many proteins [47,48]. We also identified coupled residues at the interface between TolC subunits (Figure 3a); most of these are physically close, and so are likely involved in maintaining the TolC trimeric assembly. However, one pair, Q164 and A382, was placed around 20 Å apart and so did not form a direct physical interaction, yet was still strongly evolutionarily coupled. This residue pair could potentially be involved in communicating allosteric change. Interestingly, the coupled TolC residue paired in the periplasmic helices, which have been proposed to form an intersubunit hydrogen bonding network to maintain the closed (resting) state of the channel (R367, T152, D153, and Y362 [49]) were not strongly evolutionarily coupled. We can hypothesise that through the transmission of interaction changes between the periplasmic helices upon pump activation, local residue pairs may need to be less strongly evolutionarily coupled, i.e., with various distances distribution thus making them highly dynamic, to maintain interface flexibility and allow communication of conformational changes.

In a similar manner, we found no strong ECs at the interface between AcrA subunits. This may be due to the dynamic nature of the AcrA hexamer, where interface interactions change dramatically in different conformations, thus changing distances between residue pairs [50,51]. In general, we found that whilst direct mapping of ECs onto structures can be useful in the study of protein dynamics, the throughput is too low to investigate large-scale conformational change in macromolecular complexes. One alternative to overcome this could be a vertical approach combining comparison of the modern proteins to their analogues from extinct species of the phylogenetic tree and MD simulations. This could allow to identify crucial residues involved in allosteric effects by narrowing down the number of residues to investigate. Such approach has helped to identify four residues in the β-subunit TrpB of Tryptophan synthase (TS) heterotetrameric αββα complex that are suggested to be essential for the communication between TrpB and the α-subunit TrpA [52].

3.2. Movement of TolC Periplasmic Helices and of AcrA BB Domain Appears to Be Evolutionary Conserved

Comparison of the ECs to the DDM analysis for both AcrA and TolC was found to be effective in relating sequence covariance to conformational change. Some areas of movement captured by the DDM, in particular the large outward movement of the HLH of AcrA, did not correlate with the ECs. As discussed above for the AcrA subunit interfaces, the lack of strongly coupled residues may give the domain more flexibility to change conformation, with different residue interactions and various distance distributions forming in the resting and transport states.

However, the DDM analysis also found some regions of movement that did correlate with the ECs (Figure 3b,c). The contraction of the TolC periplasmic helices correlated with some strongly coupled residues. This suggests that the movement, which likely acts to ensure the helices are optimally positioned to interact with AcrA and seal the transport channel, is conserved evolutionarily. The importance of the movement of these helices for channel opening has already been extensively reported [53,54]. In addition, the contraction of the BB domain of AcrA correlated with the ECs data, suggesting this movement is conserved as well. This finding is in accordance with previous work, which has shown that mutations in the AcrA BB domain can affect the BB domain folding in a way that impact AcrA conformational changes that influence interactions with AcrB and TolC and long-range transmission of conformational changes from AcrB to TolC [36,55]. This suggests that this domain could play a central role in long-distance communication. However, closer inspection of the correlation between the ECs and DDM data of the BB domain found the residues that move most are not themselves evolutionarily coupled. This suggests that the more static conserved residues may play a role in promoting the movement of the nearby, more mobile residues which then enable conformational change.

3.3. AcrA and TolC Show a Quaternary, Symmetrical Switch on Pump Activation

Through further investigation of the DDM analysis results, we were able to describe efflux pump conformational changes in more detail than by visual comparison of structures alone [13].

In AcrA, the BB domains, which contact AcrB, tightly contract, observed as an inward rolling movement of the BB ring (Figure 5e). This pulls the lipoyl domains to roll and contract, too. As HLH movement follows the BB and lipoyl domains movements, this suggests that the BB and lipoyl movement is communicated to the HLH domains via a flexible linker which pulls the HLH domains towards the lipoyl domains, causing the outward twisting of the HLHs. The HLH domain movement is distinct from the lipoyl domain movement, suggesting that the flexible hinge region between the domains is important, as it can enable such conformational independence. Crystal structures of AcrA indeed suggest a degree of flexibility in this region [50]. Such flexibility is suggested to also be supported by the cellular environment, as Cryo-ET structures of the AcrAB-TolC efflux pump show a rotation of AcrA protomers triggered by the lipid anchorage of AcrA in the inner membrane, leading to TolC opening [16]. In TolC, the DDM analysis showed that the periplasmic helices, in particular H7 (Figure 4b), moved outwards in the transport state while the rest of the TolC channel stayed relatively static, in accordance with previous reports [13]. Mutagenesis studies have shown the importance of the tip-to-tip interactions between the AcrA HLH domains and the TolC periplasmic helices for promoting the widening of the TolC pore [36]. The concerted movements of AcrA and TolC, discernible in detail in DDM results, act to seal the channel to prevent substrate leakage into the periplasm during drug extrusion.

Analysis of AcrA and TolC conformational changes at the tertiary and quaternary levels allowed insight into the key dynamics for pump activation. Although intrasubunit change was observed, for both AcrA and TolC, changes at the quaternary structure level were much more pronounced than at the tertiary level (Figures 4 and 5). Moreover, mutations at the tip of TolC which disrupt efflux pump function can be suppressed by mutations in AcrA, indicating the importance of the AcrA-TolC interface for efflux pump activity [36]. This suggests that changes at subunit interfaces may play an important role in communicating allostery in the pump. Interestingly, the intersubunit changes in AcrA and TolC upon pump activation are symmetrical. As AcrB adopts an asymmetric conformation on substrate binding, communication between AcrA subunits is likely important to ensure that all AcrA protomers, not just those in contact with the substrate-bound AcrB subunit, shift to a symmetrical transport state. The interdomain flexibility of AcrA may be important for this, as MD simulations have previously shown [51], but the DDM analysis in this work suggests that intersubunit flexibility may also play a role. Furthermore, as TolC is also embedded in the peptidoglycan layer, the dynamic of its interaction with the peptidoglycan and surrounding proteins may also affect both the subunits flexibility and the transmission of changes. A lipoprotein, the Braun's lipoprotein (Lpp), has been found to indirectly interact with AcrAB-TolC via the peptidoglycan layer. Lpp's role is thought to be important for pump assembly [56].

3.4. Increase in Interface Area in Transport State Indicates a Favoured Energy State

Analysis of changes in interface area in AcrA and TolC was performed to investigate energetic transitions between AcrAB-TolC states. The total interface area between AcrA subunits almost doubled from the resting to transport state and the interface area of the tip-to-tip interactions between TolC and AcrA also showed a large increase. The large changes in interface area between subunits were in accordance with the block DDM analysis, which showed contraction at a quaternary level. Taken together, these results indicate that quaternary, rather than intrasubunit, changes may be more significant in the pumping mechanism. Moreover, the increase in interface area suggests that the transport states of AcrA and TolC are at a more favourable energy than their resting states. Although a 10-Å axial contraction on pump activation had already been reported [13], this increase in interface area has not been recognised. The unexpected finding of a predicted lower energy state of AcrA and TolC on pump activation implies that energy must be provided for AcrA and TolC conformational changes to occur when switching back to the resting state.

Expansion of the interface analysis to the structures of homologous efflux pumps found similar patterns of interface changes between the resting and transport states. Of the selected homologues, MexA and AcrA are closely related, but CusB and MacB both belong to different phylogenetic clusters [57], so it is perhaps surprising that all the proteins form very similar interactions within their hexameric rings. However, these findings are in accordance with previous work which has shown the interchangeability of MFPs [14,58]. In addition, analysis of accessible surface area in the *Pseudomonas aeruginosa* pump, MexAB-OprM, found a strong affinity interaction between the MFP, MexA, and the OMP, OprM, in the transport state [28], supporting the findings shown here. The similar pattern of interface changes suggests that the lower energy transport state of the MFP and OMP may be a conserved feature of tripartite efflux pumps.

3.5. AcrAB-TolC Conformational Changes Suggest an Allosteric Transport Model

As discussed above, the open (transport) state of AcrA and TolC appeared to be at a lower energy than the closed (resting) state. This raises the paradox that the pump would be trapped in an energy minimum and not be able to efflux substrate. To reconcile this finding with the logical requirement for energetic transitions during AcrB pump cycling, we propose a model for conformational changes in AcrAB-TolC (Figure 8).

The model shown in Figure 8 shows a simplified description of the energetic transitions during the pump cycle. In the resting assembled pump (state A in Figure 8): the three AcrB subunits are all in the L conformation; AcrA is in an inactive, extended conformation, with an unsealed channel made of asymmetric protomers and the TolC channel pore is closed by the periplasmic helices [13]. Based on our interface area measurements, we propose that the pump subsequently shifts to a more stable transient intermediate state with AcrA and TolC in an open (transport) state and AcrB in an apo LTO state (state B). AcrB's shift to an LTO conformation on drug binding is thought to be coupled to proton binding to the periplasmic side of AcrB [37]. Therefore, we suggest that proton binding drives the conformational changes of AcrA and TolC to an open conformation. The conformational change of AcrB is communicated to AcrA through the extensive contacts between the docking and pore domains of AcrB and the MP and BB domains of AcrA [13].

On substrate binding to the periplasmic pocket of the AcrB O-state subunit [21], the energy of the pump is likely to be lowered (state C in Figure 8). AcrB will then cycle through conformations to extrude substrates (states C–F), whilst AcrA and TolC are predicted to stay in an open (transport) conformation [13]. Substrates are thought to bind to different AcrB subunits simultaneously and then transition through the same sequence of states: L to T to O, culminating with extrusion. Due to unfavourable energies, two AcrB subunits will never be in the O conformation at the same time [37]. The apparent fixed open conformation of AcrA and TolC during AcrB cycling may be due to the strong tip-to-tip interactions between the proteins seen in the interface analysis and due to the flexibility of AcrA. This flexibility likely acts to buffer the conformational changes of AcrB, preventing these from causing changes to the rest of the pump [50,51]. Structures of almost all of the conformational states of AcrB predicted to occur in the model during pump cycling have already been obtained, giving support to the proposed model. The only state that has not been experimentally observed is OTT. It would be useful to perform MD simulations for the OTT state to predict how energetically favourable this conformation would be.

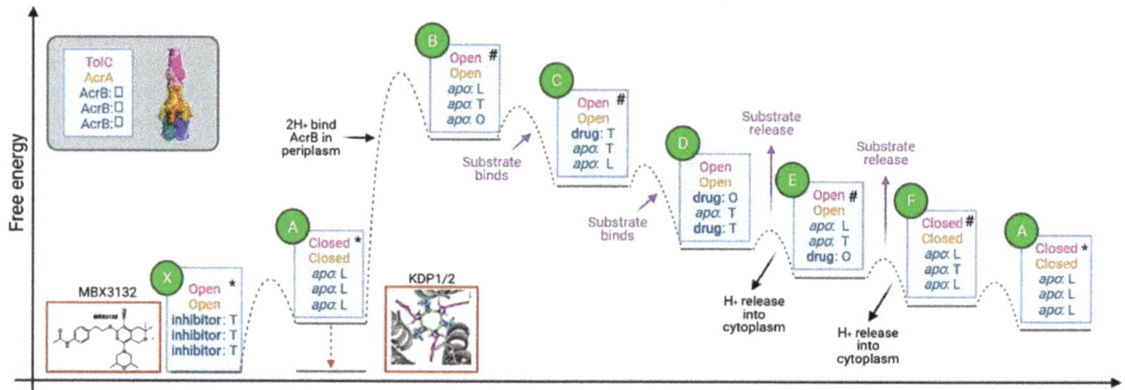

Figure 8. Proposed energy transitions during the AcrAB-TolC pump cycle. Qualitative energy diagram with relative energy levels of predicted states in the pump cycle. Text in blue boxes describes the conformations and the ligand binding states of TolC, AcrA, and the three AcrB subunits from top to bottom. Boxes outlined in red show pump inhibitors. * indicates states where high-resolution structures of the whole pump assembly exist. # indicates where high-resolution structures of AcrB in the specified conformation exist. The dotted lines indicate energy barriers to transition between the connected species, with maxima indicating transition states. AcrB subunit conformations are L = Loose, T = Tight, O = Open.

Release of substrates from AcrB through the AcrA-TolC pore (state E) is predicted to be coupled to cytoplasmic proton release to provide energy for this step [59]. Release of substrates and protons will lower the energy of the pump and so AcrA and TolC are predicted to transition back to the less favourable closed conformation with AcrB in the LLL state (state A).

This allosteric model highlights potential strategies to block pump function by trapping the pump in an energy minimum. The inhibitor MBX3132 is known to bind to the AcrB periplasmic pocket [60] in all three T state subunits with AcrA and TolC in the open (transport) state. The binding of MBX3132 to AcrB is predicted to lower the energy of the inhibitor-bound state (state X) relative to the equivalent apo state (state B). The model also highlights the central role of the PMF in promoting the conformational change needed for pump activation and so targeting the generation or maintenance of the PMF would offer an alternative strategy to inhibit the pump.

Although our diagram in Figure 8 suggests a 1:1 proton: drug stoichiometry, the model does not presuppose a specific ratio. Indeed, some studies suggest that two protons bind per cycle [37], whilst others report that only one is involved [59]. It has been proposed that changes in the pH of the periplasm due to the PMF activity of AcrB could induce AcrA to change conformation directly, and this could act instead of the conformational changes being transduced from AcrB on proton binding [55,61]. However, mutations to

AcrB residues involved in the proton-relay network abolish the activity of AcrAB-TolC, despite not causing substantial structural change, indicating the importance of proton binding for the pump's conformational changes to occur [37,62]. In addition, molecular dynamic simulations of AcrB in the presence of indole (as a substrate of protonated systems) have shown that the protonation of aspartic residues involved in the proton-relay network triggers conformational changes from binding state to extrusion state in one of the three monomers and that all three monomers return to access state in absence of indole [63]. In a similar way, the protonation of proton-relay aspartic pair of MFS transporters is likely to cause conformational changes of the N-terminal transmembrane subdomain towards the cytosolic direction [64]. The pH of the periplasm has been proposed to be buffered [65], which could also impact the extent of AcrA changes. Studies involving AcrB fluorescent substrates have shown that substrate dissociation rates are inversely related to pH, supporting the notion that protonation triggers conformational change [66].

Our model did not consider pump assembly or disassembly. Energy from the PMF is not thought to be required for pump assembly [67], but some in vitro studies suggest that the PMF may promote its disassembly [68,69]. In such case the pump may not cycle repeatedly through the energetic transitions shown in Figure 8, but instead only transition through the states once before disassembling.

3.6. The Allosteric Transport Model Might Be Common across Efflux Pumps

The shared pattern of interface changes found in this report, and the previously reported interchangeability of MFPs [14,58], suggest that the manner of energetic transitions proposed here for AcrAB-TolC may be conserved in other tripartite efflux pumps. Necessarily, the details of the conformational changes that occur will vary between pumps. For instance, the MFP of the BesABC RND pump from *Borrelia burgdorferi*, BesA, does not have a HLH domain and is predicted to form weaker interactions with the OMP BesC, with a smaller interaction interface [70]. Therefore, the active transport state of the pump will be less favourable than AcrAB-TolC and the pump may require less energy to move to the active conformation. The RND transporter AdeB from the AdeABC tripartite system of *Acinetobacter baumannii* shows structural features that are distinct from other RND trimeric pumps, as its binding, access, and extrusion states all present with two periplasmic clefts open and one cleft closed [71]. The structure of AdeB in complex with ethidium bromide shows that one protomer can simultaneously accommodate three ethidium bromide molecules and that protomers within AdeB trimer are able to independently export substrates. The substrates also tend to stabilize the trimer assembly, which may facilitate the MFP AdeA binding and substrate extrusion. Structural characterisation of more tripartite pump assemblies in different conformations would provide insight into how the allosteric mechanism of efflux pumps has been conserved and how they have diverged.

Both the ECs and DDM analyses performed here only provide information on pairwise interactions between residues, and as allostery in multi-subunit complexes is likely to be highly complex, approaches such as neural network [72] or elastic network analysis [73] may help to provide further analysis. One other approach could involve the use of a platform for allosteric analysis such as Ohm [74] to predict and map the location of allosteric sites. As with the work presented here, combining different approaches is likely to be the most fruitful way to gain insight into long-distance communication in multi-subunit efflux pumps.

4. Materials and Methods

4.1. Evolutionary Couplings

Evolutionary couplings data were calculated using the freely accessible Evolutionary Couplings Server (https://v1.evcouplings.org/, accessed on 1 April 2021) [40] queried with TolC and AcrA sequences (UniProt IDs: P02930 and P0AE06, respectively). A percentage of sequence identity is defined to down-weight redundant sequences during couplings calculation, and a value of 0.8 (at least 80% of identity) was used for the clustering threshold. Sequences taken from the UniRef90 database were used for multiple sequence alignments built and the number of sequences to include in the alignment was defined with a bit score range of 0.2 (Higher bitscores (=1) include less sequence in the alignment). Alignment positions and fraction residues (rather than gaps) were included, with a position gap threshold of 0.5 (allowing alignment positions with up to 50% gaps) and a sequence gap threshold of 0.7 (allowing sequences with up to 30% gaps). Pseudo-likelihood maximisation (plmDCA) was used during couplings calculation. Results were plotted using R/ggplot library from the R/tidyverse library used under the R studio environment. Only ECs with cn values more than or equal to 1 were plotted.

4.2. Difference Distance Matrix

A custom algorithm for the calculation of DDM was developed in Python programming language; scripts to generate DDM calculations are publicly available on GitHub (https://github.com/malitharatnaonc/differencedistance, accessed on 1 April 2021). Molecular models of the open (transport) and closed (resting) states of the pump (PDB IDs: 5V5S and 5NG5, respectively) were used as input. Outputted data were plotted using the R/reshape2 and R/tidyverse libraries in R studio environment. To perform the block analysis, the mean DDM of the domains were calculated in R studio.

4.3. Interface Surface Area

Interface surface areas were calculated using 'Protein interfaces, surfaces and assemblies' (PISA) service at the European Bioinformatics Institute, (http://www.ebi.ac.uk/pdbe/prot_int/pistart.html, accessed on 1 April 2021) [41], which measures the difference in total accessible surface areas of isolated and interfacing structures. To measure interface area changes of isolated domains, we generated structure models using Coot or PyMOL that encompass the contacting domains. The following PDB files were used: 5V5S and 5NG5 for the open (transport) and closed (resting) states of AcrAB-TolC, 3NE5 for the closed state of CusAB, 5NIK for the open state of MacAB-TolC, and 6TA6, 6IOK, 6IOL for the open state of MexAB-OprM.

5. Conclusions

The allosteric model proposed here for the mechanism of action of tripartite efflux pumps serves as a starting point for future experimental and computational research. Recent work has already found evidence of additional interactions that add complexity to the model for pump allostery. For instance, it has been found that AcrA may act as a necrosignal, binding to the external-facing BB domain of TolC to stimulate pump efflux as an adaptive resistance mechanism [75]. To gain further insight into the allostery of tripartite efflux pumps, a range of biochemical, structural, and bioinformatic approaches will be necessary, and hopefully the complex mechanism of these macromolecular machines will become clearer.

Author Contributions: Conceptualization, B.F.L.; methodology, A.W., M.R., B.F.L. and V.Y.N.E.; validation, B.F.L. and V.Y.N.E.; formal analysis, A.W. and M.R.; investigation, A.W.; resources, A.W. and M.R.; data curation, A.W. and M.R.; writing—original draft preparation, A.W. and V.Y.N.E.; writing—review and editing, A.W., M.R., A.H., B.F.L. and V.Y.N.E.; supervision, B.F.L. and V.Y.N.E.; project administration, B.F.L. and V.Y.N.E.; funding acquisition, B.F.L. All authors have read and agreed to the published version of the manuscript.

Funding: This work was funded by an ERC Advanced award to BFL (742210).

Institutional Review Board Statement: Not applicable.

Informed Consent Statement: Not applicable.

Data Availability Statement: The data presented in this study are available in the article. Structures used in this study are available in the PDB database (pdb codes: 3NE5, 5V5S, 5NG5, 6TA6, 5NIK, 1A3N and 2DN1), UniRef90 and Uniprot database (accession numbers: P02930 and P0AE06). The freely accessible Evolutionary Couplings Server (https://v1.evcouplings.org/, accessed on 1 April 2021) was used for Evolutionary couplings data calculation.

Acknowledgments: We thank Jim Smith, Kotryna Bloznelyte and Christopher Beaudoin for helpful comments on the manuscript.

Conflicts of Interest: The authors declare no conflict of interest.

References

1. No Time to Wait: Securing the Future from Drug-Resistant Infections. Available online: https://www.who.int/publications/i/item/no-time-to-wait-securing-the-future-from-drug-resistant-infections (accessed on 29 November 2021).
2. Årdal, C.; Balasegaram, M.; Laxminarayan, R.; McAdams, D.; Outterson, K.; Rex, J.H.; Sumpradit, N. Antibiotic Development-Economic, Regulatory and Societal Challenges. *Nat. Rev. Microbiol.* **2020**, *18*, 267–274. [CrossRef]
3. Aslam, B.; Wang, W.; Arshad, M.I.; Khurshid, M.; Muzammil, S.; Rasool, M.H.; Nisar, M.A.; Alvi, R.F.; Aslam, M.A.; Qamar, M.U.; et al. Antibiotic Resistance: A Rundown of a Global Crisis. *Infect. Drug Resist.* **2018**, *11*, 1645–1658. [CrossRef]
4. Li, X.-Z.; Plésiat, P.; Nikaido, H. The Challenge of Efflux-Mediated Antibiotic Resistance in Gram-Negative Bacteria. *Clin. Microbiol. Rev.* **2015**, *28*, 337–418. [CrossRef] [PubMed]
5. Blair, J.M.A.; Webber, M.A.; Baylay, A.J.; Ogbolu, D.O.; Piddock, L.J.V. Molecular Mechanisms of Antibiotic Resistance. *Nat. Rev. Microbiol.* **2015**, *13*, 42–51. [CrossRef] [PubMed]
6. Sun, J.; Deng, Z.; Yan, A. Bacterial Multidrug Efflux Pumps: Mechanisms, Physiology and Pharmacological Exploitations. *Biochem. Biophys. Res. Commun.* **2014**, *453*, 254–267. [CrossRef]
7. Du, D.; Wang-Kan, X.; Neuberger, A.; van Veen, H.W.; Pos, K.M.; Piddock, L.J.V.; Luisi, B.F. Multidrug Efflux Pumps: Structure, Function and Regulation. *Nat. Rev. Microbiol.* **2018**, *16*, 523–539. [CrossRef] [PubMed]
8. Alav, I.; Kobylka, J.; Kuth, M.S.; Pos, K.M.; Picard, M.; Blair, J.M.A.; Bavro, V.N. Structure, Assembly, and Function of Tripartite Efflux and Type 1 Secretion Systems in Gram-Negative Bacteria. *Chem. Rev.* **2021**, *121*, 5479–5596. [CrossRef]
9. McMurry, L.; Petrucci, R.E.; Levy, S.B. Active Efflux of Tetracycline Encoded by Four Genetically Different Tetracycline Resistance Determinants in Escherichia Coli. *Proc. Natl. Acad. Sci. USA* **1980**, *77*, 3974–3977. [CrossRef]
10. Hassan, K.A.; Liu, Q.; Henderson, P.J.F.; Paulsen, I.T. Homologs of the Acinetobacter Baumannii AceI Transporter Represent a New Family of Bacterial Multidrug Efflux Systems. *mBio* **2015**, *6*, e01982-14. [CrossRef]
11. Du, D.; Wang, Z.; James, N.R.; Voss, J.E.; Klimont, E.; Ohene-Agyei, T.; Venter, H.; Chiu, W.; Luisi, B.F. Structure of the AcrAB-TolC Multidrug Efflux Pump. *Nature* **2014**, *509*, 512–515. [CrossRef]
12. Jeong, H.; Kim, J.-S.; Song, S.; Shigematsu, H.; Yokoyama, T.; Hyun, J.; Ha, N.-C. Pseudoatomic Structure of the Tripartite Multidrug Efflux Pump AcrAB-TolC Reveals the Intermeshing Cogwheel-like Interaction between AcrA and TolC. *Structure* **2016**, *24*, 272–276. [CrossRef] [PubMed]
13. Wang, Z.; Fan, G.; Hryc, C.F.; Blaza, J.N.; Serysheva, I.I.; Schmid, M.F.; Chiu, W.; Luisi, B.F.; Du, D. An Allosteric Transport Mechanism for the AcrAB-TolC Multidrug Efflux Pump. *eLife* **2017**, *6*, e24905. [CrossRef] [PubMed]
14. Daury, L.; Orange, F.; Taveau, J.-C.; Verchère, A.; Monlezun, L.; Gounou, C.; Marreddy, R.K.R.; Picard, M.; Broutin, I.; Pos, K.M.; et al. Tripartite Assembly of RND Multidrug Efflux Pumps. *Nat. Commun.* **2016**, *7*, 10731. [CrossRef]
15. Shi, X.; Chen, M.; Yu, Z.; Bell, J.M.; Wang, H.; Forrester, I.; Villarreal, H.; Jakana, J.; Du, D.; Luisi, B.F.; et al. In Situ Structure and Assembly of the Multidrug Efflux Pump AcrAB-TolC. *Nat. Commun.* **2019**, *10*, 2635. [CrossRef] [PubMed]
16. Chen, M.; Shi, X.; Yu, Z.; Fan, G.; Serysheva, I.I.; Baker, M.L.; Luisi, B.F.; Ludtke, S.J.; Wang, Z. In Situ Structure of the AcrAB-TolC Efflux Pump at Subnanometer Resolution. *Structure* **2021**. [CrossRef]
17. Spengler, G.; Kincses, A.; Gajdács, M.; Amaral, L. New Roads Leading to Old Destinations: Efflux Pumps as Targets to Reverse Multidrug Resistance in Bacteria. *Molecules* **2017**, *22*, 468. [CrossRef] [PubMed]
18. Neuberger, A.; Du, D.; Luisi, B.F. Structure and Mechanism of Bacterial Tripartite Efflux Pumps. *Res. Microbiol.* **2018**, *169*, 401–413. [CrossRef]
19. Nakashima, R.; Sakurai, K.; Yamasaki, S.; Nishino, K.; Yamaguchi, A. Structures of the Multidrug Exporter AcrB Reveal a Proximal Multisite Drug-Binding Pocket. *Nature* **2011**, *480*, 565–569. [CrossRef] [PubMed]
20. Seeger, M.A.; Schiefner, A.; Eicher, T.; Verrey, F.; Diederichs, K.; Pos, K.M. Structural Asymmetry of AcrB Trimer Suggests a Peristaltic Pump Mechanism. *Science* **2006**, *313*, 1295–1298. [CrossRef]
21. Murakami, S.; Nakashima, R.; Yamashita, E.; Matsumoto, T.; Yamaguchi, A. Crystal Structures of a Multidrug Transporter Reveal a Functionally Rotating Mechanism. *Nature* **2006**, *443*, 173–179. [CrossRef]
22. Hobbs, E.C.; Yin, X.; Paul, B.J.; Astarita, J.L.; Storz, G. Conserved Small Protein Associates with the Multidrug Efflux Pump AcrB and Differentially Affects Antibiotic Resistance. *Proc. Natl. Acad. Sci. USA* **2012**, *109*, 16696–16701. [CrossRef]
23. Du, D.; Neuberger, A.; Orr, M.W.; Newman, C.E.; Hsu, P.-C.; Samsudin, F.; Szewczak-Harris, A.; Ramos, L.M.; Debela, M.; Khalid, S.; et al. Interactions of a Bacterial RND Transporter with a Transmembrane Small Protein in a Lipid Environment. *Structure* **2020**, *28*, 625–634.e6. [CrossRef]
24. Emsley, P.; Cowtan, K. Coot: Model-Building Tools for Molecular Graphics. *Acta Crystallogr. D Biol. Crystallogr.* **2004**, *60*, 2126–2132. [CrossRef] [PubMed]
25. Emsley, P.; Lohkamp, B.; Scott, W.G.; Cowtan, K. Features and Development of *Coot. Acta Crystallogr. D Biol. Crystallogr.* **2010**, *66*, 486–501. [CrossRef] [PubMed]

26. Su, C.-C.; Long, F.; Zimmermann, M.T.; Rajashankar, K.R.; Jernigan, R.L.; Yu, E.W. Crystal Structure of the CusBA Heavy-Metal Efflux Complex of Escherichia Coli. *Nature* **2011**, *470*, 558–562. [CrossRef]

27. Tsutsumi, K.; Yonehara, R.; Ishizaka-Ikeda, E.; Miyazaki, N.; Maeda, S.; Iwasaki, K.; Nakagawa, A.; Yamashita, E. Structures of the Wild-Type MexAB-OprM Tripartite Pump Reveal Its Complex Formation and Drug Efflux Mechanism. *Nat. Commun.* **2019**, *10*, 1520. [CrossRef]

28. Glavier, M.; Puvanendran, D.; Salvador, D.; Decossas, M.; Phan, G.; Garnier, C.; Frezza, E.; Cece, Q.; Schoehn, G.; Picard, M.; et al. Antibiotic Export by MexB Multidrug Efflux Transporter Is Allosterically Controlled by a MexA-OprM Chaperone-like Complex. *Nat. Commun.* **2020**, *11*, 4948. [CrossRef]

29. Fitzpatrick, A.W.P.; Llabrés, S.; Neuberger, A.; Blaza, J.N.; Bai, X.-C.; Okada, U.; Murakami, S.; van Veen, H.W.; Zachariae, U.; Scheres, S.H.W.; et al. Structure of the MacAB-TolC ABC-Type Tripartite Multidrug Efflux Pump. *Nat. Microbiol.* **2017**, *2*, 17070. [CrossRef]

30. Stegmeier, J.F.; Polleichtner, G.; Brandes, N.; Hotz, C.; Andersen, C. Importance of the Adaptor (Membrane Fusion) Protein Hairpin Domain for the Functionality of Multidrug Efflux Pumps. *Biochemistry* **2006**, *45*, 10303–10312. [CrossRef]

31. Simsir, M.; Broutin, I.; Mus-Veteau, I.; Cazals, F. Studying Dynamics without Explicit Dynamics: A Structure-Based Study of the Export Mechanism by AcrB. *Proteins* **2021**, *89*, 259–275. [CrossRef] [PubMed]

32. Wang, B.; Weng, J.; Wang, W. Substrate Binding Accelerates the Conformational Transitions and Substrate Dissociation in Multidrug Efflux Transporter AcrB. *Front. Microbiol.* **2015**, *6*, 302. [CrossRef] [PubMed]

33. Tam, H.-K.; Foong, W.E.; Oswald, C.; Herrmann, A.; Zeng, H.; Pos, K.M. Allosteric Drug Transport Mechanism of Multidrug Transporter AcrB. *Nat. Commun.* **2021**, *12*, 3889. [CrossRef]

34. López, C.A.; Travers, T.; Pos, K.M.; Zgurskaya, H.I.; Gnanakaran, S. Dynamics of Intact MexAB-OprM Efflux Pump: Focusing on the MexA-OprM Interface. *Sci. Rep.* **2017**, *7*, 16521. [CrossRef]

35. Vargiu, A.V.; Ramaswamy, V.K.; Malloci, G.; Malvacio, I.; Atzori, A.; Ruggerone, P. Computer Simulations of the Activity of RND Efflux Pumps. *Res. Microbiol.* **2018**, *169*, 384–392. [CrossRef] [PubMed]

36. Weeks, J.W.; Celaya-Kolb, T.; Pecora, S.; Misra, R. AcrA Suppressor Alterations Reverse the Drug Hypersensitivity Phenotype of a TolC Mutant by Inducing TolC Aperture Opening. *Mol. Microbiol.* **2010**, *75*, 1468–1483. [CrossRef]

37. Eicher, T.; Seeger, M.A.; Anselmi, C.; Zhou, W.; Brandstätter, L.; Verrey, F.; Diederichs, K.; Faraldo-Gómez, J.D.; Pos, K.M. Coupling of Remote Alternating-Access Transport Mechanisms for Protons and Substrates in the Multidrug Efflux Pump AcrB. *eLife* **2014**, *3*, e03145. [CrossRef]

38. Schneider, T.R. Objective Comparison of Protein Structures: Error-Scaled Difference Distance Matrices. *Acta Crystallogr. D Biol. Crystallogr.* **2000**, *56*, 714–721. [CrossRef] [PubMed]

39. Kachalova, G.S.; Popov, A.N.; Bartunik, H.D. A Steric Mechanism for Inhibition of CO Binding to Heme Proteins. *Science* **1999**, *284*, 473–476. [CrossRef]

40. Hopf, T.A.; Green, A.G.; Schubert, B.; Mersmann, S.; Schärfe, C.P.I.; Ingraham, J.B.; Toth-Petroczy, A.; Brock, K.; Riesselman, A.J.; Palmedo, P.; et al. The EVcouplings Python Framework for Coevolutionary Sequence Analysis. *Bioinformatics* **2019**, *35*, 1582–1584. [CrossRef]

41. Krissinel, E.; Henrick, K. Inference of Macromolecular Assemblies from Crystalline State. *J. Mol. Biol.* **2007**, *372*, 774–797. [CrossRef]

42. Jeon, J.; Nam, H.-J.; Choi, Y.S.; Yang, J.-S.; Hwang, J.; Kim, S. Molecular Evolution of Protein Conformational Changes Revealed by a Network of Evolutionarily Coupled Residues. *Mol. Biol. Evol.* **2011**, *28*, 2675–2685. [CrossRef]

43. Sutto, L.; Marsili, S.; Valencia, A.; Gervasio, F.L. From Residue Coevolution to Protein Conformational Ensembles and Functional Dynamics. *Proc. Natl. Acad. Sci. USA* **2015**, *112*, 13567–13572. [CrossRef] [PubMed]

44. Chen, J.; Sawyer, N.; Regan, L. Protein-Protein Interactions: General Trends in the Relationship between Binding Affinity and Interfacial Buried Surface Area. *Protein Sci.* **2013**, *22*, 510–515. [CrossRef] [PubMed]

45. Nooren, I.M.A.; Thornton, J.M. Diversity of Protein-Protein Interactions. *EMBO J.* **2003**, *22*, 3486–3492. [CrossRef] [PubMed]

46. Ovchinnikov, S.; Kamisetty, H.; Baker, D. Robust and Accurate Prediction of Residue-Residue Interactions across Protein Interfaces Using Evolutionary Information. *eLife* **2014**, *3*, e02030. [CrossRef]

47. Marks, D.S.; Colwell, L.J.; Sheridan, R.; Hopf, T.A.; Pagnani, A.; Zecchina, R.; Sander, C. Protein 3D Structure Computed from Evolutionary Sequence Variation. *PLoS ONE* **2011**, *6*, e28766. [CrossRef]

48. Anfinsen, C.B. Principles That Govern the Folding of Protein Chains. *Science* **1973**, *181*, 223–230. [CrossRef]

49. Bavro, V.N.; Pietras, Z.; Furnham, N.; Pérez-Cano, L.; Fernández-Recio, J.; Pei, X.Y.; Misra, R.; Luisi, B. Assembly and Channel Opening in a Bacterial Drug Efflux Machine. *Mol. Cell* **2008**, *30*, 114–121. [CrossRef]

50. Mikolosko, J.; Bobyk, K.; Zgurskaya, H.I.; Ghosh, P. Conformational Flexibility in the Multidrug Efflux System Protein AcrA. *Structure* **2006**, *14*, 577–587. [CrossRef]

51. Hazel, A.J.; Abdali, N.; Leus, I.V.; Parks, J.M.; Smith, J.C.; Zgurskaya, H.I.; Gumbart, J.C. Conformational Dynamics of AcrA Govern Multidrug Efflux Pump Assembly. *ACS Infect. Dis.* **2019**, *5*, 1926–1935. [CrossRef]

52. Schupfner, M.; Straub, K.; Busch, F.; Merkl, R.; Sterner, R. Analysis of Allosteric Communication in a Multienzyme Complex by Ancestral Sequence Reconstruction. *Proc. Natl. Acad. Sci. USA* **2020**, *117*, 346–354. [CrossRef]

53. Andersen, C.; Koronakis, E.; Bokma, E.; Eswaran, J.; Humphreys, D.; Hughes, C.; Koronakis, V. Transition to the Open State of the TolC Periplasmic Tunnel Entrance. *Proc. Natl. Acad. Sci. USA* **2002**, *99*, 11103–11108. [CrossRef] [PubMed]

54. Zgurskaya, H.I.; Krishnamoorthy, G.; Ntreh, A.; Lu, S. Mechanism and Function of the Outer Membrane Channel TolC in Multidrug Resistance and Physiology of Enterobacteria. *Front. Microbiol.* **2011**, *2*, 189. [CrossRef] [PubMed]
55. Zgurskaya, H.I.; Weeks, J.W.; Ntreh, A.T.; Nickels, L.M.; Wolloscheck, D. Mechanism of Coupling Drug Transport Reactions Located in Two Different Membranes. *Front. Microbiol.* **2015**, *6*, 100. [CrossRef] [PubMed]
56. Gumbart, J.C.; Ferreira, J.L.; Hwang, H.; Hazel, A.J.; Cooper, C.J.; Parks, J.M.; Smith, J.C.; Zgurskaya, H.I.; Beeby, M. Lpp Positions Peptidoglycan at the AcrA-TolC Interface in the AcrAB-TolC Multidrug Efflux Pump. *Biophys. J.* **2021**, *120*, 3973–3982. [CrossRef] [PubMed]
57. Zgurskaya, H.I.; Yamada, Y.; Tikhonova, E.B.; Ge, Q.; Krishnamoorthy, G. Structural and Functional Diversity of Bacterial Membrane Fusion Proteins. *Biochim. Biophys. Acta* **2009**, *1794*, 794–807. [CrossRef] [PubMed]
58. Krishnamoorthy, G.; Tikhonova, E.B.; Zgurskaya, H.I. Fitting Periplasmic Membrane Fusion Proteins to Inner Membrane Transporters: Mutations That Enable Escherichia Coli AcrA to Function with Pseudomonas Aeruginosa MexB. *J. Bacteriol.* **2008**, *190*, 691–698. [CrossRef] [PubMed]
59. Yue, Z.; Chen, W.; Zgurskaya, H.I.; Shen, J. Constant PH Molecular Dynamics Reveals How Proton Release Drives the Conformational Transition of a Transmembrane Efflux Pump. *J. Chem. Theory Comput.* **2017**, *13*, 6405–6414. [CrossRef] [PubMed]
60. Sjuts, H.; Vargiu, A.V.; Kwasny, S.M.; Nguyen, S.T.; Kim, H.-S.; Ding, X.; Ornik, A.R.; Ruggerone, P.; Bowlin, T.L.; Nikaido, H.; et al. Molecular Basis for Inhibition of AcrB Multidrug Efflux Pump by Novel and Powerful Pyranopyridine Derivatives. *Proc. Natl. Acad. Sci. USA* **2016**, *113*, 3509–3514. [CrossRef] [PubMed]
61. Wang, B.; Weng, J.; Fan, K.; Wang, W. Interdomain Flexibility and PH-Induced Conformational Changes of AcrA Revealed by Molecular Dynamics Simulations. *J. Phys. Chem. B* **2012**, *116*, 3411–3420. [CrossRef]
62. Liu, M.; Zhang, X.C. Energy-Coupling Mechanism of the Multidrug Resistance Transporter AcrB: Evidence for Membrane Potential-Driving Hypothesis through Mutagenic Analysis. *Protein Cell* **2017**, *8*, 623–627. [CrossRef]
63. Jewel, Y.; Van Dinh, Q.; Liu, J.; Dutta, P. Substrate-dependent Transport Mechanism in AcrB of Multidrug Resistant Bacteria. *Proteins* **2020**, *88*, 853–864. [CrossRef]
64. Zhang, X.C.; Zhao, Y.; Heng, J.; Jiang, D. Energy Coupling Mechanisms of MFS Transporters: Energy Coupling Mechanisms of MFS Transporters. *Protein Sci.* **2015**, *24*, 1560–1579. [CrossRef] [PubMed]
65. Krulwich, T.A.; Sachs, G.; Padan, E. Molecular Aspects of Bacterial PH Sensing and Homeostasis. *Nat. Rev. Microbiol.* **2011**, *9*, 330–343. [CrossRef]
66. Su, C.-C.; Nikaido, H.; Yu, E.W. Ligand-Transporter Interaction in the AcrB Multidrug Efflux Pump Determined by Fluorescence Polarization Assay. *FEBS Lett.* **2007**, *581*, 4972–4976. [CrossRef] [PubMed]
67. Tikhonova, E.B.; Zgurskaya, H.I. AcrA, AcrB, and TolC of Escherichia Coli Form a Stable Intermembrane Multidrug Efflux Complex. *J. Biol. Chem.* **2004**, *279*, 32116–32124. [CrossRef] [PubMed]
68. Janganan, T.K.; Bavro, V.N.; Zhang, L.; Borges-Walmsley, M.I.; Walmsley, A.R. Tripartite Efflux Pumps: Energy Is Required for Dissociation, but Not Assembly or Opening of the Outer Membrane Channel of the Pump. *Mol. Microbiol.* **2013**, *88*, 590–602. [CrossRef]
69. Ntsogo Enguéné, V.Y.; Verchère, A.; Phan, G.; Broutin, I.; Picard, M. Catch Me If You Can: A Biotinylated Proteoliposome Affinity Assay for the Investigation of Assembly of the MexA-MexB-OprM Efflux Pump from Pseudomonas Aeruginosa. *Front. Microbiol.* **2015**, *6*, 541. [CrossRef] [PubMed]
70. Bunikis, I.; Denker, K.; Ostberg, Y.; Andersen, C.; Benz, R.; Bergström, S. An RND-Type Efflux System in Borrelia Burgdorferi Is Involved in Virulence and Resistance to Antimicrobial Compounds. *PLoS Pathog.* **2008**, *4*, e1000009. [CrossRef] [PubMed]
71. Morgan, C.E.; Glaza, P.; Leus, I.V.; Trinh, A.; Su, C.-C.; Cui, M.; Zgurskaya, H.I.; Yu, E.W. Cryoelectron Microscopy Structures of AdeB Illuminate Mechanisms of Simultaneous Binding and Exporting of Substrates. *mBio* **2021**, *12*, e03690-20. [CrossRef] [PubMed]
72. Bravi, B.; Ravasio, R.; Brito, C.; Wyart, M. Direct Coupling Analysis of Epistasis in Allosteric Materials. *PLoS Comput. Biol.* **2020**, *16*, e1007630. [CrossRef] [PubMed]
73. Wang, B.; Weng, J.; Fan, K.; Wang, W. Elastic Network Model-Based Normal Mode Analysis Reveals the Conformational Couplings in the Tripartite AcrAB-TolC Multidrug Efflux Complex. *Proteins* **2011**, *79*, 2936–2945. [CrossRef] [PubMed]
74. Wang, J.; Jain, A.; McDonald, L.R.; Gambogi, C.; Lee, A.L.; Dokholyan, N.V. Mapping Allosteric Communications within Individual Proteins. *Nat. Commun.* **2020**, *11*, 3862. [CrossRef] [PubMed]
75. Bhattacharyya, S.; Walker, D.M.; Harshey, R.M. Dead Cells Release a "necrosignal" That Activates Antibiotic Survival Pathways in Bacterial Swarms. *Nat. Commun.* **2020**, *11*, 4157. [CrossRef] [PubMed]

Article

RND Efflux Systems Contribute to Resistance and Virulence of *Aliarcobacter butzleri*

Cristiana Mateus [1], Ana Rita Nunes [1], Mónica Oleastro [2], Fernanda Domingues [1] and Susana Ferreira [1,*]

[1] CICS-UBI-Health Sciences Research Centre, University of Beira Interior, 6200-506 Covilhã, Portugal; cristiana.lopes.mateus@ubi.pt (C.M.); anaritalmeidan@hotmail.com (A.R.N.); fcd@ubi.pt (F.D.)
[2] National Reference Laboratory for Gastrointestinal Infections, Department of Infectious Diseases, National Institute of Health Dr. Ricardo Jorge, Av. Padre Cruz, 1649-016 Lisbon, Portugal; monica.oleastro@insa.min-saude.pt
* Correspondence: susana.ferreira@fcsaude.ubi.pt

Abstract: *Aliarcobacter butzleri* is an emergent enteropathogen that can be found in a range of environments. This bacterium presents a vast repertoire of efflux pumps, such as the ones belonging to the resistance nodulation cell division family, which may be associated with bacterial resistance, as well as virulence. Thus, this work aimed to evaluate the contribution of three RND efflux systems, AreABC, AreDEF and AreGHI, in the resistance and virulence of *A. butzleri*. Mutant strains were constructed by inactivation of the gene that encodes the inner membrane protein of these systems. The bacterial resistance profile of parental and mutant strains to several antimicrobials was assessed, as was the intracellular accumulation of the ethidium bromide dye. Regarding bacterial virulence, the role of these three efflux pumps on growth, strain fitness, motility, biofilm formation ability, survival in adverse conditions (oxidative stress and bile salts) and human serum and *in vitro* adhesion and invasion to Caco-2 cells was evaluated. We observed that the mutants from the three efflux pumps were more susceptible to several classes of antimicrobials than the parental strain and presented an increase in the accumulation of ethidium bromide, indicating a potential role of the efflux pumps in the extrusion of antimicrobials. The mutant strains had no bacterial growth defects; nonetheless, they presented a reduction in relative fitness. For the three mutants, an increase in the susceptibility to oxidative stress was observed, while only the mutant for AreGHI efflux pump showed a relevant role in bile stress survival. All the mutant strains showed an impairment in biofilm formation ability, were more susceptible to human serum and were less adherent to intestinal epithelial cells. Overall, the results support the contribution of the efflux pumps AreABC, AreDEF and AreGHI of *A. butzleri* to antimicrobial resistance, as well as to bacterial virulence.

Keywords: *Aliarcobacter butzleri*; RND efflux pumps; virulence; resistance

Citation: Mateus, C.; Nunes, A.R.; Oleastro, M.; Domingues, F.; Ferreira, S. RND Efflux Systems Contribute to Resistance and Virulence of *Aliarcobacter butzleri*. *Antibiotics* **2021**, *10*, 823. https://doi.org/10.3390/antibiotics10070823

Academic Editors: Isabelle Broutin, Attilio V Vargiu, Henrietta Venter and Gilles Phan

Received: 31 May 2021
Accepted: 5 July 2021
Published: 6 July 2021

Publisher's Note: MDPI stays neutral with regard to jurisdictional claims in published maps and institutional affiliations.

Copyright: © 2021 by the authors. Licensee MDPI, Basel, Switzerland. This article is an open access article distributed under the terms and conditions of the Creative Commons Attribution (CC BY) license (https://creativecommons.org/licenses/by/4.0/).

1. Introduction

Aliarcobacter genus belongs to the *Arcobacteraceae* family, with five other genera, *Arcobacter*, *Pseudarcobacter* gen. nov., *Halarcobacter* gen. nov., *Malaciobacter* gen. nov. and *Poseidonibacter* gen. nov., and a candidate genus, *Arcomarinus* [1–3]. *Aliarcobacter* genus includes 11 species, of which 8 have been validly published, *A. butzleri*, *A. cibarius*, *A. cryaerophilus*, *A. faecis*, *A. lanthieri*, *A. skirrowii*, *A. thereius* and *A. trophiarum* [2]. Bacteria from this genus are Gram-negative, small, curved rods and present motility by a single polar flagellum [1]. They can be found in a wide range of hosts and environments and isolated from different sources [4–6]. Among these species, *A. butzleri*, *A. cryaerophilus*, *A. skirrowii*, *A. thereius* and *A. lanthieri* have been associated with human disease, such as bacteremia, enteritis, diarrhea, abdominal cramps, nausea and vomiting [7–10]. In addition, *A. butzleri* and *A. cryaerophilus* have been recognized by the International Commission on Microbiological Specifications for Food as a serious hazard to human health [11]. When considering *A. butzleri*, several virulence mechanisms have been described for this species, such as adherence, invasion and cytotoxic effects [5–7]. Diseases caused by *A. butzleri* are usually self-limiting; nonetheless, several reports have described the use of antibiotics for its treatment, mostly with β-lactams, fluoroquinolones and macrolides [5]. However, *A. butzleri* has shown resistance to major classes of antibiotics, such as fluoroquinolones, macrolides, β-lactams, aminoglycosides and tetracyclines, with a considerable portion of strains being classified as multidrug-resistant [5,7]. Therefore, this bacterium has attracted attention due to its wide distribution and antibiotic resistance, even if the underlying mechanisms are still poorly understood. The only mechanism that has been linked to *A. butzleri* antibiotic resistance is a target modification, associated with resistance to fluoroquinolones [12]. More recently, the role of an efflux pump in the erythromycin-resistance phenotype was evidenced [13,14].

Efflux pumps are relevant elements for antimicrobial resistance, contributing to antibiotic resistance and also to bacterial virulence [15,16]. These systems comprise membrane proteins that are involved in the efflux of antibiotics and other toxic substances from the cell, being classified into six families: ATP-binding cassette (ABC), multidrug and toxic compound extrusion (MATE), major facilitator superfamily (MFS), proteobacterial antimicrobial compound efflux (PACE), resistance–nodulation–division (RND) and small multidrug resistance (SMR) [17]. Among these families, the superfamily of RND efflux pumps is the most relevant among multidrug resistance efflux transporters since, when overexpressed, bacteria may exhibit significant levels of resistance [18]. In addition, these systems may have a broad substrate specificity, due to their ability to recognize substrates of different sizes and with diverse properties [19]. The RND family of efflux pumps are tripartite protein complexes found in Gram-negative bacteria, which include an inner membrane protein, an outer membrane protein, and a periplasmic membrane fusion protein [16,19]. Efflux pumps from this family, such as CmeABC of *Campylobacter jejuni*; MexAB–OprM of *Pseudomonas aeruginosa*; or AcrAB of *Enterobacter cloacae*, *Klebsiella pneumoniae*, *Salmonella enterica* and others, were shown to play multiple functions, being involved in resistance to compounds with different structures, such as bile salts, various classes of antibiotics, detergents and dyes, and also contributing to the fitness, survival and virulence of these bacteria [16,20–24].

Genetic studies conducted by Isidro et al. (2020) have identified 19 putative efflux systems in the genome of *A. butzleri*, with 8 belonging to the efflux pumps of the RND family. Among these are the AreABC (EP16), AreDEF (EP15) and AreGHI (EP4) systems, for which complete operons were found at the genomic level [13]. Nonetheless, until now, only the AreABC system has been characterized, and it was shown to play a role in resistance to several antibiotics, particularly to erythromycin [14]. Increasing the knowledge regarding the role of efflux systems in antimicrobial resistance and virulence of the emerging pathogen *A. butzleri* is, therefore, a promising line of investigation. With this work, we aimed to elucidate the role that the AreABC, AreDEF and AreGHI RND efflux pumps play in *A. butzleri* antibiotic resistance and virulence.

2. Results and Discussion
2.1. Impact of the Inactivation of Three RND Efflux Pumps on Bacterial Resistance

Bacterial resistance to various antimicrobial agents has been linked to efflux pumps that are responsible for the extrusion of these agents [18]. RND efflux pumps stand out for their role in resistance to multiple antimicrobials in Gram-negative bacteria, due to their broad substrate specificity, including (1) antibiotics, such as penicillins, cephalosporins, fluoroquinolones, macrolides, chloramphenicol and tetracyclines; (2) cationic dyes, such as crystal violet and ethidium bromide; and (3) detergents such as Triton X-100, biocides, heavy metals, organic solvents and host-derived molecules. The extruded compounds may present diverse structure charges, among other characteristics; however, their lipophilicity, or the presence of lipophilic domains, has been described as a common feature among the substrates of these systems [15,16,19,25].

AreABC, AreDEF and AreGHI are efflux pumps belonging to the RND family that were found encoded in the genome of *A. butzleri* (namely in the strain Ab_2811) [13]. A BLASTP analysis showed that the efflux pumps under study have homology with other RND efflux systems, namely from *C. jejuni*, *P. aeruginosa* and *E. coli*. The three structural components of these RND efflux systems are organized in a single operon containing outer membrane protein genes (Figure 1), similarly to the systems CmeABC and CmeDEF from *C. jejuni* or MexAB-OprM and MexEF-OprN from *P. aeruginosa*. This contrasts with other Gram-negative species, for which the RND efflux transporters operons lack genes coding for outer membrane proteins, which in turn may be located elsewhere in the chromosome, such as in the case of the AcrAB-TolC system from *Escherichia coli* [15,26–28].

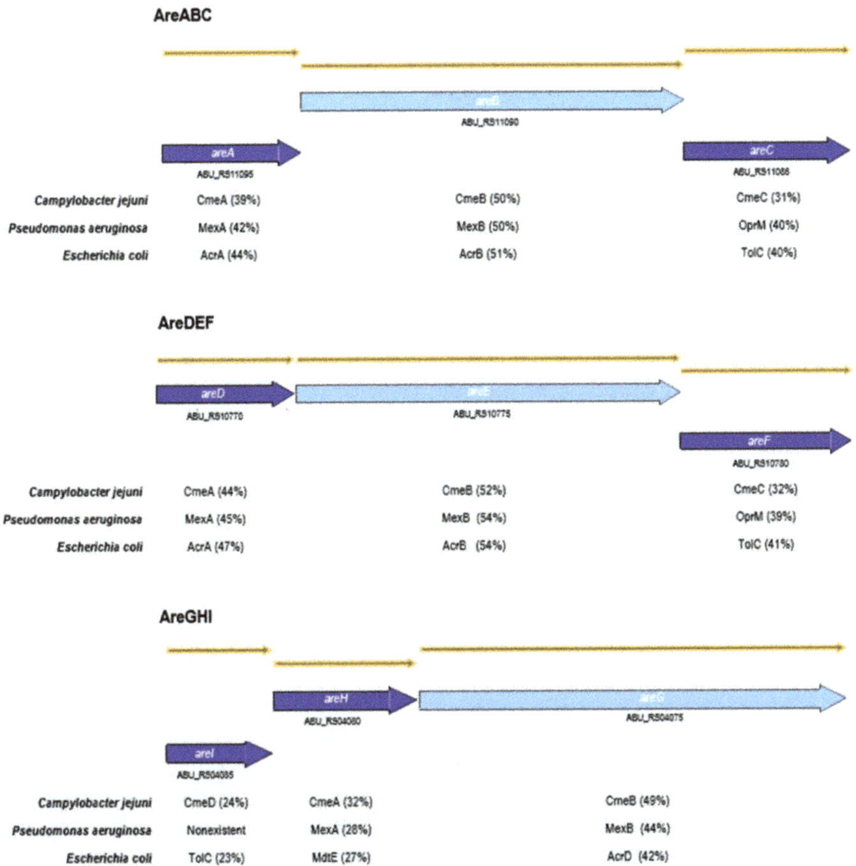

Figure 1. Genomic organization of AreABC, AreDEF and AreGHI efflux pumps. Open reading frames and their transcription directions are indicated by boxed arrows, and locus tag from *Aliarcobacter butzleri* RM4018 is identified below the arrow. Putative genes encoding the inner membrane protein of the efflux pump are marked in light blue. Amino acid identities (percentage of similarity) between protein homologs in *C. jejuni*, *P. aeruginosa* and *E. coli* are indicated for each homolog gene and were retrieved using BLASTP analysis.

Each component of these tripartite efflux pumps is necessary for the efflux, and the absence of one of these components turns the complex nonfunctional [26]. Thus, to evaluate the potential role of three RND-type efflux pumps in *A. butzleri*, a strain with a multidrug-resistant profile was selected, and three mutants of the inner membrane protein genes, *areB*, *areE* and *areG*, were generated by natural transformation. The three genes were interrupted with a kanamycin resistance marker, by insertional mutation, originating the following mutants: Ab_2811Δ*areB*, Ab_2811Δ*areE* and Ab_2811Δ*areG*. The insertion of the resistance cassette was verified by polymerase chain reaction (PCR) (Figure S1), while the inactivation of gene expression was confirmed by reverse transcription (RT) PCR (Figure S2).

The contribution of the efflux pumps was evaluated by MIC determination of several classes of antibiotics, bile salts and biocides for the parental and the mutant strains. This evaluation makes it possible to ascertain if the inactivation of the efflux systems AreABC, AreDEF and AreGHI has an influence on the resistance of the strain and to infer about potential substrates of these pumps.

All mutant strains were more susceptible than the parental strain, but different susceptibility profiles were observed for the different mutants (Table 1). The differences in MIC values obtained for the 22 antimicrobials tested suggests that the substrates of the AreABC efflux pump are cephalexin, cefotaxime, erythromycin, clarithromycin, streptomycin, ciprofloxacin and moxifloxacin, for which considerable fold-changes were observed (between 4- and 32-fold decrease in the MIC) for the mutant Ab_2811Δ*areB*. Furthermore, levofloxacin, doxycycline and acriflavine may also be potential substrates of AreABC system, despite only a 2-fold decrease in the MIC being observed. The role of AreABC in *A. butzleri* antibiotic resistance was previously described and evidenced for low to intermediate levels of erythromycin resistance [14].

Considering the AreDEF efflux pump, the impact of *areE* inactivation was less evident, with a decrease of only 4-fold for cephalexin and 2-fold for cefotaxime, erythromycin, clarithromycin, streptomycin, doxycycline and sodium cholate.

Regarding the AreGHI mutant, the results suggest that cephalexin, cefotaxime, amoxicillin, streptomycin, levofloxacin, tetracycline, doxycycline, chloramphenicol and bile salts are substrates of this efflux pump. In addition, a 2-fold decrease in the MIC for ampicillin, ciprofloxacin, sodium cholate, benzalkonium chloride and acriflavine was proved, when compared to the parental strain, suggesting that these compounds may also stand as possible substrates for this pump.

Overall, the three efflux pumps recognize a broad range of substrates, with all the efflux systems demonstrating to have a role in the extrusion of cephalexin, cefotaxime, streptomycin and doxycycline.

These results are in accordance with several studies carried out in RND efflux systems from different bacterial species, where the inactivation of genes encoding for efflux pump subunits results in increased susceptibility to different antimicrobials [15,16,27]. For example, CmeABC from *Campylobacter* spp. was described as being involved in the efflux transport of several classes of antibiotics, ethidium bromide and bile salts, proving that this system recognizes a broad range of structurally different substrates [28]. In the case of CmeDEF efflux system of *C. jejuni*, when the *cmeF* gene was interrupted, only a small decrease in MIC for ampicillin and cefotaxime was observed [29]. Therefore, the role of the CmeABC in *C. jejuni* resistance has been considered more relevant than the role of CmeDEF, with the first significantly influencing the intrinsic and acquired resistance of *Campylobacter* to various antimicrobials, such as macrolides and fluoroquinolones [28,30,31].

Accordingly, in our study, AreABC and AreGHI efflux systems demonstrated a greater influence in the extrusion of toxic compounds out of the cell than AreDEF, recognizing a different wide range of substrates. These results corroborate previous studies regarding the role of AreABC in *A. butzleri* resistance [14] and support, for the first time, the contribution of AreGHI as well. Besides that, AreDEF efflux pump has shown a selectivity for a few substrates; however, it does not seem to play such a significant role in the multidrug resistance of the *A. butzleri* strain Ab_2811.

Table 1. Minimum inhibitory concentration of several classes of antibiotics, bile salts, disinfectants and germicides for the parental and *areB*, *areE* and *areG* mutant strains of *Aliarcobacter butzleri*.

Antimicrobial	MIC (µg/mL) [Fold Reduction]			
	Ab_2811	Ab_2811Δ*areB*	Ab_2811Δ*areE*	Ab_2811Δ*areG*
β-Lactams				
Ampicillin	128	128 [ND]	128 [ND]	**64 [2]**
Cephalexin	256	**8 [32]**	**64 [4]**	**0.0625 [4096]**
Cefotaxime	64	**2 [32]**	**32 [2]**	**0.0625 [1024]**
Amoxycillin	64	64 [ND]	64 [ND]	**16 [4]**
Macrolides				
Erythromycin	16	**2 [8]**	**8 [2]**	16 [ND]
Clarithromycin	16	**1 [16]**	**8 [2]**	16 [ND]
Aminoglycosides				
Gentamycin	1	1 [ND]	1 [ND]	1 [ND]
Kanamycin *	4	>512	>512	>512
Streptomycin	8	**0.5 [16]**	**4 [2]**	**2 [4]**
Quinolones				
Nalidixic acid	>256	>256 [ND]	>256 [ND]	>256 [ND]
Ciprofloxacin	32	**8 [4]**	32 [ND]	**16 [4]**
Levofloxacin	64	**32 [2]**	64 [ND]	**16 [4]**
Moxifloxacin	32	**8 [4]**	32 [ND]	32 [ND]
Tetracyclines				
Tetracyclines	8	8 [ND]	8 [ND]	**0.5 [16]**
Doxycycline	8	**4 [2]**	**4 [2]**	**1 [8]**
Phenicol				
Chloramphenicol	32	32 [ND]	32 [ND]	**8 [4]**
Bile salts				
Sodium cholate	8000	8000 [ND]	**4000 [2]**	**4000 [2]**
Sodium deoxycholate	32,000	32,000 [ND]	32,000 [ND]	32,000 [ND]
Bile salts mix	50,000	50,000 [ND]	50,000 [ND]	**6250 [8]**
Disinfectants				
Chlorhexidine	2	2 [ND]	2 [ND]	2 [ND]
Benzalkonium chloride	32	32 [ND]	32 [ND]	**16 [2]**
Germicide				
Acriflavine	32	**16 [2]**	32 [ND]	**16 [2]**

ND, no observed MIC difference. * A kanamycin resistance marker was used to inactivate *areB*, *areE* and *areG* genes. Changes of at least 2-fold are indicated in bold type.

To evaluate if the increase in susceptibility observed for mutant strains was a consequence of efflux loss, the intracellular accumulation of ethidium bromide (EtBr) was assessed for parental and mutant strains (Figure 2). A time-dependent increase in fluorescence was observed for all strains, with the Ab_2811Δ*areB* mutant accumulating more EtBr than the parental strain and the other two mutants. When accumulation of ethidium bromide was assessed, fluorescence increased by ~5- to 9-fold after 37 min for mutant versus parental strain. Considering that a defective efflux results in dye accumulation and consequently an increase in fluorescence, these results support the functionality of the three efflux pumps of *A. butzleri* [32]. When the efflux pump inhibitor carbonyl cyanide m-chlorophenyl hydrazone was added to the parental strain, a ~6-fold increase was observed, confirming the contribution of efflux systems to the ethidium bromide accumulation, as previously described for *C. jejuni* [28].

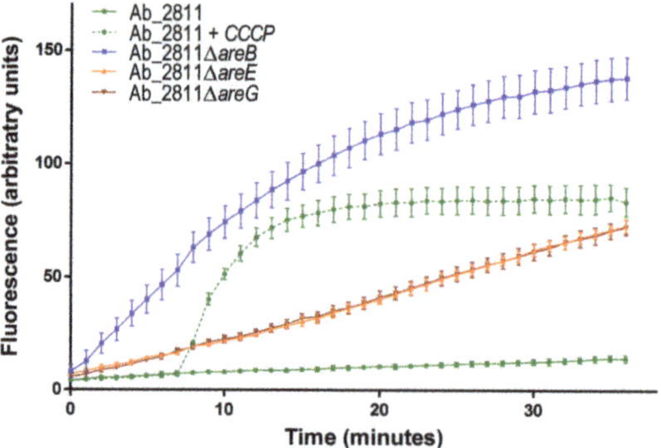

Figure 2. Accumulation of ethidium bromide after 37 min of incubation for parental *Aliarcobacter butzleri* Ab_2811 strain and derived mutants, in the absence (for mutants) or presence (for parental strain) of carbonyl cyanide m-chlorophenyl hydrazone at 50 µM. Data match to mean ± standard error of the mean (SEM) from three independent experiments.

2.2. Influence of Efflux System Impairment in Bacterial Growth and Relative Fitness

Several studies showed that the deletion of a gene from an efflux pump can affect bacterial growth and influence bacterial fitness [16,33,34]. To assess the effects, growth curves of the parental *A. butzleri* Ab_2811 strain and their mutants were constructed (Figure 3a). The mutant strains exhibited similar growth curves when compared with the parental strain. Despite the observed slight reduction in growth during the exponential phase, the mutations did not confer significant growth deficiency. This behavior is consistent with other studies showing that a mutation in the gene encoding the carrier protein of the RND-type efflux pump did not impair growth [23,24,35].

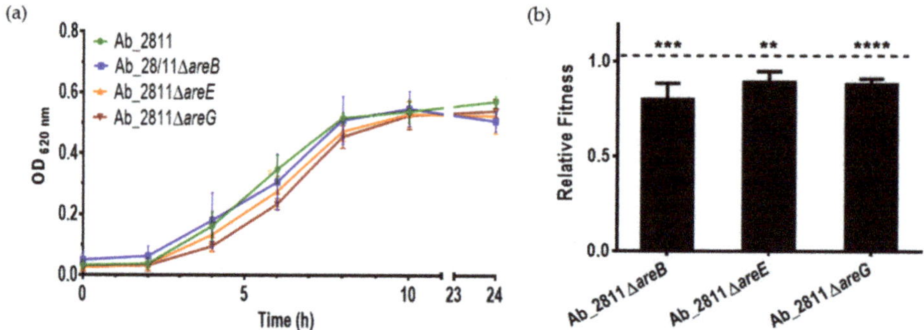

Figure 3. (**a**) Growth curves and (**b**) relative fitness of *Aliarcobacter butzleri* parental Ab_2811 strain and derived mutants. Data match to mean ± standard deviation (SD) from at least three independent experiments. ** $p < 0.01$; *** $p < 0.001$; **** $p < 0.0001$.

A fitness assessment was conducted for parental and mutant strains to determine if the inactivation of these genes could confer an advantage or disadvantage to the strain [36]. The pairwise competition assay showed that all three mutant strains underwent a fitness cost (Figure 3b), following the same trend observed by Pérez et al. (2012), where the disruption of the efflux system led to a reduction in bacterial fitness [23]. Overall, our results indicate that the efflux pumps under study participate in maintaining the biological competitiveness of *A. butzleri*. Indeed, it has been described that a variation of the expression of efflux pumps may induce specific changes in bacterial regulatory networks [16].

2.3. Effect of Inactivation of the RND Efflux Pumps in the Susceptibility to Oxidative and Bile Stress

In addition to conferring resistance to certain antibiotics, efflux pumps may have a role in resistance to oxidative stress. Oxidative stress occurs in bacteria, and in order to prevent the damage caused by the reactive oxygen species, a detoxification process occurs that involves the transport of these compounds by efflux pumps. RND efflux pumps have been associated with the extrusion of a variety of oxidative stress-inducing agents such as hydrogen peroxide and paraquat [37].

To validate this hypothesis, we tested the resistance to hydrogen peroxide and paraquat. Only the mutant Ab_2811ΔareE exhibited an increase in the inhibition diameter zone against 3% and 30% hydrogen peroxide when compared to the parental strain (Figure 4a). In contrast, all mutant strains showed an increased susceptibility to paraquat (Figure 4b), suggesting that these efflux pumps contribute to the bacterial tolerance to oxidative stress induced by paraquat.

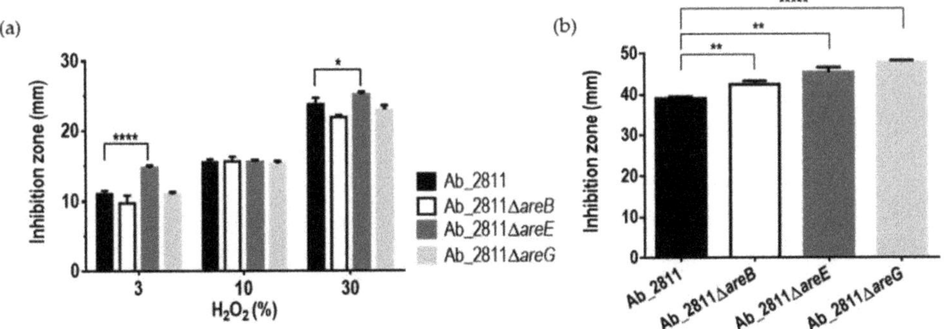

Figure 4. Oxidative stress susceptibility of *Aliarcobacter butzleri* parental Ab_2811 strain and derivate mutants, induced by (a) hydrogen peroxide and (b) paraquat. Data correspond to the mean $\pm$ SD of at least three independent experiments. * $p < 0.05$; ** $p < 0.01$; **** $p < 0.0001$; ***** $p < 0.00001$.

These results are in accordance with a previous study showing that deletion of *cmeG* gene of *C. jejuni* resulted in increased susceptibility to hydrogen peroxide. The mechanism contributing to resistance to oxidative stress in *Campylobacter* is not yet known, but it has been hypothesized that the extrusion of toxic metabolites contributes to the detoxification process and therefore to oxidative stress resistance [37]. Furthermore, in *Salmonella enterica* serovar Typhimurium, the overexpression of the RND efflux pump MdsABC resulted in an increase in the resistance to oxidative stress induced by paraquat, demonstrating a role of efflux pumps in the defense mechanism against this stress [38].

Based on our results, we suggest that the AreDEF system may present a similar function to *cmeG* in *C. jejuni*, while the three systems may alleviate stress induced by paraquat, possibly playing a critical role in the oxidative balance necessary for normal physiology.

Bacteria that infect the gastrointestinal tract are exposed to several barriers, such as bile salts [39], which can disrupt cell membranes [40]. Efflux pumps may play a role in the bacterial resistance against these compounds. As an example, *C. jejuni* uses the CmeABC efflux pump as a mechanism for resistance to bile [21].

Since *A. butzleri* is considered a pathogen capable of infecting the gastrointestinal tract [5] and the inactivation of *areG* gene led to a reduction in the MIC for bile salts, further studies were undertaken. For that, the parental and mutant strains were exposed to a physiological range of bile salt concentrations [41]. Overall, increasing concentrations of bile salts in the medium resulted in a decrease in the strains' growth, while for the Ab_2811ΔareG mutant, cellular death was observed for all the tested concentrations

(Figure 5). Several studies have shown that efflux systems are relevant for bacterial survival in the gastrointestinal tract and that this may be their ancestral function [19,27]. In *C. jejuni*, it was demonstrated that the inactivation of the CmeABC efflux pump led to a decreased resistance of the bacterium to bile salts [21,28,29,42]. The role of the efflux systems in the bacterium's survival in the presence of bile salts has also been described for other bacteria associated with human intestinal tract infections, such as *Vibrio cholerae* and *Bacteroides fragilis* [16,43]. Accordingly, our results strongly support a role of the AreGHI system in *A. butzleri* survival in the gastrointestinal tract, likely influencing the capacity of *A. butzleri* survival in response to this defense mechanism of the host, while promoting bacterial adaptation to the intestinal tract.

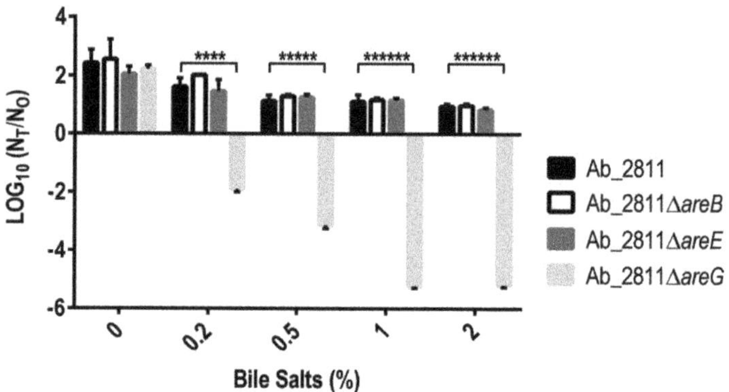

Figure 5. Bile stress susceptibility of *Aliarcobacter butzleri* parental Ab_2811 strain and derivate mutants. Data correspond to the mean ± SD of three independent experiments. **** $p < 0.0001$; ***** $p < 0.00001$; ****** $p < 0.000001$; The survival was determined as logarithm of N_T/N_0, where N_T corresponds to the colony-forming units (CFU)/mL after 6 h of incubation and N_0 corresponds to CFU/mL at time 0.

2.4. Impact of the Inactivation of RND Efflux Pumps on Motility and Biofilm Formation Ability

Motility and biofilm formation are two of the virulence mechanisms described for *A. butzleri* [5,6], being frequently associated with colonization and bacterial persistence, respectively [31,44]. A change in the expression of the efflux pumps can have an impact on the motility, which, in turn, may be associated with the formation of biofilms, influencing the virulence of the bacterium [45]. To assess whether these bacterial virulence characteristics were affected by inactivation of one of the subunits of the efflux pumps under study, motility and biofilm formation assays were performed for the parental and the mutant strains (Figure 6). Regarding motility, the migration diameter of all the mutant strains decreased; however, only the Ab_2811Δ*areG* mutant demonstrated a significant decrease when compared to the parental strain (Figure 6a). A decrease in motility has already been described for other bacteria, induced by the inactivation of an RND-type efflux pump [34,46].

The formation of biofilms was quantified by the crystal violet colorimetric assay. The inactivation of the three efflux systems led to a significant decrease in biofilm formation when compared to the parental strain (Figure 6b). Several studies described a similar behavior, where a deletion or inhibition of efflux pumps was associated with a decrease in biofilm formation [47,48]. For example, when the *adeB* gene of the AdeABC efflux system from *A. baumannii* was deleted, the mutant had a significant defect in biofilm formation. Moreover, mutants of *Salmonella* Typhimurium lacking *tolC* and *acrB* genes had compromised biofilm formation ability [49,50]. RND efflux pumps also have the ability to extrude quorum-sensing signals, with an impact on the process of intercellular communication, which in turn is associated with biofilm formation [16].

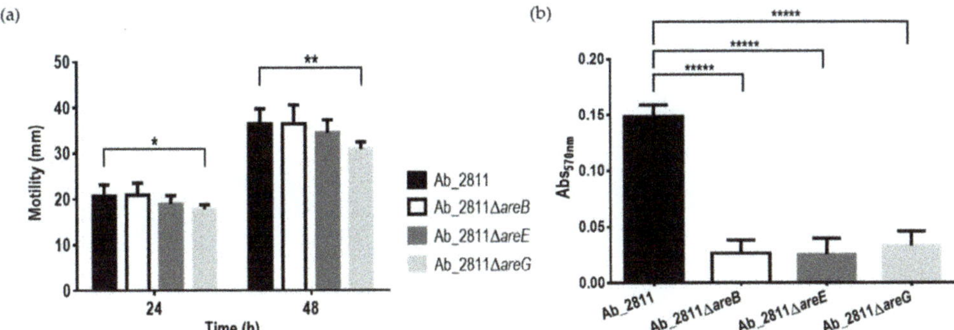

Figure 6. (**a**) Motility ability and (**b**) biofilm formation ability of *Aliarcobacter butzleri* parental Ab_2811 strain and derivate mutants. Data are the mean ± SEM of at least three independent experiments. * $p < 0.05$; ** $p < 0.01$; ***** $p < 0.00001$.

2.5. Impact of the Inactivation of RND Efflux Pumps on Human Serum Susceptibility

To verify the serum's bactericidal effects on parental and mutant strains, we performed the serum killing assay. PBS was used as control of the experiment, and no killing effects were observed for all the strains. Overall, the survival of the mutant strains in healthy human serum was shown to be significantly different from that observed for the parental strain (Figure 7). Indeed, after 180 min of exposure, all the mutants demonstrated a significant increase in their susceptibility to human serum ($p < 0.01$). These results are in agreement with a study performed in *Stenotrophomonas maltophilia*, where the deletion of the RND efflux pump SmeYZ resulted in increased susceptibility to human serum [51].

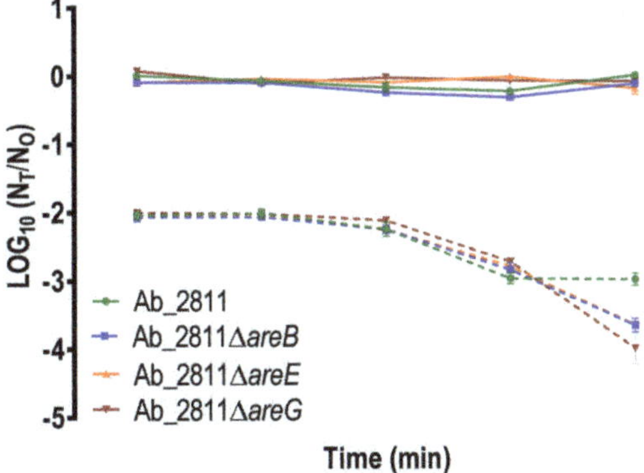

Figure 7. Susceptibility of *Aliarcobacter butzleri* parental Ab_2811 strain and derivate mutants to human serum. Death of the bacteria on PBS (solid lines) and serum (dash lines). Data correspond to the means ± SD of three independent experiments.

2.6. In Vitro Adhesion to and Invasion into Caco-2 Cells

RND-type efflux pumps are recognized for their role in resistance as well as in virulence [52]. Since capacities for adhesion to and invasion into host cells are required for successful colonization and infection [53], these in vitro assays were performed to evaluate virulence for parental and efflux pumps mutant strains. Compared to the parental strain, all mutants exhibited a significant decrease in the capacity for adhesion to Caco-2 cells (Figure 8). However, no difference was observed regarding invasion ability between the parental and mutant strains, probably due to the very low percentage of invasion observed for all the strains (Figure 8).

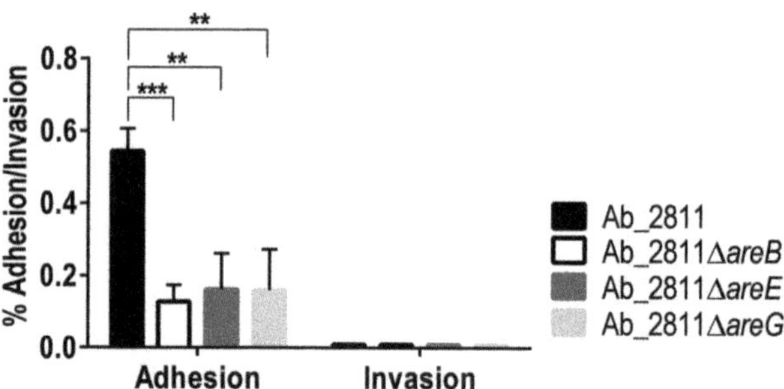

Figure 8. Adhesion and invasion assays for parental *Aliarcobacter butzleri* Ab_2811 strain and derivate mutants. Data correspond to the mean ± SEM of three independent experiments. ** $p < 0.01$; *** $p < 0.001$.

Our results are in accordance with a previous study carried out in *Salmonella enterica*, where the lack of AcrB led to a decrease in adhesion and invasion. In that study, the reduction in virulence observed for the mutant strains was caused by the loss of the efflux function [24]. Supporting this, several studies have shown that RND efflux pumps play a relevant role in the virulence of bacteria. For example, in *C. jejuni*, the CmeABC system is needed for gut colonization of 1-year-old chicks [21]; in *Moraxella catarrhalis*, a deletion of the AcrAB-oprM system led to a decrease in invasion of the mutant into pharyngeal epithelial cells [54]; and in *Vibrio cholerae*, RND efflux systems are essential for the colonization of the small intestine in mice [55]. Overall, all the *A. butzleri* efflux systems under study showed a potential role in adhesion ability, thus contributing to bacterial virulence and infection of host cells.

In summary, the results suggest that the *A. butzleri* efflux systems AreABC, AreDEF and AreGHI have a role in the extrusion of several antimicrobial agents, such as antibiotics, bile salts and biocides. In addition, results showed that these pumps are also relevant elements in the defense against oxidative stress, while AreGHI has a role in the survival of the bacterium in presence of bile salts. Finally, all the efflux systems tested showed a role in the ability to form biofilms, the ability to survive in human serum and the ability to adhere to intestinal cells. Thus, overall, the three efflux pumps studied have been shown to contribute to antimicrobial resistance and also to virulence of *A. butzleri*. Our results are in agreement with the well-known role of RND efflux systems in antibiotic resistance, while their role in other relevant physiological processes, namely those associated with bacterial virulence, has also been described for several pathogens, such as *C. jejuni*, *S. maltophilia* and *P. aeruginosa* [16]. This study produced relevant data towards a better understanding of the emerging pathogen *A. butzleri*, while allowing the identification of putative targets for new therapeutic strategies to fight this infection.

3. Materials and Methods

3.1. Bacterial Strains and Growth Conditions

The *A. butzleri* strain Ab_2811 was isolated from a poultry carcass neck skin [56] and was used as the parental strain for this study. Bacterial cells were stored at −80 °C in brain heart infusion (BHI) containing 20% glycerol and were routinely cultured on blood agar base (BA, Oxoid, Basingstoke, England) supplemented with 5% (*v/v*) of defibrinated horse blood or in tryptic soy agar (TSA, VWR, Leuven, Belgium) medium and incubated at 30 °C for 24 h, in aerobic conditions. For growth in broth culture, the *A. butzleri* strains were transferred to 10 mL of tryptic soy broth (TSB) with an initial optical density at 620 nm (OD_{620} nm) of 0.02 and incubated overnight at 30 °C under aerobic conditions.

3.2. Construction of Aliarcobacter butzleri Efflux Pump Mutants

The construction of the *A. butzleri* Ab_2811 mutants was performed using insertional mutagenesis, where the *areB*, *areE* and *areG* genes, encoding for the inner membrane protein of each of the three RND efflux systems under study, were interrupted by a kanamycin resistance cassette (*aphA-3*). The *aphA-3* cassette was obtained by enzymatic digestion of pUC18-K2 plasmid with BamHI and KpnI, followed by purification. Then, amplification of the upstream and downstream regions of each gene was performed using the primers presented in Table S1. Primers were designed with a tail sequence that allows hybridization with upstream and downstream regions of the kanamycin resistance marker. The overlap-extension PCR was used to construct the transformant DNA fragment. This fragment consisted of an *aphA-3* cassette flanked by the amplified fragments of an upstream and a downstream region of the genes under study. After purification, the transformant DNA fragment was used for the natural transformation of *A. butzleri* through a biphasic method based on a protocol previously described, with some modifications [57]. Briefly, *A. butzleri* Ab_2811 was cultured in BA at 30 °C for 24 h in a microaerophilic atmosphere (6% O_2, +/− 7.1% CO_2 and 3.6% H_2). Then, 5×10^9 CFU were resuspended in TSB and transferred to a test tube containing 1.5 mL of BA and incubated under the same conditions for 6 h. At this point, 2 µg of the transformant DNA was added and tubes were further incubated for 5 h. After incubation, the suspension was transferred to a BA plate supplemented with 50 µg/mL of kanamycin and incubated at 30 °C under microaerophilic atmosphere, for 5 days. The transformation was confirmed by PCR through analysis of the fragment size, and the gene expression was evaluated by RT-PCR technique (Table S1).

3.3. Antimicrobial Susceptibility Test

The minimum inhibitory concentration (MIC) was determined by broth microdilution method [58] for several classes of antibiotics, including β-lactams, quinolones, aminoglycosides, tetracyclines, macrolides and phenicols, and for different bile salts, disinfectants and a germicide. The MIC of compounds tested was assessed followed by spectrophotometric confirmation at 620 nm, using the value of 0.05 as a cut-off of the optical density. This test was performed in duplicate and in at least three independent assays.

3.4. Ethidium Bromide Accumulation Assay

The EtBr accumulation in parental and mutant *A. butzleri* strains was evaluated as previously described, with minor modifications [14,28]. Cultures were grown until mid-exponential phase, collected by centrifugation, washed with phosphate-buffered saline (PBS) and resuspended at OD_{620} nm of 0.2. The suspension was transferred to a black 96-well plate with a clear bottom (Greiner Bio-One, Frickenhausen, Germany). After incubation for 10 min at 30 °C, EtBr was added to each well to a final concentration of 2 µg/mL. Fluorescence was recorded with a Spectramax Gemini XS spectrofluorometer (Molecular Devices LLC, San Jose, CA, USA) at excitation and emission wavelengths of 530 and 600 nm, respectively. For the accumulation in the presence of efflux pump inhibitor, after 7 min of readings, carbonyl cyanide m-chlorophenyl hydrazone was added to each well to Ab_2811 at a final concentration of 50 µM. Each assay was performed in triplicate and in three independent assays.

3.5. Growth Curves and Competition Experiments of the Parental and Mutant Strains

The growth curves of *A. butzleri* parental and mutant strains were determined as described previously [58], and competition experiments were performed using pairwise competition assay for parental and mutant strains [58]. Briefly, a fresh culture in 10 mL of TSB was started by adding equal volumes of a preculture of parental and mutant strains. This culture was incubated at 30 °C for 24 h under aerobic condition. Samples were taken at time 0 and 24 h of incubation and the total and mutant populations were counted by plating successive dilutions on TSA and TSA supplemented with 50 µg/mL of kanamycin, respectively. The relative fitness of each mutant was determined by the ratio of the Malthusian parameter of the mutant per the parental. Both assays were repeated at least three independent times.

3.6. Stress Susceptibility Assays

The oxidative stress tolerance was determined by the disk diffusion assays, where TSA was inoculated with a cellular suspension adjusted to OD_{620} nm of 0.2 using a cotton swab, and sterile paper disks were spotted with 5 µL of 3%, 10% and 30% hydrogen peroxide (H_2O_2) and 5 µL of 250 mM methyl viologen (paraquat). After incubation at 30 °C in aerobic atmosphere for 24 h, the zone of growth inhibition diameter was measured. To evaluate bile stress, an overnight culture of the parental and mutant strains was used to inoculate a culture in the absence and presence of bile salts, with a range from 0 to 2% (Sigma-Aldrich, St. Louis, MO, USA). At 0 and 6 h, the drop-plate method was used for determination of CFU/mL. These assays were performed three independent times.

3.7. Motility Assay

The motility of parental and mutant strains was evaluated as previously described [58]. The center of a plate of TSA medium with 0.4% of agar was stabbed with 5 µL of mid-exponential phase culture. Motility was measured at 24 and 48 h, after incubation at 30 °C in aerobic atmosphere. This assay was performed at least three independent times.

3.8. Biofilm Formation Ability

The ability of *A. butzleri* to form biofilm was evaluated by the protocol described by Reeser et al. (2007), with some modifications [59]. An overnight culture was diluted to an OD_{620} nm of 0.2, and 100 µL of the bacterial suspension was applied in nonuplicate on a round-bottom polystyrene 96-well microtiter plate. Wells with only medium were used as negative control. After 48 h of incubation at 30 °C in aerobic atmosphere, the medium was removed, and the plate was incubated for 1 h at 55 °C. Then, the biofilm was stained with 100 µL of crystal violet at 0.1% (w/V) in deionized water and incubated for 15 min at room temperature, followed by washing of the wells three times with distilled water and incubation again at 55 °C for 15 min. After drying, the biofilm was destained with 120 µL of a 30% methanol and 10% acetic acid solution. One hundred microliters was moved to a new microtiter plate and the absorbance at 570 nm was determined. The assay was performed at least three independent times.

3.9. Serum Killing Assays

Serum bactericidal assays were performed according to the protocol previously described by O'Shaughnessy et al. (2012), with some modifications [60]. Blood was collected from a healthy adult, and serum was separated by centrifugation at 2000 rpm for 10 min at 4 °C and frozen in aliquots at −80 °C in sterile cryogenic tubes. For the assay, the parental and mutant strains at mid-exponential phase were washed and resuspended in PBS at a final concentration of ~10^7 CFU/mL with 90% of serum. As control, serum was replaced with PBS. The assay was performed at 30 °C, and viable cells were counted by successive dilution at the time points of 0, 15, 30, 45, 90 and 180 min. This assay was done three independent times.

3.10. Adhesion and Invasion Assays

Adhesion and invasion assays were performed in the Caco-2 human intestinal epithelial cell line. The Caco-2 cells were grown in tissue culture flasks in Dulbecco's modified Eagle medium supplemented with 10% fetal bovine serum, 1% nonessential amino acids, 100 µg/mL streptomycin and 100 U/mL penicillin and incubated at 37 °C in 5% of CO_2. For assays, 24-well plates were seeded with 1×10^5 cells/ well and incubated for 48 h. To determine adhesion, monolayers were infected at an MOI of 100 and incubated for 2 h in the same atmospheric conditions. Cells were washed three times and lysed with 0.1% Triton X-100. For invasion of Caco-2 cells, a gentamicin protection assay was performed.

After 2 h of infection, extracellular bacteria were killed by incubating with 125 μg/mL of gentamicin for 1 h. Then, cells were washed to remove residual antibiotic and lysed with 0.1% Triton X-100. Percentage of adherent and invading bacteria was determined by successive dilution of lysates and plating by drop-plate method. Each assay was performed at least three independent times, in triplicate.

Supplementary Materials: The following are available online at https://www.mdpi.com/article/10.3390/antibiotics10070823/s1, Table S1: Primers used in PCR in this study, Figure S1. Confirmation of constructed mutants by PCR, Figure S2. Confirmation of gene expression by RT-PCR.

Author Contributions: Conceptualization, S.F.; formal analysis, C.M.; investigation, C.M. and A.R.N.; writing—original draft preparation, C.M.; writing—review and editing, M.O., F.D. and S.F.; funding acquisition, F.D., M.O. and S.F. All authors have read and agreed to the published version of the manuscript.

Funding: This work was financed by the Foundation for Science and Technology (FCT), through funds from the State Budget, and by the European Regional Development Fund (ERDF), under the Portugal 2020 Program, through the Regional Operational Program of the Center (Centro2020), through the Project with the reference UIDB/00709/2020.

Institutional Review Board Statement: Not applicable.

Informed Consent Statement: Not applicable.

Data Availability Statement: Data are contained within the text and the Supplementary Materials.

Acknowledgments: Cristiana Mateus is recipient of a doctoral fellowship (UI/BD/151023/2021) under the scope of the CICS-UBI Programmatic Funding (UIDP/00709/2020). Susana Ferreira acknowledges UBI and FCT by the contract of Scientific Employment according to DL57/2016.

Conflicts of Interest: The authors declare no conflict of interest.

References

1. Pérez-Cataluña, A.; Salas-Massó, N.; Diéguez, A.L.; Balboa, S.; Lema, A.; Romalde, J.L.; Figueras, M.J. Revisiting the Taxonomy of the Genus Arcobacter: Getting Order from the Chaos. *Front. Microbiol.* **2018**, *9*, 2077. [CrossRef]
2. Oren, A.; Garrity, G.M. List of New Names and New Combinations Previously Effectively, but Not Validly, Published. *Int. J. Syst. Evol. Microbiol.* **2020**, *70*, 1–5. [CrossRef]
3. Oren, A.; Garrity, G.M. List of New Names and New Combinations Previously Effectively, but not Validly, Published. *Int. J. Syst. Evol. Microbiol.* **2020**, *70*, 2960–2966. [CrossRef] [PubMed]
4. Whiteduck-Léveillée, K.; Whiteduck-Léveillée, J.; Cloutier, M.; Tambong, J.T.; Xu, R.; Topp, E.; Arts, M.; Chao, J.; Adam, Z.; Lévesque, C.A.; et al. Arcobacter Lanthieri sp. nov., Isolated from Pig and Dairy Cattle Manure. *Int. J. Syst. Evol. Microbiol.* **2015**, *65*, 2709–2716. [CrossRef] [PubMed]
5. Ferreira, S.; Queiroz, J.; Oleastro, M.; Domingues, F.C. Insights in the Pathogenesis and Resistance of *Arcobacter*: A Review. *Crit. Rev. Microbiol.* **2015**, *42*, 1–20. [CrossRef] [PubMed]
6. Chieffi, D.; Fanelli, F.; Fusco, V. *Arcobacter butzleri*: Up-to-Date Taxonomy, Ecology, and Pathogenicity of An Emerging Pathogen. *Compr. Rev. Food Sci. Food Saf.* **2020**, *19*, 2071–2109. [CrossRef]
7. Collado, L.; Figueras, M.J. Taxonomy, Epidemiology, and Clinical Relevance of the Genus Arcobacter. *Clin. Microbiol. Rev.* **2011**, *24*, 174–192. [CrossRef]
8. Figueras, M.J.; Levican, A.; Pujol, I.; Ballester, F.; Quilez, M.J.R.; Gomez-Bertomeu, F. A Severe Case of Persistent Diarrhoea Associated with *Arcobacter cryaerophilus* but Attributed to *Campylobacter* sp. and A Review of the Clinical Incidence of *Arcobacter* Spp. *New Microbes New Infect.* **2014**, *2*, 31–37. [CrossRef]
9. Kerkhof, P.-J.; Abeele, A.-M.V.D.; Strubbe, B.; Vogelaers, D.; Vandamme, P.; Houf, K. Diagnostic Approach for Detection and Identification of Emerging Enteric Pathogens Revisited: The (*Ali*)arcobacter Lanthieri Case. *New Microbes New Infect.* **2021**, *39*, 100829. [CrossRef]
10. Abeele, A.-M.V.D.; Vogelaers, D.; Van Hende, J.; Houf, K. Prevalence of *Arcobacter* Species among Humans, Belgium, 2008–2013. *Emerg. Infect. Dis.* **2014**, *20*, 1746–1749. [CrossRef]
11. ICMSF International Commission on Microbiological Specifications for Foods. *Microorganisms in Foods 7: Microbiological Testing in Food Safety Management*; Kluwer Academic/Plenum Publishers: New York, NY, USA, 2002; p. 171.
12. Abdelbaqi, K.; Mãnard, A.; Prouzet-Mauleon, V.; Bringaud, F.; Lehours, P.; Mãgraud, F. Nucleotide Sequence of the *gyrA* Gene of *Arcobacter* species and Characterization of Human Ciprofloxacin-Resistant Clinical Isolates. *FEMS Immunol. Med. Microbiol.* **2007**, *49*, 337–345. [CrossRef]

13. Isidro, J.; Ferreira, S.; Pinto, M.; Domingues, F.; Oleastro, M.; Gomes, J.P.; Borges, V. Virulence and Antibiotic Resistance Plasticity of *Arcobacter butzleri*: Insights on the Genomic Diversity of An Emerging Human Pathogen. *Infect. Genet. Evol.* **2020**, *80*, 104213. [CrossRef]

14. Ferreira, S.; Silva, A.L.; Tomás, J.; Mateus, C.; Domingues, F.; Oleastro, M. Characterization of AreABC, An RND-type Efflux System Involved in Antimicrobial Resistance of *Aliarcobacter butzleri*. *Antimicrob. Agents Chemother.* **2021**, *65*, AAC0072921. [CrossRef]

15. Blair, J.; Piddock, L.J. Structure, Function and Inhibition of RND Efflux Pumps in Gram-Negative Bacteria: An Update. *Curr. Opin. Microbiol.* **2009**, *12*, 512–519. [CrossRef]

16. Alcalde-Rico, M.; Hernando-Amado, S.; Blanco, P.; Martínez, J.L. Multidrug Efflux Pumps at the Crossroad between Antibiotic Resistance and Bacterial Virulence. *Front. Microbiol.* **2016**, *7*, 1483. [CrossRef] [PubMed]

17. Hassan, K.A.; Liu, Q.; Elbourne, L.D.; Ahmad, I.; Sharples, D.; Naidu, V.; Chan, C.L.; Li, L.; Harborne, S.; Pokhrel, A.; et al. Pacing across the Membrane: The Novel PACE Family of Efflux Pumps is Widespread in Gram-Negative Pathogens. *Res. Microbiol.* **2018**, *169*, 450–454. [CrossRef]

18. Blair, J.; Webber, M.A.; Baylay, A.J.; Ogbolu, D.O.; Piddock, L.J.V. Molecular Mechanisms of Antibiotic Resistance. *Nat. Rev. Microbiol.* **2015**, *13*, 42–51. [CrossRef] [PubMed]

19. Kumar, A.; Schweizer, H.P. Bacterial Resistance to Antibiotics: Active Efflux and Reduced Uptake. *Adv. Drug Deliv. Rev.* **2005**, *57*, 1486–1513. [CrossRef] [PubMed]

20. Hirakata, Y.; Srikumar, R.; Poole, K.; Gotoh, N.; Suematsu, T.; Kohno, S.; Kamihira, S.; Hancock, R.; Speert, D.P. Multidrug Efflux Systems Play an Important Role in the Invasiveness of *Pseudomonas aeruginosa*. *J. Exp. Med.* **2002**, *196*, 109–118. [CrossRef]

21. Lin, J.; Sahin, O.; Michel, L.O.; Zhang, Q. Critical Role of Multidrug Efflux Pump CmeABC in Bile Resistance and In Vivo Colonization of *Campylobacter jejuni*. *Infect. Immun.* **2003**, *71*, 4250–4259. [CrossRef]

22. Padilla, E.; Llobet, E.; Doménech-Sánchez, A.; Martínez-Martínez, L.; Bengoechea, J.; Albertí, S. *Klebsiella pneumoniae* AcrAB Efflux Pump Contributes to Antimicrobial Resistance and Virulence. *Antimicrob. Agents Chemother.* **2010**, *54*, 177–183. [CrossRef]

23. Pérez, A.; Poza, M.; Fernández, A.; Fernández, M.D.C.; Mallo, S.; Merino, M.; Rumbo-Feal, S.; Cabral, M.P.; Bou, G. Involvement of the AcrAB-TolC Efflux Pump in the Resistance, Fitness, and Virulence of *Enterobacter cloacae*. *Antimicrob. Agents Chemother.* **2012**, *56*, 2084–2090. [CrossRef] [PubMed]

24. Wang-Kan, X.; Blair, J.M.A.; Chirullo, B.; Betts, J.; La Ragione, R.M.; Ivens, A.; Ricci, V.; Opperman, T.J.; Piddock, L.J.V. Lack of AcrB Efflux Function Confers Loss of Virulence on *Salmonella enterica* Serovar Typhimurium. *mBio* **2017**, *8*, e00968-17. [CrossRef]

25. Nikaido, H. Structure and Mechanism of RND-Type Multidrug Efflux Pumps. *Adv. Enzymol. Relat. Areas Mol. Biol.* **2011**, *77*, 1–60. [CrossRef]

26. Nikaido, H.; Takatsuka, Y. Mechanisms of RND Multidrug Efflux Pumps. *Biochim. Biophys. Acta Proteins Proteom.* **2009**, *1794*, 769–781. [CrossRef] [PubMed]

27. Ealvarez-Ortega, C.; Eolivares, J.; Martínez, J.L. RND Multidrug Efflux Pumps: What are They Good for? *Front. Microbiol.* **2013**, *4*, 7. [CrossRef]

28. Lin, J.; Michel, L.O.; Zhang, Q. CmeABC Functions as a Multidrug Efflux System in *Campylobacter jejuni*. *Antimicrob. Agents Chemother.* **2002**, *46*, 2124–2131. [CrossRef] [PubMed]

29. Akiba, M.; Lin, J.; Barton, Y.-W.; Zhang, Q. Interaction of CmeABC and CmeDEF in Conferring Antimicrobial Resistance and Maintaining Cell Viability in *Campylobacter jejuni*. *J. Antimicrob. Chemother.* **2005**, *57*, 52–60. [CrossRef]

30. Pumbwe, L.; Piddock, L.J.V. Identification and Molecular Characterisation of CmeB, a *Campylobacter jejuni* Multidrug Efflux Pump. *FEMS Microbiol. Lett.* **2002**, *206*, 185–189. [CrossRef]

31. Luo, N.; Pereira, S.; Sahin, O.; Lin, J.; Huang, S.; Michel, L.; Zhang, Q. Enhanced In Vivo Fitness of Fluoroquinolone-Resistant *Campylobacter jejuni* in the Absence of Antibiotic Selection Pressure. *Proc. Natl. Acad. Sci. USA* **2005**, *102*, 541–546. [CrossRef]

32. Opperman, T.J.; Kwasny, S.M.; Kim, H.-S.; Nguyen, S.T.; Houseweart, C.; D'Souza, S.V.; Walker, G.C.; Peet, N.P.; Nikaido, H.; Bowlin, T.L. Characterization of A Novel Pyranopyridine Inhibitor of the AcrAB Efflux Pump of *Escherichia coli*. *Antimicrob. Agents Chemother.* **2013**, *58*, 722–733. [CrossRef] [PubMed]

33. Srinivasan, V.B.; Venkataramaiah, M.; Mondal, A.; Rajamohan, G. Functional Characterization of AbeD, an RND-Type Membrane Transporter in Antimicrobial Resistance in *Acinetobacter baumannii*. *PLoS ONE* **2015**, *10*, e0141314. [CrossRef] [PubMed]

34. Srinivasan, V.B.; Vaidyanathan, V.; Rajamohan, G. AbuO, a TolC-Like Outer Membrane Protein of *Acinetobacter baumannii*, Is Involved in Antimicrobial and Oxidative Stress Resistance. *Antimicrob. Agents Chemother.* **2014**, *59*, 1236–1245. [CrossRef]

35. Blanco, P.; Corona, F.; Martínez, J.L. Involvement of the RND Efflux Pump Transporter SmeH in the Acquisition of Resistance to Ceftazidime in *Stenotrophomonas maltophilia*. *Sci. Rep.* **2019**, *9*, 4917. [CrossRef]

36. Morgan, J. Pairwise Competition and the Replicator Equation. *Bull. Math. Biol.* **2003**, *65*, 1163–1172. [CrossRef]

37. Jeon, B.; Wang, Y.; Hao, H.; Barton, Y.-W.; Zhang, Q. Contribution of CmeG to Antibiotic and Oxidative Stress Resistance in *Campylobacter jejuni*. *J. Antimicrob. Chemother.* **2010**, *66*, 79–85. [CrossRef]

38. Song, S.; Lee, B.; Yeom, J.-H.; Hwang, S.; Kang, I.; Cho, J.-C.; Ha, N.-C.; Bae, J.; Lee, K.; Kim, Y.-H. MdsABC-Mediated Pathway for Pathogenicity in *Salmonella enterica* Serovar Typhimurium. *Infect. Immun.* **2015**, *83*, 4266–4276. [CrossRef] [PubMed]

39. Fox, E.; Raftery, M.; Goodchild, A.; Mendz, G.L. *Campylobacter jejuni* Response to Ox-Bile Stress. *FEMS Immunol. Med. Microbiol.* **2007**, *49*, 165–172. [CrossRef] [PubMed]

40. Urdaneta, V.; Casadesús, J. Interactions between Bacteria and Bile Salts in the Gastrointestinal and Hepatobiliary Tracts. *Front. Med.* **2017**, *4*, 163. [CrossRef]

41. Gunn, J.S. Mechanisms of Bacterial Resistance and Response to Bile. *Microbes Infect.* **2000**, *2*, 907–913. [CrossRef]
42. Lin, J.; Cagliero, C.; Guo, B.; Barton, Y.-W.; Maurel, M.-C.; Payot, S.; Zhang, Q. Bile Salts Modulate Expression of the CmeABC Multidrug Efflux Pump in *Campylobacter jejuni*. *J. Bacteriol.* **2005**, *187*, 7417–7424. [CrossRef] [PubMed]
43. Pumbwe, L.; Skilbeck, C.A.; Nakano, V.; Avila-Campos, M.J.; Piazza, R.M.; Wexler, H.M. Bile Salts Enhance Bacterial Co-Aggregation, Bacterial-Intestinal Epithelial Cell Adhesion, Biofilm Formation and Antimicrobial Resistance of *Bacteroides fragilis*. *Microb. Pathog.* **2007**, *43*, 78–87. [CrossRef]
44. Donlan, R.M. Biofilms: Microbial Life on Surfaces. *Emerg. Infect. Dis.* **2002**, *8*, 881–890. [CrossRef]
45. Houry, A.; Gohar, M.; Deschamps, J.; Tischenko, E.; Aymerich, S.; Gruss, A.; Briandet, R. Bacterial Swimmers that Infiltrate and Take Over the Biofilm Matrix. *Proc. Natl. Acad. Sci. USA* **2012**, *109*, 13088–13093. [CrossRef] [PubMed]
46. Webber, M.A.; Bailey, A.M.; Blair, J.M.A.; Morgan, E.; Stevens, M.P.; Hinton, J.C.D.; Ivens, A.; Wain, J.; Piddock, L.J.V. The Global Consequence of Disruption of the AcrAB-TolC Efflux Pump in *Salmonella enterica* Includes Reduced Expression of SPI-1 and Other Attributes Required to Infect the Host. *J. Bacteriol.* **2009**, *191*, 4276–4285. [CrossRef] [PubMed]
47. Kvist, M.; Hancock, V.; Klemm, P. Inactivation of Efflux Pumps Abolishes Bacterial Biofilm Formation. *Appl. Environ. Microbiol.* **2008**, *74*, 7376–7382. [CrossRef]
48. Baugh, S.; Phillips, C.R.; Ekanayaka, A.S.; Piddock, L.; Webber, M.A. Inhibition of Multidrug Efflux as A Strategy to Prevent Biofilm Formation. *J. Antimicrob. Chemother.* **2014**, *69*, 673–681. [CrossRef]
49. Yoon, E.-J.; Chabane, Y.N.; Goussard, S.; Snesrud, E.; Courvalin, P.; Dé, E.; Grillot-Courvalin, C. Contribution of Resistance-Nodulation-Cell Division Efflux Systems to Antibiotic Resistance and Biofilm Formation in *Acinetobacter baumannii*. *mBio* **2015**, *6*, e00309-15. [CrossRef]
50. Baugh, S.; Ekanayaka, A.S.; Piddock, L.; Webber, M.A. Loss of or Inhibition of all Multidrug Resistance Efflux Pumps of *Salmonella enterica* Serovar Typhimurium Results in Impaired Ability to Form a Biofilm. *J. Antimicrob. Chemother.* **2012**, *67*, 2409–2417. [CrossRef]
51. Lin, Y.-T.; Huang, Y.-W.; Chen, S.-J.; Chang, C.-W.; Yang, T.-C. The SmeYZ Efflux Pump of *Stenotrophomonas maltophilia* Contributes to Drug Resistance, Virulence-Related Characteristics, and Virulence in Mice. *Antimicrob. Agents Chemother.* **2015**, *59*, 4067–4073. [CrossRef]
52. Poole, K. Bacterial Multidrug Efflux Pumps Serve Other Functions. *Microbe* **2008**, *3*, 179–185. [CrossRef]
53. Pizarro-Cerdá, J.; Cossart, P. Bacterial Adhesion and Entry into Host Cells. *Cell* **2006**, *124*, 715–727. [CrossRef]
54. Spaniol, V.; Bernhard, S.; Aebi, C. *Moraxella catarrhalis* AcrAB-OprM Efflux Pump Contributes to Antimicrobial Resistance and Is Enhanced during Cold Shock Response. *Antimicrob. Agents Chemother.* **2015**, *59*, 1886–1894. [CrossRef] [PubMed]
55. Bina, X.R.; Provenzano, D.; Nguyen, N.; Bina, J.E. *Vibrio cholerae* RND Family Efflux Systems Are Required for Antimicrobial Resistance, Optimal Virulence Factor Production, and Colonization of the Infant Mouse Small Intestine. *Infect. Immun.* **2008**, *76*, 3595–3605. [CrossRef]
56. Ferreira, S.; Fraqueza, M.J.; Queiroz, J.; Domingues, F.C.; Oleastro, M. Genetic Diversity, Antibiotic Resistance and Biofilm-Forming Ability of *Arcobacter butzleri* Isolated from Poultry and Environment from a Portuguese Slaughterhouse. *Int. J. Food Microbiol.* **2013**, *162*, 82–88. [CrossRef] [PubMed]
57. Bonifácio, M.; Mateus, C.; Alves, A.; Maldonado, E.; Duarte, A.; Domingues, F.; Oleastro, M.; Ferreira, S. Natural Transformation as A Mechanism of Horizontal Gene Transfer in *Aliarcobacter butzleri*. *Pathogens*. submitted.
58. Ferreira, S.; Correia, D.R.; Oleastro, M.; Domingues, F.C. *Arcobacter butzleri* Ciprofloxacin Resistance: Point Mutations in DNA Gyrase A and Role on Fitness Cost. *Microb. Drug Resist.* **2018**, *24*, 915–922. [CrossRef]
59. Reeser, R.J.; Medler, R.T.; Billington, S.J.; Jost, B.H.; Joens, L.A. Characterization of *Campylobacter jejuni* Biofilms under Defined Growth Conditions. *Appl. Environ. Microbiol.* **2007**, *73*, 1908–1913. [CrossRef] [PubMed]
60. O'Shaughnessy, C.M.; Cunningham, A.F.; MacLennan, C.A. The Stability of Complement-Mediated Bactericidal Activity in Human Serum against *Salmonella*. *PLoS ONE* **2012**, *7*, e49147. [CrossRef] [PubMed]

Communication

Exploring the Contribution of the AcrB Homolog MdtF to Drug Resistance and Dye Efflux in a Multidrug Resistant *E. coli* Isolate

Sabine Schuster [1,*], Martina Vavra [1], Ludwig Greim [1] and Winfried V. Kern [1,2]

[1]   Division of Infectious Diseases, Department of Medicine II, University Hospital and Medical Center, 79106 Freiburg, Germany; martina.vavra@uniklinik-freiburg.de (M.V.); ludwig.greim@med.uni-duesseldorf.de (L.G.); winfried.kern@uniklinik-freiburg.de (W.V.K.)

[2]   Faculty of Medicine, Albert-Ludwigs-University, 79106 Freiburg, Germany

*   Correspondence: sabine.schuster@uniklinik-freiburg.de

Abstract: In *Escherichia coli*, the role of RND-type drug transporters other than the major efflux pump AcrB has largely remained undeciphered (particularly in multidrug resistant pathogens), because genetic engineering in such isolates is challenging. The present study aimed to explore the capability of the AcrB homolog MdtF to contribute to the extrusion of noxious compounds and to multidrug resistance in an *E. coli* clinical isolate with demonstrated expression of this efflux pump. An *mdtF/acrB* double-knockout was engineered, and susceptibility changes with drugs from various classes were determined in comparison to the parental strain and its *acrB* and *tolC* single-knockout mutants. The potential of MdtF to participate in the export of agents with different physicochemical properties was additionally assessed using accumulation and real-time efflux assays with several fluorescent dyes. The results show that there was limited impact to the multidrug resistant phenotype in the tested *E. coli* strain, while the RND-type transporter remarkably contributes to the efflux of all tested dyes. This should be considered when evaluating the efflux phenotype of clinical isolates via dye accumulation assays. Furthermore, the promiscuity of MdtF should be taken into account when developing new antibiotic agents.

Keywords: MdtF (YhiV); multidrug resistance; RND-type efflux pump; dye accumulation; real-time efflux

Citation: Schuster, S.; Vavra, M.; Greim, L.; Kern, W.V. Exploring the Contribution of the AcrB Homolog MdtF to Drug Resistance and Dye Efflux in a Multidrug Resistant *E. coli* Isolate. *Antibiotics* **2021**, *10*, 503. https://doi.org/10.3390/antibiotics10050503

Academic Editor: Isabelle Broutin

Received: 7 April 2021
Accepted: 28 April 2021
Published: 28 April 2021

Publisher's Note: MDPI stays neutral with regard to jurisdictional claims in published maps and institutional affiliations.

Copyright: © 2021 by the authors. Licensee MDPI, Basel, Switzerland. This article is an open access article distributed under the terms and conditions of the Creative Commons Attribution (CC BY) license (https://creativecommons.org/licenses/by/4.0/).

1. Introduction

Knowledge about resistance mechanisms is crucial for future drug development in particular against Gram-negative pathogens. Many of them rank within the uppermost levels of the WHO priority list for the urgent requirement for new antibiotics [1]. The importance of RND (resistance nodulation cell division)-type efflux pumps for the emergence of multidrug resistance (MDR) in Gram-negative pathogens has been demonstrated, among others, for *Escherichia coli* [2–4], *Klebsiella* [5], *Enterobacter* [6] *Acinetobacter* [7], and *Pseudomonas aeruginosa* strains [4]. As shown recently with the latter, in isolates lacking carbapenemases, the overexpression of efflux pumps appeared as the predominant reason for carbapenem and multidrug resistance [8]. In contrast to *Pseudomonas aeruginosa* strains (in which several different transporters of the RND superfamily contributes to MDR), AcrB has been the only multidrug efflux transporter with a proven impact in *E. coli*. The relevance of further efflux pumps encoded in the *E. coli* chromosome has been less explored. One reason is that they are not found significantly expressed in laboratory strains [9] (in which most of the experiments were carried out), whereas results from MDR clinical isolates have remained underrepresented.

In recent studies with such isolates, a significant percentage of the collection revealed *mdtF* expression in addition to *acrB* expression [10,11]. MdtF (formerly YhiV), exhibiting a sequence similarity of 79% compared with AcrB [12], is complexed with the membrane fusion protein MdtE (formerly YhiU) and with TolC, the ubiquitous outer membrane channel in *E. coli* not only working together with RND-type efflux pumps but also with transporters of other families, such as the ABC-transporter MacB and the MFS-transporter EmrB [13]. The principal capability of MdtF to efflux drugs from different classes, with the exception of the oxazolidinone linezolid, has been shown [9], but little is known about a potential role in MDR development. In the present study, we aimed to explore the MdtF functionality and its contribution to drug resistance in an MDR *E. coli* isolate with verified expression of this RND-type efflux pump [3,11].

2. Results and Discussion

2.1. Impact of mdtF Inactivation on Drug Susceptibility

In order to evaluate the contribution of MdtF to total TolC-dependent efflux in MDR *E. coli* strains, we aimed to compare an *mdtF/acrB* double-knockout of the patient isolate KUN9180 (besides *acrB* also showing *mdtF* expression) with the *acrB* (KUNΔ*acrB*) and *tolC* single-knockout (KUNΔ*tolC*) mutants from previous studies [3,11]. Because selection options are limited in MDR strains, the reutilization of the kanamycin/neomycin selection cassette used for switching off *acrB* was an option. Thus, the first step was its removal from *acrB* and the second step was its reintroduction in *mdtF* (see Section 3.3). The resulting double-knockout was subjected to susceptibility testing, and the results were compared with those from the KUNΔ*acrB* and KUNΔ*tolC* single-knockout mutants. Between the latter, significant differences in the susceptibilities to nadifloxacin, zoliflodacin [11], and novobiocin (Table 1) had been shown, but not to several other proven AcrB substrates, including the more hydrophilic fluoroquinolones levofloxacin and moxifloxacin and the non-fluoroquinolone gyrase inhibitor gepotidacin [11], to tetracycline, chloramphenicol, linezolid, clindamycin, and rifaximin (Table S1), all of which were demonstrably effluxed in isolate KUN9180 [3,11]. Regarding those results, only nadifloxacin, zoliflodacin, and novobiocin are potentially extruded from a TolC-dependent transporter other than AcrB.

Table 1. Drug susceptibilities for which significant difference between *acrB* and *tolC* mutants of the MDR *E. coli* isolate KUN9180 were reported.

E. coli **Strain/Mutant**	MIC in μg/mL [1]		
	Nadifloxacin	**Zoliflodacin**	**Novobiocin**
KUN9180	>512 (0)	4 (0)	128 (64)
KUNΔ*acrB*	32 (16)	0.25 (0)	4 (2)
KUNΔ*acrB*Δ*mdtF*	16 (0)	0.125 (0)	2 (0)
KUNΔ*tolC*	4 (2)	0.045 (0.02)	1 (0)

[1] MIC, minimal inhibitory concentration; the median of ≥7 independent assays is shown and the MAD (median absolute deviation) is given in parenthesis.

Indeed, we detected susceptibility increases for these three agents with the KUNΔ*acrB*Δ*mdtF* double-mutant in comparison with the *acrB* single-knockout, but the changes were small, with significance only confirmed for nadifloxacin (*p* value 0.01). Remarkably, the MICs remain one or two dilution steps above those shown for the *tolC* deletion mutant (Table 1). This could be due to another TolC-dependent efflux pump. However, we did not find any significant expression of such an additional transporter in the KUN9180 isolate [11]. Residual differences in the susceptibilities of the Δ*acrB*Δ*mdtF* and the *tolC* inactivated mutants could be explained by further physiological functions of TolC [13]. Among others, a role in enterobactin export has been reported [14]. An increasing accumulation of the siderophore in the periplasm due to *tolC* deletion could impair the bacterial fitness and thereby, to some extent, drug susceptibilities [15]. We have to outline that we did not include investigations of a putative physiological impact of MdtF under differing environmental conditions other than drug exposure in this study. A previous report had detected a protective role of MdtF against nitrosative damage under anaerobic growth conditions [16].

2.2. Impact of mdtF Inactivation on Intracellular Dye Accumulation

To gain further insights regarding the functionality and specificity of the MdtF transporter, we carried out accumulation assays with six chemically diverse fluorescent dyes which are known substrates of AcrB. In a previous study with the MDR clinical isolate KUN9180, we had observed residual efflux of dyes not fully prevented after AcrB inactivation. That was in contrast to findings with derivatives of laboratory *E. coli* strains, such as the *acrB* overexpressing K-12 derivative 3-AG100 [2,3]. A recent study with new powerful pyranopyridine efflux pump inhibitors showed an increase in the accumulation of the fluorescent dye Hoechst 33342 exceeding that caused by *acrB* inactivation, which also suggests the existence of at least one more active efflux transporter [17].

The accumulation assays with the MDR *E. coli* isolate reveal a remarkable contribution of MdtF to the total efflux of all dyes tested, because only the *acrB*/*mdtF* double-knockout (but not KUNΔ*acrB*) reached the intracellular accumulation comparable with that of the *tolC* knockout (Figure 1). Moreover, most of these dyes seemed to be even better substrates of MdtF than of AcrB. The latter contributes most efficiently to total TolC-dependent efflux only in the case of Pyronine Y, whereas all other dyes appeared more successfully expelled by MdtF.

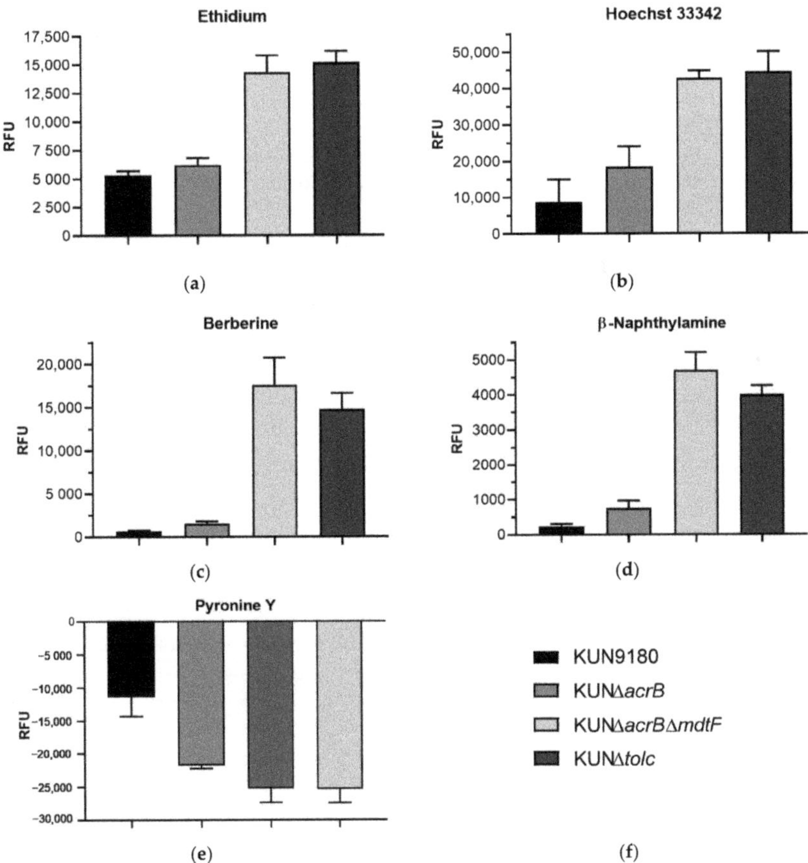

Figure 1. Accumulation of dyes in isolate KUN9180 and derived mutants (RFU, relative fluorescence unit; bars represent means of ≥3 assays ±SD from values at 30 min, for Pyronine Y at 20 min of incubation): (**a**) ethidium; (**b**) Hoechst 33342; (**c**) berberine; (**d**) β-naphthylamine; (**e**) quenching of fluorescence by intracellular accumulated Pyronine Y (lower fluorescence values reflect higher accumulation of Pyronine Y). (**f**) Graph legend for panels (**a**–**e**).

Regarding berberine and β-naphthylamine, the latter being the intracellular fluorescent cleavage product of the efflux pump inhibitor PAβN (phenylalanine arginine β-naphthylamide) [18], the accumulation seemed even higher within the *acrB/mdtF* double-knockout versus the KUNΔ*tolC* mutant, but the differences were not significant (*p* values > 0.05).

The results may help to resolve the previous discussion about ethidium as a substrate of AcrB in different *E. coli* isolates [19]. In some *E. coli* strains (including the clinical strain investigated in the present study), the inactivation of *acrB* does not remarkably affect the efflux of ethidium. We here show that this is associated with the presence of MdtF which obviously strongly reduced ethidium accumulation in a Δ*acrB* background (Figure 1).

2.3. Impact of mdtF Inactivation on Real-Time Efflux

Since intracellular compound accumulation might also be impaired by altered influx or uptake, we conducted real-time efflux assays with three different dyes suitable for this application. We used Nile red [20], 1,2′-dinaphthylamine (1,2′-DNA) [21], and the piperazine arylideneimidazolone BM-27 [22] to assess the extrudability of these substances in the KUN9180 clinical isolate and its mutants. Obviously, MdtF remarkably contributes to the efflux of these dyes, too (Figure 2). Moreover, it appears as major efflux transporter in the case of BM-27. This is unlike the findings with the *acrB* overexpressing K-12 derivative 3-AG100, in which the inactivation of *acrB* caused an efflux breakdown reaching the level of the *tolC*-knockout for all dyes tested [11], but in accordance with the fact that *mdtF* is not expressed in 3-AG100 and its Δ*acrB* derivative [3]. Some residual efflux with Nile red and possibly also with 1,2′-DNA occurred in the *acrB/mdtF* double-knockout mutant (Figure 2) suggesting the putative activity of another TolC-dependent transporter. However, as just mentioned, we could not detect any significant expression of such an efflux pump besides AcrB and MdtF in the KUN9180 isolate [11].

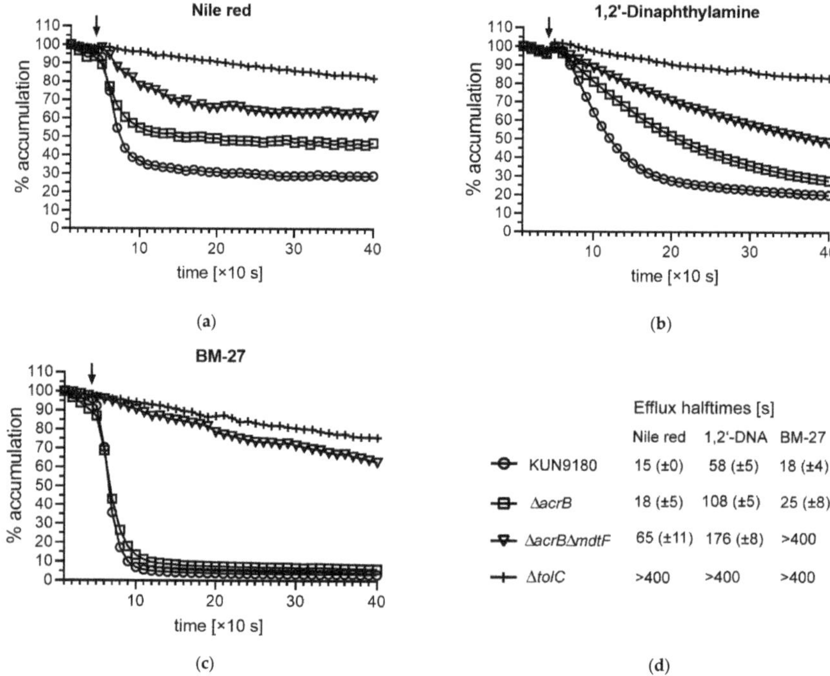

Figure 2. Real-time efflux with fluorescent dyes (the arrow indicates the addition of glucose for the re-energization of the efflux-arrested cells loaded with dye): (**a**) Nile red; (**b**) 1,2′-dinaphthylamine (1,2′-DNA); (**c**) BM-27. (**d**) Graph legend for panels (**a–c**) and efflux halftimes (means with SD, *n* ≥ 3) detected with the KUN9180 isolate and its mutants.

Differences in the substrate compatibility with MdtF regarding the dyes used in the assays are probably due to different physicochemical properties of these compounds. They all are highly lipophilic, but differ remarkably for instance in size and shape (Figure S1). BM-27 is the largest molecule with a molecular weight of 425 [22], whereas that of Nile red is 318 and that of 1,2′-DNA is only 269. (https://pubchem.ncbi.nlm.nih.gov/compound/; accessed on 30 March 2021) [23]. Of course, many other determinants (such as the polarity of the molecules or available hydrogen bond donors and acceptors) might play a role.

3. Materials and Methods

3.1. Strains, Growth Conditions, Chemicals

The MDR clinical isolate KUN9180 was kindly provided from Yasufumi Matsumura from the Department of Clinical Laboratory Medicine, Kyoto University Graduate School of Medicine, Japan. The *acrB* and the *tolC* knockout mutants of the MDR *E. coli* isolate KUN9180 had been engineered within the scope of previous studies [3,11]. The KUNΔ*acrB*Δ*mdtF* double-knockout was constructed as described in Section 3.3. All strains and mutants were cultured in cation-adjusted Mueller Hinton broth (MH2) or on MH2 agar plates at 37 °C overnight.

Chemicals were purchased from Merck (Darmstadt, Germany) with the exception of 1,2′-DNA, which was purchased from TCI (Eschborn, Germany), and BM-27, which was a kind gift from Jadwiga Handzlik from the Department of Technology and Biotechnology of Drugs, Jagiellonian University Medical College, Faculty of Pharmacy, (Kraków; Poland).

3.2. Susceptibility Testing

Minimal inhibitory concentrations (MICs) of antibiotics were determined by broth microdilution assays according to EUCAST guidelines (https://www.eucast.org/; accessed on 30 March 2021).

3.3. Engineering of Mutant KUNΔacrBΔmdtF

Mutant KUNΔ*acrB* (KUN9180Δ*acrB*:FRT-PGK-gb2-neo-FRT [3]) was used to engineer the double-mutant KUNΔ*acrB*Δ*mdtF*. In the first step, the FRT-PGK-gb2-neo-FRT cassette was removed from *acrB* with the assistance of an FLPe (708-FLPe) expression plasmid (Gene Bridges, Heidelberg, Germany), as described in the manual of the "Quick & Easy *E. coli* Gene Deletion Kit" (Gene Bridges). The resulting *acrB* deficient mutant without any resistance marker was cured from the FLPe plasmid (with a temperature sensitive origin) by cultivating at 37 °C and then transformed with a curable Red/ET plasmid (Gene Bridges) harboring a chloramphenicol resistance marker. An FRT-PGK-gb2-neo-FRT cassette was PCR-amplified using oligonucleotides with flanks homologous to the desired substitution region in *mdtF* (Table 2). Red/ET recombination with the purified PCR product was carried out as described in the "Quick & Easy *E. coli* Gene Deletion Kit" protocol (Gene Bridges). Recombinants were selected on agar plates containing 100 µg/ml kanamycin (cross-resistance with neomycin), and successful insertion of the FRT-PGK-gb2-neo-FRT cassette was verified by PCR with check-primers binding up- and downstream from the replacement site in *mdtF* (Table 2).

Table 2. Oligonucleotides used in this study.

Oligonucleotide	Sequence (5′-3′) [1]	Application
upOl-*mdtF*-FRT-PGKgb2neo	*gtcactcaggtgattgagcaaaatatgaatgggcttgatggcctgatgta-*aattaaccctcactaaagggcg	FRT-PGK-gb2neo-FRTcassette amplification
lowOl-*mdtF*-FRT-PGKgb2neo	*gcggtgccatcgtgccagaggcgttgcgtacataccattggttgatgtta-*taatacgactcactatagggctc	FRT-PGK-gb2neo-FRTcassette amplification
Check-*mdtF*-fw	ggcgatcatgaacttaccgg	Check-primer for *mdtF* insertions
Check-*mdtF*-rv	ggatgccgttgtagcgttc	Check-primer for *mdtF* insertions

[1] Underlined letters represent the primer sequence for the FRT-PGKgb2neo-FRT template, letters in italic the homology flanks fitting to *mdtF*.

3.4. Dye Accumulation Assays

The intracellular accumulation of fluorescent dyes was determined as described previously [24–26], with minor modifications. Briefly, bacteria from an overnight cultivated MH2 agar plate were suspended in PBS (phosphate buffered saline, pH 7.4) supplemented with $MgCl_2$ and with glucose to final concentrations of 1 mM and 0.4%, respectively. The suspensions were adjusted to an OD_{600} of 1 and immediately used for determining the intracellular accumulation of dyes, which were added to a final concentration of 2.5 µM for ethidium bromide, 2.5 µM for Hoechst 33342, 30 µg/mL for berberine, 200 µM for PAβN (β-napththylamine determination), and 22.5 µM for Pyronine Y. Fluorescence was measured in the TECAN Infinite M200 PRO plate reader (Crailsheim, Germany) with an incubation temperature of 37 °C. For ethidium, the excitation and emission wavelengths were 518 and 605 nm, respectively, for Hoechst 33343 350 and 461 nm, for berberine 355 and 517 nm, for β-naphthylamine 320 and 460 nm, and for Pyronine Y 545 and 570 nm. Fluorescence values were corrected by subtracting the values detected from the bacterial suspensions without dye.

3.5. Real-Time Efflux Assays

Real-time efflux assays were conducted with Nile red, 1,2′-DNA, and BM-27 according to protocols published earlier [20–22] with slight modifications. Briefly, 20 mL MH2 broth was inoculated with a colony from a freshly grown MH2 agar plate and cultivated overnight at 37 °C with shaking (200 rpm). Bacterial cells were harvested by centrifugation (3220 g, 10 min) and washed twice with PBS. The pellets were suspended in PBS containing 1 mM $MgCl_2$ and adjusted to an OD_{600} of 1. Efflux arrest was induced by incubating with the proton gradient uncoupling agent CCCP (carbonyl cyanide 3-chlorophenylhydrazone, final concentration 5 µM) at 37 °C for 20 min followed by dye loading with Nile red or BM-27 to final concentrations of 10 µM at 37 °C for 2 h. 1,2′-DNA was added to a final concentration of 32 µM (4 h incubation, 37 °C). To monitor the real-time efflux, 180 µL aliquots of cells were washed by centrifugation (5800 g, 2 min) and resuspended in the same volume of PBS containing 1 mM $MgCl_2$. After 40 s of fluorescence recording with the TECAN plate reader M200 Pro, cells were re-energized by the addition of glucose to a final concentration of 1 mM and the measurement was continued for 360 s (at 37 °C). For Nile red the excitation and emission wavelengths were 544 and 650 nm, respectively, for BM-27 400 and 457 nm, and for 1,2′-DNA 370 and 420 nm.

3.6. Statistical Data Analysis

Standard deviations (SD) were calculated from the mean of n experiments as indicated and the statistical significance of differences was analyzed by two-tailed t-tests using the software GraphPad Prism (San Diego, CA, USA) version 8.4.2 (p values < 0.05 represent significance).

4. Conclusions

We found limited contribution of the RND-type transporter MdtF to the antibiotic resistance profile of an MDR *E. coli* isolate, but a remarkable capacity to export dyes (including ethidium) from different chemical substance classes suggesting a potential risk that new compounds including drugs could be substrates, too. Furthermore, it should be kept in mind that dye assays established to evaluate the efflux competence of bacteria do not necessarily allow conclusions about the activity of the major drug exporter AcrB in MDR *E. coli* isolates.

Supplementary Materials: The following are available online at https://www.mdpi.com/article/10.3390/antibiotics10050503/s1, Figure S1: Dyes used in the study of the KUNΔacrB/*mdtF* mutant. Table S1: Further drug susceptibilities of the MDR *E. coli* isolate KUN9180 and derived knockout mutants.

Author Contributions: Conceptualization, S.S. and W.V.K.; methodology, S.S. and L.G.; validation, S.S. and W.V.K.; formal analysis, S.S.; investigation, S.S., M.V. and L.G.; resources, W.V.K.; data curation, S.S.; writing—original draft preparation, S.S.; writing—review and editing, W.V.K.; visualization, S.S.; supervision, W.V.K.; project administration, W.V.K.; funding acquisition, W.V.K. All authors have read and agreed to the published version of the manuscript.

Funding: This research was funded in part by the Innovative Medicines Initiative (IMI) Joint Undertaking project no. 115525 ND4BB Translocation (http://www.translocation.eu/; accessed on 30 March 2021, contributions from the European Union seventh framework program and EFPIA companies).

Acknowledgments: We thank Yasufumi Matsumura from the Department of Clinical Laboratory Medicine, Kyoto University Graduate School of Medicine, Japan, for kindly providing *E. coli* isolate KUN9180, and Jadwiga Handzlik from the Department of Technology and Biotechnology of Drugs, Jagiellonian University Medical College, Faculty of Pharmacy, Kraków, Poland, for compound BM-27.

Conflicts of Interest: The authors declare no conflict of interest.

References

1. Tacconelli, E.; Carrara, E.; Savoldi, A.; Harbarth, S.; Mendelson, M.; Monnet, D.L.; Pulcini, C.; Kahlmeter, G.; Kluytmans, J.; Carmeli, Y.; et al. Discovery, research, and development of new antibiotics: The WHO priority list of antibiotic-resistant bacteria and tuberculosis. *Lancet Infect. Dis.* **2018**, *18*, 318–327. [CrossRef]
2. Kern, W.V.; Oethinger, M.; Jellen-Ritter, A.S.; Levy, S.B. Non-Target Gene Mutations in the Development of Fluoroquinolone Resistance in *Escherichia coli*. *Antimicrob. Agents Chemother.* **2000**, *44*, 814–820. [CrossRef]
3. Schuster, S.; Vavra, M.; Schweigger, T.M.; Rossen, J.W.; Matsumura, Y.; Kern, W.V. Contribution of AcrAB-TolC to multidrug resistance in an *Escherichia coli* sequence type 131 isolate. *Int. J. Antimicrob. Agents* **2017**, *50*, 477–481. [CrossRef] [PubMed]
4. Cunrath, O.; Meinel, D.M.; Maturana, P.; Fanous, J.; Buyck, J.M.; Auguste, P.S.; Seth-Smith, H.M.; Körner, J.; Dehio, C.; Trebosc, V.; et al. Quantitative contribution of efflux to multi-drug resistance of clinical *Escherichia coli* and *Pseudomonas aeruginosa* strains. *EBioMedicine* **2019**, *41*, 479–487. [CrossRef]
5. Bialek-Davenet, S.; Lavigne, J.-P.; Guyot, K.; Mayer, N.; Tournebize, R.; Brisse, S.; Leflon-Guibout, V.; Nicolas-Chanoine, M.-H. Differential contribution of AcrAB and OqxAB efflux pumps to multidrug resistance and virulence in *Klebsiella pneumoniae*. *J. Antimicrob. Chemother.* **2015**, *70*, 81–88. [CrossRef] [PubMed]
6. Gravey, F.; Cattoir, V.; Ethuin, F.; Fabre, L.; Beyrouthy, R.; Bonnet, R.; Le Hello, S.; Guérin, F. *ramR* deletion in an *Enterobacter hormaechei* isolate as a consequence of therapeutic failure to key antibiotics in a long-term hospitalized patient. *Antimicrob. Agents Chemother.* **2020**, *64*. [CrossRef]
7. Ruzin, A.; Immermann, F.W.; Bradford, P.A. RT-PCR and Statistical Analyses of *adeABC* Expression in Clinical Isolates of *Acinetobacter calcoaceticus–Acinetobacter baumannii* Complex. *Microb. Drug Resist.* **2010**, *16*, 87–89. [CrossRef]
8. Gajdács, M. Carbapenem-Resistant but Cephalosporin-Susceptible *Pseudomonas aeruginosa* in Urinary Tract Infections: Opportunity for Colistin Sparing. *Antibiotics* **2020**, *9*, 153. [CrossRef]
9. Bohnert, J.A.; Schuster, S.; Fähnrich, E.; Trittler, R.; Kern, W.V. Altered spectrum of multidrug resistance associated with a single point mutation in the *Escherichia coli* RND-type MDR efflux pump YhiV (MdtF). *J. Antimicrob. Chemother.* **2006**, *59*, 1216–1222. [CrossRef]
10. Camp, J.; Schuster, S.; Vavra, M.; Schweigger, T.; Rossen, J.W.A.; Reuter, S.; Kern, W.V. Limited Multidrug Resistance Efflux Pump Overexpression among Multidrug-Resistant *Escherichia coli* Strains of ST131. *Antimicrob. Agents Chemother.* **2021**, *65*, 01735-20. [CrossRef]
11. Schuster, S.; Vavra, M.; Köser, R.; Rossen, J.W.A.; Kern, W.V. New Topoisomerase Inhibitors: Evaluating the Potency of Gepotidacin and Zoliflodacin in Fluoroquinolone-Resistant *Escherichia coli* upon *tolC* Inactivation and Differentiating Their Efflux Pump Substrate Nature. *Antimicrob. Agents Chemother.* **2020**, *65*. [CrossRef] [PubMed]
12. Nishino, K.; Yamaguchi, A. Analysis of a Complete Library of Putative Drug Transporter Genes in *Escherichia coli*. *J. Bacteriol.* **2001**, *183*, 5803–5812. [CrossRef] [PubMed]
13. Zgurskaya, H.I.; Krishnamoorthy, G.; Ntreh, A.; Lu, S. Mechanism and Function of the Outer Membrane Channel TolC in Multidrug Resistance and Physiology of Enterobacteria. *Front. Microbiol.* **2011**, *2*, 189. [CrossRef] [PubMed]
14. Bleuel, C.; Grosse, C.; Taudte, N.; Scherer, J.; Wesenberg, D.; Krauß, G.J.; Nies, D.H.; Grass, G. TolC Is Involved in Enterobactin Efflux across the Outer Membrane of *Escherichia coli*. *J. Bacteriol.* **2005**, *187*, 6701–6707. [CrossRef]
15. Vega, D.E.; Young, K.D. Accumulation of periplasmic enterobactin impairs the growth and morphology of *Escherichia coli tolC* mutants. *Mol. Microbiol.* **2013**, *91*, 508–521. [CrossRef]
16. Zhang, Y.; Xiao, M.; Horiyama, T.; Zhang, Y.; Li, X.; Nishino, K.; Yan, A. The Multidrug Efflux Pump MdtEF Protects against Nitrosative Damage during the Anaerobic Respiration in *Escherichia coli*. *J. Biol. Chem.* **2011**, *286*, 26576–26584. [CrossRef]
17. Sjuts, H.; Vargiu, A.V.; Kwasny, S.M.; Nguyen, S.T.; Kim, H.-S.; Ding, X.; Ornik, A.R.; Ruggerone, P.; Bowlin, T.L.; Nikaido, H.; et al. Molecular basis for inhibition of AcrB multidrug efflux pump by novel and powerful pyranopyridine derivatives. *Proc. Natl. Acad. Sci. USA* **2016**, *113*, 3509–3514. [CrossRef]

18. Lomovskaya, O.; Warren, M.S.; Lee, A.; Galazzo, J.; Fronko, R.; Lee, M.; Blais, J.; Cho, D.; Chamberland, S.; Renau, T.; et al. Identification and Characterization of Inhibitors of Multidrug Resistance Efflux Pumps in *Pseudomonas aeruginosa*: Novel Agents for Combination Therapy. *Antimicrob. Agents Chemother.* **2001**, *45*, 105–116. [CrossRef]
19. Wang-Kan, X.; Rodríguez-Blanco, G.; Southam, A.D.; Winder, C.L.; Dunn, W.B.; Ivens, A.; Piddock, L.J.V. Metabolomics Reveal Potential Natural Substrates of AcrB in *Escherichia coli* and *Salmonella enterica* Serovar Typhimurium. *mBio* **2021**, *12*. [CrossRef]
20. Bohnert, J.A.; Karamian, B.; Nikaido, H. Optimized Nile Red Efflux Assay of AcrAB-TolC Multidrug Efflux System Shows Competition between Substrates. *Antimicrob. Agents Chemother.* **2010**, *54*, 3770–3775. [CrossRef]
21. Bohnert, J.A.; Schuster, S.; Szymaniak-Vits, M.; Kern, W.V. Determination of Real-Time Efflux Phenotypes in *Escherichia coli* AcrB Binding Pocket Phenylalanine Mutants Using a 1,2′-Dinaphthylamine Efflux Assay. *PLoS ONE* **2011**, *6*, e21196. [CrossRef]
22. Bohnert, J.A.; Schuster, S.; Kern, W.V.; Karcz, T.; Olejarz, A.; Kaczor, A.; Handzlik, J.; Kieć-Kononowicz, K. Novel Piperazine Arylideneimidazolones Inhibit the AcrAB-TolC Pump in *Escherichia coli* and Simultaneously Act as Fluorescent Membrane Probes in a Combined Real-Time Influx and Efflux Assay. *Antimicrob. Agents Chemother.* **2016**, *60*, 1974–1983. [CrossRef] [PubMed]
23. Kim, S.; Chen, J.; Cheng, T.; Gindulyte, A.; He, J.; He, S.; Li, Q.; A Shoemaker, B.; A Thiessen, P.; Yu, B.; et al. PubChem in 2021: New data content and improved web interfaces. *Nucleic Acids Res.* **2021**, *49*, D1388–D1395. [CrossRef]
24. Schumacher, A.; Steinke, P.; Bohnert, J.A.; Akova, M.; Jonas, D.; Kern, W.V. Effect of 1-(1-naphthylmethyl)-piperazine, a novel putative efflux pump inhibitor, on antimicrobial drug susceptibility in clinical isolates of Enterobacteriaceae other than *Escherichia coli*. *J. Antimicrob. Chemother.* **2005**, *57*, 344–348. [CrossRef]
25. Schuster, S.; Kohler, S.; Buck, A.; Dambacher, C.; König, A.; Bohnert, J.A.; Kern, W.V. Random Mutagenesis of the Multidrug Transporter AcrB from *Escherichia coli* for Identification of Putative Target Residues of Efflux Pump Inhibitors. *Antimicrob. Agents Chemother.* **2014**, *58*, 6870–6878. [CrossRef] [PubMed]
26. Schuster, S.; Bohnert, J.A.; Vavra, M.; Rossen, J.W.; Kern, W.V. Proof of an Outer Membrane Target of the Efflux Inhibitor Phe-Arg-β-Naphthylamide from Random Mutagenesis. *Molecules* **2019**, *24*, 470. [CrossRef]

Review

Role of RND Efflux Pumps in Drug Resistance of Cystic Fibrosis Pathogens

Viola Camilla Scoffone [1], Gabriele Trespidi [1], Giulia Barbieri [1], Samuele Irudal [1], Elena Perrin [2,*] and Silvia Buroni [1,*]

[1] Department of Biology and Biotechnology "Lazzaro Spallanzani", University of Pavia, 27100 Pavia, Italy; viola.scoffone@unipv.it (V.C.S.); gabriele.trespidi01@universitadipavia.it (G.T.); giulia.barbieri@unipv.it (G.B.); samuele.irudal@iusspavia.it (S.I.)

[2] Department of Biology, University of Florence, 50019 Sesto Fiorentino, Italy

[*] Correspondence: elena.perrin@unifi.it (E.P.); silvia.buroni@unipv.it (S.B.); Tel.: +39-055-4574731 (E.P.); +39-0382-985571 (S.B.)

Abstract: Drug resistance represents a great concern among people with cystic fibrosis (CF), due to the recurrent and prolonged antibiotic therapy they should often undergo. Among Multi Drug Resistance (MDR) determinants, Resistance-Nodulation-cell Division (RND) efflux pumps have been reported as the main contributors, due to their ability to extrude a wide variety of molecules out of the bacterial cell. In this review, we summarize the principal RND efflux pump families described in CF pathogens, focusing on the main Gram-negative bacterial species (*Pseudomonas aeruginosa*, *Burkholderia cenocepacia*, *Achromobacter xylosoxidans*, *Stenotrophomonas maltophilia*) for which a predominant role of RND pumps has been associated to MDR phenotypes.

Keywords: RND efflux pumps; drug resistance; cystic fibrosis

Citation: Scoffone, V.C.; Trespidi, G.; Barbieri, G.; Irudal, S.; Perrin, E.; Buroni, S. Role of RND Efflux Pumps in Drug Resistance of Cystic Fibrosis Pathogens. *Antibiotics* **2021**, *10*, 863. https://doi.org/10.3390/antibiotics10070863

Academic Editors: Isabelle Broutin, Attilio V. Vargiu, Henrietta Venter and Gilles Phan

Received: 28 May 2021
Accepted: 13 July 2021
Published: 15 July 2021

Publisher's Note: MDPI stays neutral with regard to jurisdictional claims in published maps and institutional affiliations.

Copyright: © 2021 by the authors. Licensee MDPI, Basel, Switzerland. This article is an open access article distributed under the terms and conditions of the Creative Commons Attribution (CC BY) license (https://creativecommons.org/licenses/by/4.0/).

1. Introduction

According to the Cystic Fibrosis Foundation Patient Registry, worldwide more than 70,000 people suffer from Cystic Fibrosis (CF) [1]. Mutations in the Cystic Fibrosis Transmembrane conductance Regulator (CFTR) gene are responsible for the insurgence of a pathological condition, with different severities based on the type of mutation [2]. The CFTR channel is required for the homeostatic control of chloride and bicarbonate ions in the lung. Its malfunctioning leads to mucin overproduction along airways and disruption of the regular mucociliary activity [3,4]. Together, these defects promote polymicrobial proliferation in the respiratory tract, where bacteria are trapped in the mucus and their clearance becomes harder and harder [4]. Moreover, their presence stimulates an exaggerated inflammatory response, making CF pathology characterized by a progressive loss of lung function.

It is noteworthy that the microbial community in CF lungs changes during the lifetime: in 3–5 year-old children, one or a few CF pathogens are detected [1], while in adolescents and adults a polymicrobial community or the prevalence of one typical CF bacterium (e.g., *Pseudomonas*, *Staphylococcus*, *Stenotrophomonas*, or *Burkholderia*) has been reported [5]. The introduction of CFTR modulator therapy has greatly improved the general health conditions of CF people; however, the effects of lumacaftor-ivacaftor, tezacaftor-ivacaftor, and elexacaftor-tezacaftor-ivacaftor therapy in patients with diverse genetic backgrounds, as well as their effects on the airway microbiota, need to be addressed [6].

A major concern regards the Multi Drug Resistance (MDR) phenotype of CF lung-associated pathogens. Beside the classical drug resistance mechanisms (i.e., drug modification and inactivation, decreased membrane permeability, modification of antibiotic targets, target protection, drug efflux), during the progression of infection, *Pseudomonas aeruginosa* may switch to the mucoid phenotype, which is very difficult to eradicate [7]. In addition, the highly resistant small-colony variant phenotype of *Staphylococcus aureus* and *P. aeruginosa* may be induced by repetitive antibiotic therapy [8,9]. Also, the proportion of methicillin-resistant *S. aureus* (MRSA), together with metallo-β-lactamase-producing *P. aeruginosa* strains is worrisome [10,11]. Moreover, during the COVID-19 pandemic, the increased usage of antibiotics to control secondary bacterial infections may further accelerate the spread of antibiotic resistance among nosocomial pathogens [12].

Indeed, while early infections by CF pathogens can be intermittent and involve different strains with multiple levels of antibiotic resistance (AR) profiles, subsequently, people with CF are chronically colonized with well adapted strains with properties (among which high levels of MDR) that differ significantly from those exhibited by the isolate which gave rise to the infection [13,14].This change is related to the adaptation of bacteria to the fluctuating and heterogeneous conditions of the CF lung environment, which exerts a high selective pressure [15]. CF lung is indeed an ecological niche characterized by several selective elements, including the host immune response, the oxidative stresses especially derived from the liberation of reactive oxygen species (ROS) by neutrophils, the interactions among different microorganisms, the nutrient availability, the modified acidity and salinity of the surrounding environment, and the oxygen deprivation in mucus [14–16]. Moreover, a strong selective pressure is exerted by the high levels of antibiotics used to treat the infections caused by CF pathogens (a summary of the antibiotic treatment used for the CF pathogens described in this review is reported in Supplementary Table S1) [14–16].

Among the consequences of this high selective pressure, there is the emergence of hypermutable strains, whose presence has been strongly associated with bacteria adaptation to the lung environment [12,13]. Hypermutable strains, together with the characteristic transition from the planktonic to the biofilm lifestyle of CF pathogens during chronic infections, lead to the development of high levels of AR in strains adapted to the CF lung. Together, all these factors increase the rate of AR through horizontal gene transfer [12,13]. Although no single mutations can lead to MDR profiles, the use of all antibiotics is prone to be compromised by the acquisition of mutations that can lead to overexpression of efflux pumps, hyperproduction of antibiotic degrading enzymes, porin loss or altered antibiotic targets [13]. Among efflux pumps, those belonging to the Resistance-Nodulation-cell Division (RND) family are able to translocate different molecules (including drugs) out of the bacterial cell in an aspecific manner, thus increasing the ability of bacteria to resist a wide range of treatments [17] RND efflux systems are tripartite complexes composed of an inner membrane protein, a periplasm associated subunit (membrane fusion protein or MFP), and an outer membrane protein (OMP), that span the inner and outer Gram-negative membranes. These pumps are activated by a proton motive force to export compounds into the extracellular environment. The best-described members of this family are the AcrAB-TolC and the MexAB-OprM of *Escherichia coli* and *P. aeruginosa*, respectively [18,19].

In this review, we will describe the principal RND efflux pump families which have been found in CF pathogens, then we will focus on the main Gram-negative bacterial species (*P. aeruginosa*, *Burkholderia cenocepacia*, *Achromobacter xylosoxidans*, *Stenotrophomonas maltophilia*) for which a predominant role of RND pumps has been associated to MDR phenotypes.

2. RND Efflux Pump Families in CF Pathogens

The RND superfamily is a ubiquitous group of efflux pumps conserved in all domains of life (for a recent review see [17]). This superfamily is divided into nine functionally recognized families, six of which have representatives in Gram-negative bacteria [17,20].

In particular, the SecDF efflux pumps are involved in the general secretion (Sec) pathway and members of this family are present in both Bacteria and Archea [17,20]. However, most of the characterized RND proteins of Gram-negative bacteria belong to the Hydrophobe/Amphiphile Efflux 1 (HAE-1) and Heavy Metal Efflux (HME) families, involved in the export of multiple drugs and heavy metals respectively [17,20]. In addition, three other families with few representatives have been found in Gram-negative bacteria that are less known and characterized: (i) the Nodulation Factor Exporter (NFE) family that was identified as a probable nodulation factor exporter, although recently added members of this family are drug exporters; (ii) the Aryl Polyene Pigment Exporters (APPEs), that have been found in *Xanthomonas oryzae* where they are involved in exporting a pigment [17,20]; (iii) the Hydrophobe/Amphiphile Efflux 3 (HAE-3) family that included some Archea transporters but also HpnN proteins, a group of Gram-negative pumps apparently involved in the transport of hopanoids to the outer membrane [17,20]. The RND proteins of the HAE-1, HME and NFE families are generally associated with an MFP and an OMP protein to form a complex that allows the extrusion of substrates directly out of the cells. The genes coding for these three proteins are usually associated in an operon [17,20].

Most of the RND systems identified and experimentally characterized in cystic fibrosis pathogens belong to the HAE-1 family and are involved in antibiotic efflux. In *P. aeruginosa*, twelve different RND operons have been found (*mexAB-oprM*, *mexCD-oprJ*, *mexEF-oprN*, *mexXY*, *mexJK*, *mexGHI-opmD*, *mexPQ-opmE*, *mexMN*, *muxABC-ompB*, *mexVW*, *triABC* and *czcABC*) [21]. The CzcABC system belongs to the HME family, while all the others belong to the HAE-1 family [22]. All the twelve systems have been experimentally characterized and most of them are conserved among different strains (in particular, MexAB-OprM, MexCD-OprJ, MexEF-OprN, MexXY and MexJK) [22–26].

In the *Burkholderia cepacia* complex at least 19 different putative HAE-1 RND efflux pumps are present, four of which (operon RND-4 or *bpeAB-oprB*, operon RND-6 RND-7, operon RND-10 or *ceoAB-opcM* or *bpeEF-oprC* and operon RND-13) are being conserved among several different strains [20,27,28]. Most of these proteins belong to the HAE-1 family and for several of them, the role in antibiotic efflux have been experimentally confirmed in several *Burkholderia* species (RND-1, RND-3 or AmrAB-OprA, RND-4 or BpeAB-OprB, RND-8 and RND-9, RND-10 or CeoAB-OpcM or BpeEF-OprC) [29]. Moreover, for two systems (RND-11 or CusABC and RND-12 or CzcABC) identified as belonging to the HME family [20,30], the role in heavy-metal efflux has been experimentally validated [31]. The genes coding for putative SecDF, HpnN/HAE-3 and APPE proteins have been found but not experimentally confirmed [20]. Finally, in this genus, a group of operons which appear not to belong to any of the recognized RND families have been identified and defined as Uncertain Function (UF) [20,30].

In the genome of the type strain of *A. xylosoxidans*, ATCC 27061, the genes coding for 9 different RND efflux pumps have been identified [32]. Three of these efflux pumps have been functionally characterized: AxyABM (homolog of MexAB-OprM) [33], AxyXY-OprZ (with homology to MexXY-OprM) [34] and AxyEF-OprN [35]. All these systems are involved in the transport of several different antibiotics and belong to the HAE1 family of RND transporters [33–35]. The substrates and the family of the other six pumps have yet to be determined. A comparative genomic analysis showed that the genes coding for most of these nine systems are conserved among different *A. xylosoxidans* strains [34], with one of them, *axyABM*, conserved in all the sequenced *Achromobacter* genomes [36], while *axyXY–oprZ* has been found also in *Achromobacter ruhlandii* [37]. Regarding proteins belonging to the HME family, RND transport systems homologous of CzcABC and CusABC are present in the genomes of other *Achromobacter* strains [38,39].

Finally, in *S. maltophilia*, the genes coding for fifteen putative HAE-1 RND systems have been found, seven of which (*smeVWX*, *smeYZ*, *smeGH*, *smeMN*, *smeOP*, *smeDEF*, *smeIJK*) seem to be conserved among different strains [40,41]. Eight out of these fifteen pumps (SmeVWX, SmeYZ, SmeOP, SmeDEF, SmeIJK, SmeABC, SmeGH, SmeMN) have also been experimentally characterized, confirming that they are actually involved in AR [40]. In addition, the genes coding for six others putative HME RND efflux pumps have been found in the genome of the K279a strain [41], but none of them have been experimentally validated.

3. RND in *Pseudomonas aeruginosa*

3.1. Pseudomonas aeruginosa Infections in CF

P. aeruginosa is a Gram-negative bacterium that belongs to the family of *Pseudomonadaceae*. Thanks to its metabolic versatility it is able to colonize many different environments and to establish opportunistic infections [42]. The World Health Organization classified as a priority one *P. aeruginosa* carbapenem resistant [43]. *P. aeruginosa* is the most common causative agent of Gram-negative nosocomial infections and lung infection in CF patients [44]. MDR *P. aeruginosa* is responsible for over 72,000 infections and 4800 deaths annually in Europe and the majority of these cases are attributed to carbapenem and colistin-resistant strains [45].

P. aeruginosa has a relatively large genome of 5.5–7 million base pairs, encoding a large number of regulatory enzymes involved in metabolism, transport and efflux [46]. During childhood, CF patients are colonized by both *P. aeruginosa* and *S. aureus*, while in adulthood *P. aeruginosa* is predominant and induces lung function decline [47]. The interaction between *P. aeruginosa* and its hosts is still poorly understood and its persistence in the airways is due to highly complex and multifactorial reasons [48]. The CF airways environment helps *P. aeruginosa* colonization over other bacteria (*S. aureus*) and the consequence of this is the prevalence of *P. aeruginosa* in adults, ranging from 31 to 47% [49]. One possible reason for this prevalence is that the physiological defects linked to CFTR mutations (such as mucus viscosity, production of reactive oxygen species, impaired autophagy, reduced airway acidity and accumulation of ceramides) induce advantages to *P. aeruginosa* [50].

During the course of the infection, the genetic and phenotypic traits of *P. aeruginosa* strains in CF airways are subjected to evolutionary changes in response to the selective pressure of the environment [51]. Chronic *P. aeruginosa* infections are recalcitrant to antibiotic treatment, which are extremely challenging due to the ability of the bacterium to resist the commonly used compounds thanks to its numerous mechanisms of resistance (efflux pumps, ability to form biofilm, persistence) [52]

P. aeruginosa is resistant to numerous antibiotics belonging to the aminoglycosides, quinolones, and β-lactams families [53]. Mechanisms of AR of *P. aeruginosa* are classified into intrinsic, acquired, and adaptive. Mechanisms of intrinsic AR are encoded by the core genome of the organism, adaptive resistance is induced by environmental stimuli, while acquired resistance depends on the gain of resistance genes derived from other organisms or those which originated after the selection of mutations [54]. Among intrinsic resistance mechanisms there are: the low outer membrane permeability, the expression of efflux pumps, lipopolysaccharides modification, and the production of enzymes that inactivate antibiotics. The adaptive resistance is related to biofilm formation that limits antibiotic access to bacterial cells, decreases bacterial motility and promotes the formation of persister cells [55]. Acquired resistance is the result of horizontal transfer of resistance-related genes or of mutational changes [56].

3.2. Pseudomonas aeruginosa RND Efflux Systems

Antibiotic extrusion and resistance in *P. aeruginosa* can be closely related to tripartite RND efflux pumps [57]. Efflux pumps are also involved in cellular stress response. Stress signals such as host factors, detergents and endogenous inducers of bacterial stress could help to select mutants, which over-express efflux systems [58]. The constant inflammation of CF lungs exposes *P. aeruginosa* to reactive oxygen species (ROS), which might induce the prevalence of strains over-expressing efflux pumps (MexAB-OprM and MexXY-OprM) [59]. Moreover, Fraud and colleagues showed that ROS over-exposure selects resistant mutants expressing the RND MexXY-OprM [60].

Among the 12 RND efflux pumps identified in *P. aeruginosa*, six contribute to AR [61]. These RNDs are: MexAB-OprM, MexCD-OprJ, MexEF-OprN, MexXY-OprM, MexJK-OprM and MexVW-OprM (Table 1) [52,62]. MexAB-OprM and MexXY-OprM are constitutively expressed at the basal level in wild type strains and are induced by antibiotic substrates, while the other systems are not expressed in wild type strains [52,63]. The genes encoding these tripartite efflux pumps are organized in operons, but in certain cases the operon does not contain the OMF gene, such as in the case of MexXY, MexJK and MexVW.

Table 1. RND efflux pumps in *P. aeruginosa*.

RND Efflux Pump	Systematic ID	Family	Identified Regulator(s)	Substrates
MexAB-OprM	PA0425-PA0427	HAE-1	MexR, repressor (MarR-type regulator)	β-Lactams (except imipenem), β-lactam inhibitors, fluoroquinolones, tetracycline, chloramphenicol, novobiocin, macrolides, trimethoprim, triclosan (irgasan), ethidium bromide, SDS, aromatic hydrocarbons, thiolactomycin, cerulenin, acylated homoserine lactones
MexCD-OprJ	PA4599- PA4597	HAE-1	NfxB, repressor (TetR/AcrR-type regulator)	β-Lactams, fluoroquinolones, chloramphenicol, tetracycline, novobiocin, trimethoprim, macrolides, crystal violet, ethidium bromide, acriflavine, SDS, aromatic hydrocarbons, triclosan
MexEF-OprN	PA2493-PA2495	HAE-1	MexT, activator (LysR-type regulator)	Fluoroquinolones, chloramphenicol, trimethoprim, aromatic hydrocarbons, triclosan, *Pseudomonas* quinolone signal
MexXY	PA2019-PA2018	HAE-1	MexZ, repressor (TetR-type regulator	Fluoroquinolones, aminoglycosides, tetracycline, erythromycin
MexJK	PA3677-PA3676	HAE-1	MexL, repressor (TetR/AcrR-type regulator)	Tetracycline, erythromycin, triclosan
MexVW	PA4374-PA4375	HAE-1	N.D.	Norfloxacin, ofloxacin, chloramphenicol, cefpirome, tetracycline, ethidium bromide and acriflavine

MexAB-OprM extrudes carbapenems, chloramphenicol, fluoroquinolones, lincomycin, macrolides, novobiocin, tetracyclines, and all β-lactams except imipenem. It is also involved in the efflux of triclosan (antiseptic compound) and of sodium dodecyl sulfate (surfactant). While deletion of *mexAB-oprM* results in a *P. aeruginosa* strains sensitive to all the above-mentioned antibiotics, a mutant overexpressing MexAB-OprM is characterized by a significant level of resistance [64,65]. The efflux pump MexAB-OprM is composed of an inner membrane protein MexA, a fusion protein MexB and the outer membrane protein OprM [66]. Genes encoding these proteins constitute an operon which is controlled by the transcriptional regulator MexR [67]. The *mexR* gene is localized upstream of the *mexAB-oprM* operon and encodes a transcriptional repressor which binds the intergenic region between *mexA* and *mexR*, in proximity to their promoters [68]. When MexR is not functional, there is MexAB-OprM overexpression. *P. aeruginosa* clinical isolates showed different types of *mexR* mutations, leading to the production of a protein unable to dimerize, to bind the DNA and to repress *mexAB-oprM* operon or mutations that result in the complete absence of a functional MexR (such as peptide premature termination) [69,70]. A recent study focused on the evolution of resistance during infections showed that the frequency of mutations (frameshift in either *mexA* or *oprM*) in *mexAB-oprM* rises rapidly during infection, providing evidence that the loss of this pump is adaptive [71]. Mutants have a low meropenem resistance, suggesting that these mutations arise in a sub-population of cells of the ancestral strain that are protected from meropenem by physical barriers, such as biofilm, or by phenotypic resistance (tolerance or persistence) [72,73].

The RND efflux pump MexCD-OprJ is expressed in *nfxB P. aeruginosa* mutants only. NfxB is the negative regulator of MexCD-OprJ and clinical isolates with diverse mutations in *nfxB* gene were isolated. These mutants showed different levels of resistance to the antibiotics effluxed by MexCD-OprJ, such as chloramphenicol, erythromycin, fluoroquinolones, and tetracyclines [74].

Another RND efflux pump is MexEF-OprN that, unlike the other efflux systems, is positively regulated by the transcriptional activator MexT [64]. This efflux pump extrudes chloramphenicol, fluoroquinolones, tetracycline, and trimethoprim [75]. In most laboratory strains deriving from reference *P. aeruginosa* strain PAO1, the *mexT* gene is frequently unfunctional, causing the suppression of *mexEF-oprN* operon [76]. On the other hand, when MexT is active, it also works as a repressor of the OprD porin, inducing an increase of resistance to carbapenem [77]. *P. aeruginosa* mutants in the *nfxC* gene (norfloxacin resistance gene) are characterized by the over-expression of *mexEF-oprN* operon and are more resistant to chloramphenicol, fluoroquinolones, tetracycline, trimethoprim, and imipenem [78].

One of the most studied RND of *P. aeruginosa* is the efflux system MexXY-OprM, which contributes to intrinsic resistance to aminoglycosides, tetracyclines, erythromycin, and cefepime [79]. The MexXY can form functional complexes with two different outer membrane proteins, OprM and OprA, in *P. aeruginosa* PA7 [80]. Recently, it has been shown that the substrate specificities of MexXY can change depending on which OM protein it complexes with [81]. Both OprM and OprA are involved in aminoglycosides efflux, while carbenicillin and sulbenicillin are substrates only of the MexXY-OprA complex [81]. The regulator of this RND is the repressor MexZ and mutations in its gene, or in the regulatory region, lead to overexpression of MexXY [82,83]. In *P. aeruginosa* CF clinical isolates, the most common mutations are localized in the *mexZ* gene, inducing MexXY-OprM overproduction. These *mexZ* mutations arise during chronic infections in CF patients, contributing to tobramycin resistance, one of the first-line antibiotics used in CF [84]. The expression of *mexY* and *mexZ* was found to be higher in adults with chronic infection than in children with new or chronic infections, suggesting that these mutations are subjected to positive selection [85].

Another RND efflux pump, MexJK, was identified using triclosan (biocide) as selective agent in *mexL* mutants in a Δ*mexAB-oprM* and Δ*mexCD-oprJ* strains [86]. Furthermore, MexCD-OprJ expression is selected by triclosan and could be considered an interesting selective tool to study efflux systems [86] MexJK expression is controlled by the product of an upstream regulatory gene, *mexL*, similar to what has been described in other RND efflux pumps. MexJK lacks its own outer membrane protein and requires OprM for the efflux of antibiotics [86].

Using a *P. aeruginosa* mutant lacking *mexAB*, *mexCD–oprJ*, *mexEF–oprN* and *mexXY*, the RND efflux pump MexVW was characterized [87]. In the proximity of the *mexVW* genes, no ORFs are present that could encode a regulatory protein; similarly, no genes coding for an outer membrane protein are present in the downstream region. MexVW works as a multidrug efflux pump and uses OprM as OMP. Overexpression of *mexVW* was demonstrated to confer resistance to norfloxacin, ofloxacin, chloramphenicol, cefpirome, tetracycline, and ethidium bromide [87].

3.3. P. aeruginosa RND Efflux Pumps Inhibitors

Among the *P. aeruginosa* efflux pump inhibitors, the most studied is Phe-Arg-β- naphthylamide (PAβN), a broad spectrum peptidomimetic compound. PAβN was shown to interfere with the four RND systems of *P. aeruginosa*: MexAB-OprM, MexCD-OprJ, MexEF-OprN, MexXY-OprM. The association of chloramphenicol, fluoroquinolones, macrolides, ketolides, oxazolidinones, and rifampicin with PAβN increases their effects [88]. PAβN functions as substrate of Mex efflux pumps and competes with antibiotics, preventing their extrusion [89]. Unfortunately, PAβN and its derivatives during phase 1 clinical trials showed adverse toxicity and pharmacokinetic profile [90].

Another efflux pump inhibitor is the pyridopyrimidine derivative D13-9001 [89]. It blocks MexAB-OprM in vivo and in vitro and it showed low toxicity profiles [91]. The mechanism of action of this compound relies on a tight interaction with the hydrophobic trap of the pump, preventing its conformational changes. At the same time, D13-9001 blocks the substrate binding to MexB [92]. The limit of this molecule is its specificity for MexAB-OprM: in fact, efflux pump inhibitors should be broad spectrum compounds in order to be used as adjuvants together with antibiotics that are substrates of several efflux pumps. Moreover, different mechanisms of resistance were identified when the compound was administered with carbenicillin. The resistance occurred due to a mutation in the residue F628 of MexB, a site involved in inhibitor binding [93,94].

A polyamine scaffold was identified as an efflux pump inhibitor by Fleeman and co-workers [95]. Polyamines are essential organic polycations ubiquitous in all forms of life and are composed of an aliphatic carbon chain with numerous amino groups. Five polyamine derivatives were demonstrated to potentiate the effect of aztreonam, chloramphenicol, and tetracycline, inducing an MIC_{90} decrease of 5- to 8-fold. These compounds have limited toxicity and no inhibitory effects on the eukaryotic Ca^{2+} channel of human kidney cells [95].

Among the natural products that target MDR efflux pumps, there are EA-371α and EA-371δ, identified by screening a library of 78,000 microbial fermentation extracts [96]. These compounds are the products of a *Streptomyces* strain and are potent MexAB-OprM inhibitors, with a MPC_8 (minimum potentiation concentration decreasing the MIC of 8-fold) values of 4.29 µM (EA-371α) and 2.15 µM (EA-371δ) for levofloxacin against strain PAM103. Unfortunately, EA-371α could not be considered a lead compound because of its moderate cytotoxicity [96].

Another type of RND efflux inhibition relies on the application of phage therapy. While the traditional phage therapy is based on the administration of phages to block bacterial cell growth, another approach used phages to steer AR evolution. An example is the lytic *Myoviridae* bacteriophage OMKO1 that uses OprM as a receptor binding site. Bacteria resistant to OMKO1, lacking OprM, are more sensitive to ciprofloxacin, tetracycline, ceftazidime, and erythromycin due to the counterselection of MDR *P. aeruginosa* and, possibly, to a change in the efflux pump mechanism [97].

4. RND in *Burkholderia cenocepacia*

4.1. *Burkholderia cenocepacia* Infections in CF

Burkholderia cepacia complex (Bcc) species are abundant in the polymicrobial communities inhabiting the lungs of adult CF patients [98]. Within this group of 24 phenotypically related but genetically distinct bacterial species, *Burkholderia cenocepacia* and *Burkholderia multivorans* are responsible for approximately 70–85% of all Bcc infections in this cohort of patients [99,100]. The wide variety of potential virulence factors (e.g., catalases, proteases and siderophores) produced by these bacteria to evade host defenses, their innate resistance to many antibiotics and disinfectants, their ability to adhere and invade epithelial cells and to survive inside macrophages, render *B. cenocepacia* infections very difficult to treat [101–104]. Clinical effects vary from transient carriage to chronic lung infection, which can rapidly deteriorate to necrotizing pneumonia and sepsis, the so-called "cepacia syndrome", resulting in a significant decrease in patients' survival [105,106]. Moreover, the poor post lung transplant outcomes of individuals affected by *B. cenocepacia* renders chronic infection as a contraindication for lung transplantation [107]. In this scenario, despite the relatively low and stable prevalence of *B. cenocepacia* infections, affecting around 3% of CF patients in Europe [108], this opportunistic pathogen represents a serious burden for the management of people affected by CF.

The main challenges in the treatment of *B. cenocepacia* infections are represented by the intrinsic resistance of this species to clinically relevant antibiotics and by their tolerance to antibiotic exposure, typically associated with a biofilm lifestyle [109,110]. In the absence of evidence-based guidelines for treatment [111], various therapeutic protocols based on the use of single or multiple antibiotics administered by different routes (intravenous, oral, inhaled, or combined) for varying periods of time have been employed in clinics. However, complete eradication of the infection is difficult to achieve [112,113]. Strategies based on compounds that improve the activity of antibiotics (helper compounds) by blocking the main resistance mechanisms or altering the physiological state of antibiotic-tolerant cells are in clinical trials or under study [114–117]. These molecules generally act by impairing bacterial growth, permeabilizing bacteria through the alteration of the structure of the outer membrane, inhibiting biofilm formation and eradicating established biofilms [114,116]. Alternative approaches based on molecules used for other diseases, natural products, quorum-sensing inhibitors and antimicrobial peptides are under investigation [118–121]. Finally, interest in the design of *B. cenocepacia* vaccines has recently risen [122].

4.2. Burkholderia cenocepacia RND Efflux Systems

The ability to produce a variety of efflux pumps significantly contributes to the inherent multidrug resistance of *B. cenocepacia* [112,123]. After the identification of the gene cluster encoding the conserved salicylate-regulated RND-10 efflux pump responsible for chloramphenicol, trimethoprim, and ciprofloxacin resistance [124], sixteen genes encoding transporters of the RND family, organized in 14 operons, have been identified in the genome of the reference *B. cenocepacia* strain J2315 (Table 2) [20,125,126]. This CF isolate, belonging to the highly transmissible epidemic ET12 lineage, was used for the preparation of a collection of mutant strains, each carrying a marker-less deletion of a single RND operon, thus allowing the investigation of the role of these systems in *B. cenocepacia* physiology and antibiotic susceptibility [127–129]. While the RND-deleted strains did not show any defect in their growth characteristics, the absence of a few specific RND-systems resulted in increased antibiotic susceptibility and, in some cases, alterations in the production of biofilm matrix compared to their parental strain [129]. In particular, when grown in planktonic cultures, mutants lacking the RND-3 and RND-4 efflux systems displayed a higher susceptibility to both ciprofloxacin and tobramycin and a reduced secretion of quorum-sensing molecules [127,129]. Interestingly, lifestyle specific effects could be observed for the different mutants. While the contribution of the RND-3 system to the intrinsic AR of *B. cenocepacia* J2315 was exerted both in planktonic and sessile cells, the RND-4 efflux pump played a major role in the efflux of ciprofloxacin, tobramycin, minocycline, and chloramphenicol only in planktonic cells. On the contrary, the RND-8 and RND-9 efflux systems were demonstrated to confer protection against tobramycin only in biofilms, but not in planktonic cultures [129]. The lifestyle-specific activity of these pumps appears as a cellular response to regulatory signals governing the physiology of the cell. In fact, besides contributing to the extrusion of antibiotics and of a variety of compounds toxic for cellular metabolism, RND systems play a role in the control of physiological processes and virulence of *B. cenocepacia* [128]. Deletion of RND-efflux pumps was reported to affect motility-related phenotypes and biofilm formation, with RND-4 and RND-9 mutant deletion strains showing an enhanced biofilm formation ability and an increased and reduced swimming motility, respectively [128]. As revealed by transcriptomic analysis, while the motility phenotype could be easily correlated to a differential expression of motility genes in the mutant strains compared to wild type, the increased ability to form a biofilm could not be linked to an altered expression of genes involved in biofilm formation, suggesting indirect regulatory mechanisms, possibly activated by altered concentrations of toxic compounds or metabolic signals that accumulate in the cell as a consequence of efflux pump inactivation.

Table 2. RND efflux pumps in *B. cenocepacia*.

RND-Efflux Pump	Systematic ID	Family	Identified Regulator(s)	Antibiotic Substrates
RND-1	BCAS0591-BCAS0593	HAE-RND	N.A.	EO
RND-2	BCAS0766- BCAS0764	HAE-RND	LysR family transcriptional regulator (BCAS0767) AraC family transcriptional regulator (BCAS0768)	Fluoroquinolones, tetracycline, rifampicin, novobiocin, EO
RND-3	BCAL1674-BCAL1676	HAE-RND	Tet-R type regulator(BCAL1672)	Nalidixic acid, ciprofloxacin, tobramycin, meropenem, chlorhexidine
RND-4	BCAL2820-BCAL2822	HAE-RND	Tet-R type regulator(BCAL2823)	Aztreonam, chloramphenicol, fluoroquinolones, tobramycin, tetracycline, rifampicin, novobiocin, essential oils, ethidium bromide, 2-thiocyanatopyridine derivative (11026103)
RND-6-7	BCAL1079-BCAL1081	HAE-RND	N.A.	EO
RND-8	BCAM0925-BCAM0927	HAE-RND	N.A.	Tobramycin
RND-9	BCAM1945-BCAM1947	HAE-RND	Mer-R type regulator(BCAM1948)	Tobramycin, chlorhexidine, EO, 2-thiocyanatopyridine derivative (11026103), 2,1,3-benzothiadiazol-5-yl family compound (10126109)
RND-10	BCAM2549-BCAM2551	HAE-RND	Tet-R type regulator (BCAM2548)	Chloramphenicol, fluoroquinolones, Trimethoprim, EO
RND-11	BCAM0711-BCAM0713	HME-RND	N.A.	Divalent cations (Zn^{2+}, Co^{2+}, Cd^{2+} and Ni^{2+})
RND-12	BCAM0433-BCAM0435	HME-RND	N.A.	Monovalent cations (Cu^+ and Ag^+), EO
RND-16	BCAL2134-BCAL2136	U.F.-RND	N.A.	Minocycline, meropenem ciprofloxacin

HAE: Hydrophobe/Amphiphile Efflux-1; HME = Heavy-Metal Efflux; U.F. = Uncertain Function. N.A. Not available; EO: Essential oils.

The presence of multiple operons encoding RND-efflux pumps in the *B. cenocepacia* genome suggests a functional redundancy and synergistic activity, accounting for the lack of alterations in the phenotype and in the antibiotic susceptibility of the majority of single RND deletion mutants [127,129]. Interestingly, the high level of conservation of the RND-4 operon in the genomes of *Burkholderia* species is consistent with the multiple functions in which this system is involved and with the effects of its inactivation on the increased susceptibility to different antimicrobial compounds, including essential oils and disinfectants [127,130–132]. On the other hand, RNDs with a narrow phylogenetic distribution, like RND-9, show a more specific activity, with consequent milder phenotypic changes observed in the corresponding J2315 deletion strain [128,129]. Noteworthy, when the conserved RND-4 efflux pump is missing or inactivated, overexpression of the RND-9 system can compensate for its function. For example, in a *B. cenocepacia* RND-4 deletion strain, mutations in a gene (*bcam1948*) encoding a transcriptional repressor of the RND-9 operon were demonstrated to confer resistance to a new antitubercular thiopyridine compound whose antimicrobial activity was previously demonstrated to be impaired by RND-4 mediated extrusion [130,133]. Interestingly, mutations in the same regulator confer resistance to a 2,1,3-benzothiadiazol-5-yl family compound and to multiple antibiotics (chloramphenicol, ciprofloxacin, levofloxacin, norfloxacin, sparfloxacin and nalidixic acid) [134]. It is noteworthy that, despite the important contribution of RND-4 in facilitating multiple AR in the *B. cenocepacia* J2315 laboratory strain, no significant differences in the expression of the *RND-4* gene (*bcal2822*) was detected in multidrug-resistant clinical isolates which, on the contrary, displayed a high expression level of *RND-3* (*bcal1674*) and *RND-9* (*bcal1947*) [135].

However, the upregulation of the genes encoding RND-6 and RND-4 were found to be involved in conferring resistance to different classes of antimicrobials (aminoglycosides, β-lactams, fluoroquinolones, folate-pathway inhibitors) in a clonal variant of *B. cenocepacia* isolated during long-term infection in CF lungs [136].

Phylogenetic analysis revealed a high degree of sequence similarity between *RND-4* and the functionally distinct *RND-2* operon, encoding a system present in only some Bcc species [27]. RND-2 is not expressed in bacteria growing in LB medium and its ability to confer resistance to fluoroquinolones, tetraphenylphosphonium, streptomycin and ethidium bromide could be identified only by overexpression experiments in *E. coli* [125]. Noteworthy, RND-2 overexpression is able to restore resistance to some antibiotics in an RND-4 deletion mutant, supporting the hypothesis that this operon originated from an RND-4 duplication event that led to the creation of a system maintaining the ancestral substrate specificity but subjected it to different regulatory mechanisms [28].

5. RND in *Achromobacter xylosoxidans*

5.1. Achromobacter Infections in CF

The *Achromobacter* genus consists of 19 species [137] of motile, non-lactose fermenting Gram-negative environmental bacilli isolated from soil and water sources. Even though they are not intrinsically pathogenic bacteria, they can represent a threat for critically ill, immunocompromised and CF patients. *A. xylosoxidans* has been known to cause pulmonary infections in CF patients since the 1980s [138], but only recently has it been recognized as one of the main CF pathogens. There is a high regional variability in its infection rate [139] but different reports highlight a worldwide rise in prevalence [140–142]. This increase could be due both to the selective antimicrobial pressure present on the CF lung bacterial community, and to the recent improvement of the detection methods, which allow the unequivocal identification of *Achromobacter* isolates at the species level [143]. This highlighted the presence in CF of different species aside from *A. xylosoxidans*, which still remains the most prevalent, such as *Achromobacter ruhlandii*, *Achromobacter dolens*, and *Achromobacter insuavis* [143]. Although the impact of these infections on lung function is not fully understood yet [144,145], it is known that these bacterial species, so closely related to the pathogenic *Bordetella* genus, have a high host adaptation potential, possessing several virulence-associated genes [146].

The treatment of *Achromobacter* spp. infections is extremely challenging since they show inherent resistance to most penicillins and cephalosporins, as well as to aztreonam, fluoroquinolones, and aminoglycosides [147]. Besides the intrinsic resistance mechanisms, *Achromobacter* often exhibits acquired resistances, especially towards β-lactams, but also to aminoglycosides and trimethoprim, achieved by horizontal gene transfer [148]. This array of resistance determinants makes these bacteria potentially resistant to every class of antibiotics, and cases of pan-drug-resistant *Achromobacter* spp. have been already reported [149]. For this reason, the optimal antibiotic therapy for these infections is patient-specific, even if piperacillin–tazobactam, trimethoprim–sulfamethoxazole, and meropenem are usually the most active agents [147]. Concerning the innate resistance mechanisms, initially some β-lactamases were biochemically characterized [150–153], but the class D β-lactamases OXA-114, in *A. xylosoxidans*, and OXA-258, in *A. ruhlandii*, are nowadays the best characterized enzymes, although their role in the β-lactams resistance profile is likely secondary [37,154]. To better study the resistance potential of *A. xylosoxidans*, Hu and colleagues performed a genome-wide analysis, predicting the presence of 50 drug resistance genes, 38 of which were efflux pump genes [32].

5.2. Achromobacter spp. RND Efflux Systems

The genome of *Achromobacter* spp. contains a significantly higher number of efflux pump-related genes compared with other genera [155]. Only three of nine RND efflux systems have been studied so far (Table 3), and a lot of work is still needed to have a comprehensive overview of the intrinsic and acquired antimicrobial resistance patterns in the *Achromobacter* genus.

Table 3. Characterized RND efflux pumps in *A. xylosoxidans*.

RND Efflux-Pump	PAO1 Orthologous (% of Identity)	Identified Regulator(s)	Antibiotic Substrates
AxyABM	MexAB-OprM (60-72-60%)	AxyR (putative LysR-type regulator)	Cephalosporins, aztreonam, nalidixic acid, fluoroquinolones, chloramphenicol, trimethoprim/sulfamethoxazole
AxyXY-OprZ	MexXY-OprM (62-74-48%)	AxyZ (TetR-type regulator)	Aminoglycosides, carbapenems, cefepime, ceftazidime, fluoroquinolones, tetracyclines, erythromycin
AxyEF-OprN	MexEF-OprN (50-65-31%)	AxyT (LysR-type regulator)	Fluoroquinolones, carbapenems, tetracyclines

The first RND-type multidrug efflux system described in *A. xylosoxidans* (even though the strain used in this work was later reclassified as *A. insuavis*) was the AxyABM [33]. This RND system is the ortholog of the MexAB-OprM system of *P. aeruginosa* (60–72% protein identity) and shares with it the same operon organization. Indeed, the genes composing the multiprotein complex are grouped in a cluster of three open reading frames, *axyA* (the MFP), *axyB* (the RND transporter protein), and *axyM* (*oprM*; the OMP). Moreover, upstream of the operon a gene coding for a transcriptional regulator, namely *axyR*, is present, as already seen in *P. aeruginosa* for *mexR*, although the two genes do not share any homology [33]. By inactivation of *axyB*, it was also demonstrated that the spectrum of activity of AxyABM is comparable, even if not identical, to the one of MexAB-OprM, being involved in the innate resistance to a broad spectrum of antibiotics, in particular most cephalosporins and aztreonam, but also nalidixic acid, fluoroquinolones, and chloramphenicol [33]. This RND system is present in all the sequenced *Achromobacter* genomes [146], but it was better characterized only in *A. ruhlandii*, where it seems to have a narrower spectrum of activity. Indeed, by cloning the *axyABM* operon in *E. coli*, it was demonstrated to be only involved in the extrusion of chloramphenicol, nalidixic acid and trimethoprim/sulfamethoxazole [37]. Finally, besides the innate antibiotic tolerance, AxyABM is probably involved also in persistence and biofilm metabolism of *A. xylosoxidans*, since the gene *axyA* was found to be 21-fold upregulated upon the establishment of chronic infections in CF lungs [156]. Moreover, in the same strain, the expression of *axyA* increased more than 7-fold in sessile cells, highlighting the importance of this efflux system in biofilm formation [156].

To identify the mechanism(s) responsible for the high-levels of innate resistance of *A. xylosoxidans* towards aminoglycosides, a genomic comparison with *P. aeruginosa* was performed. This approach led to the characterization of the AxyXY-OprZ efflux pump, the ortholog of the MexXY-OprM RND system of *P. aeruginosa* [34]. AxyXY-OprZ is encoded by an operon conserved in many *Achromobacter* species, predominantly in those often recovered from CF patients, and it is described as the major resistance mechanism to aminoglycosides, since its presence is always associated with a resistant phenotype, whereas its absence leads to a sensitive phenotype [157].

The *axyXY-oprZ* operon is under the negative control of AxyZ, a TetR-type transcriptional repressor homolog of the *P. aeruginosa* MexZ and is encoded by the gene *axyZ*, found upstream of the cluster [158]. Surprisingly, this transcription factor plays a role also in the regulation of a novel carbapenemase, Axc, highly expressed in meropenem-resistant *A. xylosoxidans* clinical isolates [159]. Indeed, loss of function mutations in the *axyZ* sequence, and especially the V29G substitution localized in the DNA-binding domain of the protein, lead to the overexpression of AxyXY-OprZ, but also of the Axc carbapenemase, increasing the MICs of antibiotic substrates of these proteins [158,159]. This demonstrates that AxyZ is involved in a wide regulatory pathway controlling the activation of disparate AR mechanisms. The AxyZ mutations can be quite easily selected in vitro by exposure of the bacterium to aminoglycosides [158], a class of antibiotic extensively used for CF infections treatment, and consequently these are reported to be frequently associated with the pathoadaptive process of *A. xylosoxidans*, *A. ruhlandii* and *A. insuavis* in CF lung [160].

The AxyXY-OprZ possesses the ability to extrude a broad spectrum of antibiotics, since its inactivation leads to a drastic decrease in the MICs of aminoglycosides and, to a lesser extent, of carbapenems, cefepime (the only cephalosporin not extruded by AxyABM), ceftazidime, some fluoroquinolones, tetracyclines, and erythromycin. Moreover, this RND pump seems to be partially involved in the *Achromobacter* spp. acquired resistance to carbapenems, since its impairment leads to a significant decrease of carbapenem MICs in a resistant clinical isolate [34]. However, the MIC value results higher than the carbapenem-sensitive *Achromobacter* strains, suggesting the presence of additional resistance mechanisms, such as the recently described carbapenemase Axc. Despite the high similarity between AxyXY-OprZ and its *P. aeruginosa* counterpart, the *Achromobacter* efflux pump confers up to a 32-fold higher level of resistance to aminoglycosides. It was hypothesized that this difference is probably due to the different Opr protein associated with the RND complex, since OprZ is the homolog of OprA (not OprM), the outer membrane protein coupled with MexXY in some *P. aeruginosa* genetic lineages [34].

The last RND efflux pump characterized in *Achromobacter* spp. was the AxyEF-OprN, the ortholog of the *P. aeruginosa* MexEF-OprN [35]. In contrast to the other two RND systems, this pump has a narrow spectrum of activity and was initially demonstrated to have a role in the *Achromobacter* innate resistance to few fluoroquinolones, carbapenems, and tetracyclines. Indeed, by analyzing the effect of *axyE* deletion in the AX08 clinical isolate, Nielsen and collaborators showed a decrease of the MIC of levofloxacin, making this strain susceptible to this antibiotic according to the EUCAST interpretative criterion for *Pseudomonas* spp. Moreover, a 2-fold decrease in the MIC of ertapenem, ciprofloxacin, and doxycycline was reported [35]. Surprisingly, in the same paper they also described an increase in the MICs of some β-lactams as a consequence of the pump inactivation, but this aspect was not further investigated [35]. AxyEF-OprN was also characterized as the main mechanism responsible for acquired fluoroquinolone resistance in *Achromobacter* [161]. Indeed, it was demonstrated that, different to many Gram-negative bacilli, the fluoroquinolones-resistant phenotype is not due to amino acid substitutions within the Quinolone Resistance Determining Regions (QRDRs) of the targets (DNA gyrase and topoisomerase IV), but it is mainly due to AxyEF-OprN overexpression. In particular, the overproduction of the efflux pump in *Achromobacter*-resistant clinical isolates is often caused by gain-of-function mutations of AxyT, the transcriptional activator of the *axyEF-oprN* operon, although the big difference found in fold change in strains owning the same mutation suggests an interplay between different regulatory pathways [161].

5.3. Achromobacter spp. RND Efflux Pumps Inhibitors

Until now, despite the prominent role of RND efflux pumps in *Achromobacter* innate and acquired AR, no specific inhibitors have been studied. The only active compound present in the literature is berberine, a benzylisoquinoline alkaloid isolated from many medicinal plants, and its derivatives, characterized as specific inhibitors of the *P. aeruginosa* MexXY system, but tested also against *A. xylosoxidans* [162,163]. Indeed, in this bacterium berberine significantly reduced the tolerance to aminoglycosides, decreasing the MICs of amikacin, arbekacin, gentamicin, and tobramycin (the substrates of the AxyXY-OprZ efflux pump) up to 32-fold [162]. Moreover, among eleven berberine derivatives, the 13-(2-methylbenzyl) berberine (13-o-MBB) showed the best activity against *P. aeruginosa* and thus it was tested against *A. xylosoxidans*. The presence of 13-o-MBB resulted in a further increased sensitivity to aminoglycosides, and the most impressive result was obtained in combination with gentamicin, reducing its MIC of more than 512-fold [163]. However, even low concentrations (30 µg/mL) of this molecule are cytotoxic to human cells in vitro [163], making the development of less toxic derivatives fundamental for future application in humans.

6. RND in *Stenotrophomonas maltophilia*

6.1. Stenotrophomonas maltophilia Infections in CF

Stenotrophomonas maltophilia is a Gram-negative, aerobic, non-fermentative bacillus, belonging to the class of gammaproteobacteria. It is an ubiquitous contaminant in soil, water, food, and hospital settings [164]. Its major presence in healthcare centers, after *Acinetobacter* spp. and *Pseudomonas aeruginosa*, is linked to opportunistic infections with relevant morbidity among patients with underlying pathologies, such as cystic fibrosis, or immunocompromised subjects, with an incidence in USA intensive care units of 4.3% of all Gram-negative infections [41,165]. Risk factors include malignancy, chronic respiratory diseases, and long-term hospitalization. In CF patients, *S. maltophilia* isolation in the respiratory tract is linked to intravenous antibiotic use and oral quinolone administration, as for the use of anti-pseudomonal antibiotics; approximately 11% of CF patients are colonized by this bacterium, even if its role in such condition is not clear [164]. *S. maltophilia* chronic infection is correlated to a lower mean percent predicted Forced Expiratory Volume in the 1st second (FEV1) compared to the uninfected control, with a significantly higher risk of pulmonary exacerbation [164]. Combinatorial treatments are efficient in avoiding clone selection, e.g., with trimethoprim-sulfamethoxazole and ticarcillin-clavulanate, doxycycline and ticarcillin-clavulanate, trimethoprim-sulfamethoxazole and piperacillin-tazobactam, ciprofloxacin and ticarcillin-clavulanate. Nevertheless, MDR strains were isolated from topical antiseptic, hand-washing soap, bottled water, and intravenous cannulae, nebulizers and prosthetic devices, showing how hazardous direct-contact transmission and how tolerant this pathogen can be [164]. Such persistence in the environment is adjuvanted by a broad array of intrinsic AR determinants against β-lactams, macrolides, aminoglycosides, cephalosporins, polymyxins, tetracyclines, chloramphenicol, fluoroquinolones, carbapenems, and trimethoprim-sulfamethoxazole [164]. Such phenotype results from the interaction of different layers, as poor membrane permeability, the presence of chromosomally encoded L1 and L2 β-lactamases [166], AAC(6′)-Iz and APH(3′)-IIc aminoglycoside-modifying enzymes [167], and multidrug resistance efflux pumps [164].

6.2. Stenotrophomonas RND Efflux Systems

In the *S. maltophilia* K279a strain genome eight pumps have been annotated, while seven (*smeABC, smeDEF, smeGH, smeIJK, smeOP, smeU1VWU2X, smeYZ*) of them have been characterized as hydrophobic and amphiphilic efflux (HAE)-RND pumps (Table 4) [167–173].

Table 4. RND efflux pumps in *S. maltophilia*.

RND-Efflux Pump	Systematic ID	Family	Identified Regulator(s)	Antibiotic Substrates
SmeABC	Smlt4474-4476	HAE-RND	Two-component regulator SmeSR	trimethoprim; third-generation β-lactams; aminoglycosides; fluoroquinolones
SmeDEF	Smlt4070-4072	HAE-RND	Tet-R type regulator SmeT; Two component regulator SmeRySy	chloramphenicol; ceftazidime; amikacin; aztreonam; novobiocin; fosfomycin; quinolones
SmeGH	Smlt3170-3171	HAE-RND	Tet-R type regulator	ceftazidime; tetracycline; polymyxin B; β-lactams; quinolones; fluoroquinolones
SmeIJK	Smlt4279/4281	HAE-RND	N.D.	tetracyclines; fluoroquinolones; aminoglycosides
SmeMN	Smlt3788-3787	HAE-RND	N.D.	N.D.
SmeOP	Smlt3925-3924	HAE-RND	Tet-R type regulator SmeRo	nalidixic acid; doxocycline; aminoglycosides; macrolides
SmeYZ	Smlt2201-2202	HAE-RND	Two-component regulator SmeRySy	trimethoprim-sulfamethoxazole; leucomycin; aminoglycosides
SmeU$_1$VWU$_2$Z	Smlt1829-1833	HAE-RND	Lys-R type regulator SmeRv	chloramphenicol; tetracycline; quinolones

N.D. Not Determined.

One of the first identified HAE-RND pumps has been SmeABC, which shows similarities to different efflux pumps, such as MexAB-OprM in *P. aeruginosa*, TtgABC in *P. putida* and AcrAB in *E. coli* [172]. This tripartite efflux pump, whose operon is controlled by the SmeSR sensor proteins, confers resistance to third-generation β-lactams, aminoglycosides, and fluoroquinolones and leads to trimethoprim susceptibility once overexpressed [166,172], while physiologically it does not confer intrinsic resistance due to its low-basal expression level. The determinants involved in MDR are being identified as *smeC* and *smeR*, whose deletion leads to the reversal of the resistance phenotype [172].

A similar quiescent behaviour is provided by the *smeU1VWU2X* operon, whose encoded SmeVWX proteins show 51%, 56%, and 48% amino acid identity with *P. aeruginosa* MexEF-OprN, respectively. The SmeRv protein, a LysR-type regulator, negatively regulates the operon in the *S. maltophilia* KJ strain, but it acts as a positive regulator in the *S. maltophilia* MDR KJ09C strain [170]. No mutations have been identified in *smeRv*, so the presence of an activator ligand could be able to switch on the expression of the entire operon. Differently from the other RND-efflux pumps, it possesses two additional sensor proteins, SmeU1 and SmeU2, belonging to the Short-chain Dehydrogenase/Reductase (SDR) family. The latter has been shown to mediate alleviation from environmental oxidative stress, which is found to trigger the expression of *smeU1VWU2X* [170,174]. KJ09C mutant overexpressing this operon shows increased resistance to chloramphenicol, quinolones, and tetracycline, with the MICs of aminoglycosides unexpectedly decreased [170]. Interestingly, the *smeX* deletion of KJ09C mutant reverts both the resistance and the susceptibility patterns, while *smeU2* deletion in the same strain leads only to a slight decrease in the resistance, up to a 2-fold MIC decrease in the case of aminoglycosides, suggesting an additional control exerted by SmeU2 on SmeX overexpression [174].

SmeDEF intrinsically confers a two- to eight-fold increase in the MICs of quinolones, tetracycline, chloramphenicol, and novobiocin [175]. Its components show several homologies with different Gram-negative bacterial efflux pumps: SmeD and SmeE share the highest similarities to *E. coli* AcrA and AcrE (48%) and AcrB and AcrF (61% and 58%), while SmeF is similar to SmeC (42%) [168]. The *smeDEF* operon is directly regulated by the SmeT protein, which acts as a negative regulator [176]. Different mutations in *smeT* have been linked to the acquisition of the resistance phenotype, such as L166Q and T197P, allowing tigecycline, aztreonam, and quinolones tolerance, but also fosfomycin susceptibility [169]. This pattern is reasonable, as overexpression of *smeD* and *ameF* has been linked to levofloxacin, moxifloxacin, ceftazidime, and tetracycline resistance and amikacin resistance, respectively; in addition, deletion of the *smeF* gene in K1385 and K1439 MDR strains leads to the reversion of the MDR phenotype [175]. Indirectly, the expression of this efflux pump is influenced by the SmeRySy two-component regulatory system, the main regulator of the *smeYZ* operon [176]. The deletion of these particular sensor proteins is linked with *smeDEF* up-regulation and to subsequent chloramphenicol, ciprofloxacin, tetracycline, and macrolide resistance. Counterintuitively, such deletion also increases also *smeT* expression: a possible explanation involves the presence of an intermediary modulator, whose expression is altered by *smeRySy* deletion and which mediates the interaction between SmeT and its operator, resulting in the derepression of both *smeDEF* and *smeT* [176]. Interestingly, a biocide called triclosan acts as a SmeT inactivator, consequently leading to *smeDEF* overexpression and MDR strain selection [177,178].

Two highly expressed efflux pumps, SmeYZ, and SmeIJK, play a major role in the intrinsic resistance to antimicrobials [171,173]. The *smeYZ* operon, sharing 44% and 59% amino acid identity with *Acinetobacter baumanii* AdeAB [41], confers resistance to amikacin, gentamicin, kanamycin, and leucomycin. Parallelly, its deletion leads to both aminoglycosides and trimethoprim-sulfamethoxazole susceptibility [173]. As previously stated, the operon is controlled by SmeRySy, with *smeRy* deletion downregulating *smeZ* expression and conferring aminoglycoside susceptibility, in addition to the acquired resistances involving *smeDEF* pump expression [176]. Celastrol, an anti-inflammatory natural terpenoid compound, can down-regulate *smeYZ* expression, thus proposing a possible candidate to control virulence in *S. maltophilia* [179].

smeIJK has a particular genetic organization, as it is the only efflux pump in *S. maltophilia* coding for two inner membrane proteins, SmeJ and SmeK, both showing high similarity (59%) among them [171]. The smeIJK operon shares 41%, 50%, and 44% amino acid identity, respectively, to MtdABC of *E. coli* [41]. SmeIJK confers intrinsic resistance to tetracycline and, to a lesser extent, to aminoglycosides; overexpression can be found in *S. maltophilia* KJ and KM5 strains leads to an up to 16-fold increase in aminoglycosides MICs and to an increase in fluoroquinolones and tetracyclines resistance, phenotypes reverted after *smeJK* deletion [167,171]. In addition, deletion of the entire operon in the KJ mutant is linked to polymyxin E susceptibility, thus suggesting a role for SmeI in membrane integrity and permeability [171].

SmeOP proteins are not conserved in other Gram-negative bacteria, as they share less than 30% of the amino acid identity of other antimicrobial efflux pumps [41]. In the strain KJ, this efflux pump is involved in the extrusion of nalidixic acid, doxycycline, macrolides, and more relevantly aminoglycosides, and in the elimination of some toxic compounds such as carbonyl cyanide 3-chlorophenylhydrazone (CCCP) and tetrachlorosalicylanilide (TCS) [180]. The operon is controlled by a TetR-type transcription regulator SmeRo, which represses the expression of the genes [180]. Its deletion only produces a slight increase in the MICs of chloramphenicol, quinolones, and tetracyclines. To properly work, the pump requires the cognate OMP TolCSm: deletion of the corresponding gene has been associated with higher decreases in the MICs than those caused by *smeOP* inactivation, suggesting the involvement of this OMP in the function of another uncharacterized efflux system [180].

Finally, *smeGH* is the last operon characterized, whose components share 39% and 49% amino acid identity to *Morganella morganii* AcrAB [41,181]. It is controlled by a TetR-type regulator, which acts as a repressor. In the *S. maltophilia* D457 strain, a *smeH* deletion mutant shows an increased susceptibility to ceftazidime, β-lactams, quinolones, and fluoroquinolones, and polymyxin B, suggesting the role of this pump in intrinsic resistance. In the same mutant, a wide variety of other noxious compounds are identified as substrates, such as menadione, benzalkonium chloride and naringenin [181]. Overexpression of *smeH* in the *S. maltophilia* clinical strain L1301 is linked to quinolones, macrolides, chloramphenicol, and tetracycline resistance; a similar effect is observed in another strain, named C2206, except for macrolide MIC, which remains unchanged [182]. Through an approach of laboratory experimental evolution, where the D457 strain was exposed to increased ceftazidime concentrations, two subsequent mutations in *smeH* were identified as linked to MDR [181]: the first to be acquired was P326Q, which confers a 5-fold and 2-fold increase in the MICs of ceftazidime and cefazolin and for aztreonam, respectively; the second acquired mutation was Q663R, which further increased the resistance to ceftazidime, cefazolin, and aztreonam, and conferred a 2-fold and 3-fold increase in the MICs of cefotaxime and norfloxacin, respectively. Finally, the role of Q663R mutation alone was explored, resulting only in a 4-fold increase in MIC of tetracycline: this suggests how relevant the order of mutation acquisition for the final phenotypic outcome is [181].

7. Conclusions

Multidrug-resistant strains represent a major threat for cystic fibrosis patients, who undergo heavy antibiotic therapies to face the recurrent bacterial infections that damage their lungs especially.

Major contributors to the MDR phenotype are the efflux pumps belonging to the Resistance-Nodulation-cell Division family. These transporters are able to translocate a lot of unrelated compounds out of the bacterial cell, thus impairing the effect of the antibiotic therapy, even when a new molecule is administered for the first time [134]. Although nine families of RND proteins have been described, the Hydrophobe/Amphiphile Efflux 1 (HAE-1) is the most represented among CF bacteria, mainly being involved in the extrusion of drugs.

The contribution of this RND family in MDR has been particularly highlighted in *P. aeruginosa*, *B. cenocepacia*, *A. xylosoxidans*, and *S. maltophilia*.

In *P. aeruginosa*, six RND systems have been demonstrated to be related to the insurgence of drug resistance in clinical isolates. These pumps are involved in the extrusion of drugs belonging to different categories and were all used for the treatment of CF infections (beta-lactams, tetracyclines, fluoroquinolones, aminoglycosides, etc., Supplementary Table S1), but also detergents, dyes, and quorum-sensing signal molecules. Whole-genome sequencing of *P. aeruginosa* clinical isolates derived from CF patients revealed that, among the gene-encoding efflux pumps or their regulators, MexZ presents a high rate of mutation [183]). Indeed, a study by Henrichfreise and collaborators [184] reported that the 82% of multidrug-resistant *P. aeruginosa* strains overproduced MexXY-OprM. However, another work revealed mutations also in the efflux regulator genes *mexR*, *mexT*, and *nfxB* [185]. Non-synonymous mutations have been reported also in the transcriptional regulator of MexAB-OprM, *nalC* [186]. The same clinical isolates have mutations which lead to the activation of MexT, the positive regulator of MexEF-OprN [186]. A high mutation rate was identified also in the genes encoding the components of RND efflux pumps, such as *mexA*, *mexY*, *oprM* [187] and *mexB* [188].

In *B. cenocepacia*, sixteen genes encoding RND pumps have been identified, although a differential contribution to drug resistance has been reported when bacterial cells grow as planktonic or sessile ones [129]. Also in this case, their major role has been described for unrelated compounds, such as antimicrobial compounds, essential oils, disinfectants, and new molecules [20,130–132,189]. A study aimed at dissecting the mechanisms responsible for antibiotic resistance in clinical *B. cepacia* complex isolates revealed that the majority of them exhibited efflux pump activity, which correlated with resistance to various antimicrobial agents, including those used for the treatment of infections in CF patients (e.g., meropenem, ceftazidime, trimethoprim/sulfamethoxazole, Supplementary Table S1) [135]. In particular, RND-3 and RND-9 overexpression was observed in all clinical isolates, with RND-3 being the most up-regulated among the RND pumps tested [135].

During chronic infections, the long-term colonization of the lungs of CF patients is accompanied by an adaptive remodeling of the *B. cenocepacia* transcriptome. Adaptive changes include the overexpression of various genes encoding drug efflux pumps, like RND-6 and RND-4. As a consequence, the higher active drug export capacity of clinical isolates from the lungs of CF patients affected by long-term chronic infections is accompanied by an increased resistance to clinically relevant antibiotics with very different biological targets [136].

In *A. xylosoxidans*, seventeen predicted efflux systems have been reported [32]. Only three of these efflux systems have been fully characterized so far, showing the ability to confer resistance to CF used drugs, such as fluoroquinolones andtrimethoprim/sulfamethoxazole (Supplementary Table S1). As an example, Gabrielaite and collaborators [160], performing a genomic analysis on 101 clinical strains isolated within a time span of 20 years in a Denmark CF center, found that in 38% of the analyzed lineages mutations in the gene *axyZ* (*axyXY-oprZ* transcriptional repressor) were present. The presence of *axyZ* mutations led to an overall increase of tolerance to antibiotics since AxyXY-OprZ has a broad spectrum of activity.

Finally, in the *S. maltophilia* K279a genome eight pumps belonging to the HAE family have been annotated [168]. Their involvement in the resistance has been assessed in 102 clinical isolates, where 70%, 77%, 59% and 61% overexpressed *smeB, smeC, smeD, and smeF*, respectively [190]. In particular, as regarding the drugs currently used to treat *S. maltophilia* infections in CF (Supplementary Table S1), *smeD* overexpression was responsible for levofloxacin and minocycline resistance, *smeC* for ceftazidime and ticarcillin-clavulonate-nonsusceptibility, while *smeF* overexpression was significantly correlated with ceftazidime and levofloxacin resistance [190].

Another interesting point is that all the described pathogens are able to chronically colonize the CF airway. This implies their ability to adapt to the host environment, characterized by peculiar nutrient and oxygen availabilities, to interact with the host immune response and to deal with the presence of drugs administered to try to clear the infections. In order to understand this phenomenon, different papers reported results achieved through transcriptomics, which analyzed differential gene expression of strains isolated from CF patients, or genomic analyses which evaluated the presence of mutations in clinical isolates. Interestingly, efflux pump encoding genes were listed among those in which altered level of expression or mutations were reported as contributors to CF lung adaptation in *P. aeruginosa* [191], *B. cenocepacia* [136], *Achromobacter* sp. [146] and *S. maltophilia* [16]. This has been mainly ascribed both to their role in biofilm formation and in bacterial virulence [61], highlighting a wider role of efflux systems. Indeed, the role of RND efflux pumps in drug resistance can be demonstrated in vitro, where the amount of antibiotic can be measured, while in the clinical environment it is much more complicated to evaluate the achieved antibiotic concentrations and the consequent contribution of efflux to MDR, which might allow the acquisition of other resistance mechanisms.

Despite the recognized role in drug resistance of RND efflux transporters, more efforts are necessary to find efflux inhibitors to be administered to patients. As some molecules were shown to be effective against *P. aeruginosa* and *A. xylosoxidans*, the high degree of similarity found among the RND systems of all the described CF colonizing bacteria could lead to the discovery of new inhibitors effective against a broad range of pathogens. These molecules could be used in combination with antibiotics to avoid extrusion and MDR insurgence. Indeed, given the important contribution of specific efflux systems in the insurgence of MDR, the combined use of antibiotics and specific efflux inhibitors could represent a promising therapeutic strategy for CF patients. Interestingly, phage therapy has been shown to target specific efflux pumps in *P. aeruginosa*: this also represents a new route in the fight against drug resistance.

Supplementary Materials: The following are available online at https://www.mdpi.com/article/10.3 390/antibiotics10070863/s1, Table S1: Antibiotoiotics used for the treatment of *P. aeruginosa, B. cenocepacia, A. xylosoxidans* ans *S. maltophilia* infections in CF patients and efflux pumps involved in resistance.

Funding: This research was funded by the Italian Ministry of Education, University and Research (MIUR) (Dipartimenti di Eccellenza, Program 2018–2022) to Department of Biology and Biotechnology, "L. Spallanzani", University of Pavia (to G.B. and S.B.).

Institutional Review Board Statement: Not applicable.

Informed Consent Statement: Not applicable.

Data Availability Statement: Not applicable.

Conflicts of Interest: The authors declare no conflict of interest.

References

1. Cystic Fibrosis Foundation. *Cystic Fibrosis Foundation Patient Registry: 2018 Annual Data Report*; Cystic Fibrosis Foundation: Bethesda, MD, USA, 2019.
2. Burgener, E.B.; Moss, R.B. Cystic fibrosis transmembrane conductance regulator modulators: Precision medicine in cystic fibrosis. *Curr. Opin. Pediatr.* **2018**, *30*, 372–377. [CrossRef] [PubMed]
3. Cribbs, S.K.; Beck, J.M. Microbiome in the pathogenesis of cystic fibrosis and lung transplant-related disease. *Transl. Res.* **2017**, *179*, 84–96. [CrossRef]
4. Françoise, A.; Héry-Arnaud, G. The microbiome in cystic fibrosis pulmonary disease. *Genes* **2020**, *11*, 536. [CrossRef]
5. Zemanick, E.T.; Wagner, B.D.; Robertson, C.E.; Ahrens, R.C.; Chmiel, J.F.; Clancy, J.P.; Gibson, R.L.; Harris, W.T.; Kurland, G.; Laguna, T.A.; et al. Airway microbiota across age and disease spectrum in cystic fibrosis. *Eur. Respir. J.* **2017**, *50*, 1700832. [CrossRef]
6. Yi, B.; Dalpke, A.H.; Boutin, S. Changes in the cystic fibrosis airway microbiome in response to CFTR Modulator therapy. *Front. Cell Infect. Microbiol.* **2021**, *11*, 548613. [CrossRef]
7. Li, Z.; Kosorok, M.R.; Farrell, P.M.; Laxova, A.; West, S.E.; Green, C.G.; Collins, J.; Rock, M.J.; Splaingard, M.L. Longitudinal development of mucoid *Pseudomonas aeruginosa* infection and lung disease progression in children with cystic fibrosis. *JAMA* **2005**, *293*, 581–588. [CrossRef]
8. Besier, S.; Smaczny, C.; von Mallinckrodt, C.; Krahl, A.; Ackermann, H.; Brade, V.; Wichelhaus, T.A. Prevalence and clinical significance of *Staphylococcus aureus* small-colony variants in cystic fibrosis lung disease. *J. Clin. Microbiol.* **2007**, *45*, 168–172. [CrossRef]
9. Haussler, S.; Ziegler, I.; Lottel, A.; Götz, F.V.; Rohde, M.; Wehmhöhner, D.; Saravanamuthu, S.; Tümmler, B.; Steinmetz, I. Highly adherent small-colony variants of *Pseudomonas aeruginosa* in cystic fibrosis lung infection. *J. Med. Microbiol.* **2003**, *52*, 295–301. [CrossRef] [PubMed]
10. Nadesalingam, K.; Conway, S.P.; Denton, M. Risk factors for acquisition of methicillin-resistant *Staphylococcus aureus* (MRSA) by patients with cystic fibrosis. *J. Cyst. Fibros.* **2005**, *4*, 49–52. [CrossRef]
11. Senda, K.; Arakawa, Y.; Ichiyama, S.; Nakashima, K.; Ito, H.; Ohsuka, S.; Shimokata, K.; Kato, N.; Ohta, M. PCR detection of metallo-beta-lactamase gene (blaIMP) in gram-negative rods resistant to broad-spectrum beta-lactams. *J. Clin. Microbiol.* **1996**, *34*, 2909–2913. [CrossRef] [PubMed]
12. Langendonk, R.F.; Neill, D.R.; Fothergill, J.L. The Building Blocks of Antimicrobial Resistance in *Pseudomonas aeruginosa*: Implications for Current Resistance-Breaking Therapies. *Front. Cell Infect. Microbiol.* **2021**, *11*, 665759. [CrossRef]
13. López-Causapé, C.; Rojo-Molinero, E.; Macià, M.D.; Oliver, A. The problems of antibiotic resistance in cystic fibrosis and solutions. *Expert Rev. Respir. Med.* **2015**, *9*, 73–88. [CrossRef] [PubMed]
14. Coutinho, C.P.; Dos Santos, S.C.; Madeira, A.; Mira, N.P.; Moreira, A.S.; Sá-Correia, I. Long-term colonization of the cystic fibrosis lung by *Burkholderia cepacia* complex bacteria: Epidemiology, clonal variation, and genome-wide expression alterations. *Front. Cell. Infect. Microbiol.* **2011**, *1*, 12. [CrossRef]
15. Hogardt, M.; Heesemann, J. Microevolution of *Pseudomonas aeruginosa* to a chronic pathogen of the cystic fibrosis lung. *Curr. Top. Microbiol. Immunol.* **2013**, *358*, 91–118.
16. Menetrey, Q.; Sorlin, P.; Jumas-Bilak, E.; Chiron, R.; Dupont, C.; Marchandin, H. *Achromobacter xylosoxidans* and *Stenotrophomonas maltophilia*: Emerging Pathogens Well-Armed for Life in the Cystic Fibrosis Patients' Lung. *Genes* **2021**, *12*, 610. [CrossRef]
17. Nikaido, H. RND transporters in the living world. *Res. Microbiol.* **2018**, *169*, 363–371. [CrossRef]
18. Du, D.; Wang, Z.; James, N.R.; Voss, J.E.; Klimont, E.; Ohene-Agyei, T.; Venter, H.; Chiu, W.; Luisi, B.F. Structure of the AcrAB-TolC multidrug efflux pump. *Nature* **2014**, *509*, 512–515. [CrossRef]
19. Glavier, M.; Puvanendran, D.; Salvador, D.; Decossas, M.; Phan, G.; Garnier, C.; Frezza, E.; Cece, Q.; Schoehn, G.; Picard, M.; et al. Antibiotic export by MexB multidrug efflux transporter is allosterically controlled by a MexA-OprM chaperone-like complex. *Nat. Commun.* **2020**, *11*, 4948. [CrossRef] [PubMed]

20. Perrin, E.; Fondi, M.; Papaleo, M.C.; Maida, I.; Emiliani, G.; Buroni, S.; Pasca, M.R.; Riccardi, G.; Fani, R. A census of RND superfamily proteins in the Burkholderia genus. *Future Microbiol.* **2013**, *8*, 923–937. [CrossRef] [PubMed]

21. Schweizer, H.P. Efflux as a mechanism of resistance to antimicrobials in *Pseudomonas aeruginosa* and related bacteria: Unanswered questions. *Genet. Mol. Res.* **2003**, *2*, 48–62. [PubMed]

22. Milton, H.S., Jr.; Vamsee, S.R.; Gabriel, M.H.; Kevin, J.H.; Yichi, Z.; Vasu, I.; Katie, J.K.L.; Nuo, T.; Steven, R.; Jianing, W.; et al. The Transporter Classification Database (TCDB): 2021 update. *Nucleic Acids Res.* **2021**, *49*, D461–D467.

23. Teixeira, P.; Tacão, M.; Alves, A.; Henriques, I. Antibiotic and metal resistance in a ST395 *Pseudomonas aeruginosa* environmental isolate: A genomics approach. *Mar. Pollut. Bull.* **2016**, *110*, 75–81. [CrossRef] [PubMed]

24. McFarland, A.G.; Bertucci, H.K.; Littman, E.; Shen, J.; Huttenhower, C.; Hartmann, E.M. Triclosan Tolerance Is Driven by a Conserved Mechanism in Diverse *Pseudomonas* Species. *Appl. Environ. Microbiol.* **2021**, *87*, e02924–e03020. [CrossRef]

25. Sood, U.; Hira, P.; Kumar, R.; Bajaj, A.; Rao, D.L.N.; Lal, R.; Shakarad, M. Comparative Genomic Analyses Reveal Core-Genome-Wide Genes under Positive Selection and Major Regulatory Hubs in Outlier Strains of *Pseudomonas aeruginosa*. *Front. Microbiol.* **2019**, *10*, 53. [CrossRef]

26. Jeukens, J.; Kukavica-Ibrulj, I.; Emond-Rheault, J.G.; Freschi, L.; Levesque, R.C. Comparative genomics of a drug-resistant *Pseudomonas aeruginosa* panel and the challenges of antimicrobial resistance prediction from genomes. *FEMS Microbiol. Lett.* **2017**, *364*. [CrossRef]

27. Perrin, E.; Fondi, M.; Papaleo, M.C. Exploring the HME and HAE1 efflux systems in the genus *Burkholderia*. *BMC Evol. Biol.* **2010**, *10*, 164. [CrossRef] [PubMed]

28. Perrin, E.; Fondi, M.; Bosi, E.; Mengoni, A.; Buroni, S.; Scoffone, V.C.; Valvano, M.; Fani, R. Subfunctionalization influences the expansion of bacterial multidrug antibiotic resistance. *BMC Genom.* **2017**, *18*, 834. [CrossRef] [PubMed]

29. Podnecky, N.L.; Rhodes, K.A.; Schweizer, H.P. Efflux pump-mediated drug resistance in *Burkholderia*. *Front. Microbiol.* **2015**, *6*, 305. [CrossRef]

30. Zhang, J.; Li, Q.; Zeng, Y.; Zhang, J.; Lu, G.; Dang, Z.; Guo, C. Bioaccumulation and distribution of cadmium by *Burkholderia cepacia* GYP1 under oligotrophic condition and mechanism analysis at proteome level. *Ecotoxicol. Environ. Saf.* **2019**, *176*, 162–169. [CrossRef] [PubMed]

31. Wang, X.; Zhang, X.; Liu, X.; Huang, Z.; Niu, S.; Xu, T.; Zeng, J.; Li, H.; Wang, T.; Gao, Y.; et al. Physiological, biochemical and proteomic insight into integrated strategies of an endophytic bacterium *Burkholderia cenocepacia* strain YG-3 response to cadmium stress. *Metallomics* **2019**, *11*, 1252–1264. [CrossRef]

32. Hu, Y.; Zhu, Y.; Ma, Y.; Liu, F.; Lu, N.; Yang, X.; Luan, C.; Yi, Y.; Zhu, B. Genomic insights into intrinsic and acquired drug resistance mechanisms in *Achromobacter xylosoxidans*. *Antimicrob. Agents Chemother.* **2015**, *59*, 1152–1161. [CrossRef]

33. Bador, J.; Amoureux, L.; Duez, J.M.; Drabowicz, A.; Siebor, E.; Llanes, C.; Neuwirth, C. First description of an RND-type multidrug efflux pump in *Achromobacter xylosoxidans*, AxyABM. *Antimicrob. Agents Chemother.* **2011**, *55*, 4912–4914. [CrossRef]

34. Bador, J.; Amoureux, L.; Blanc, E.; Neuwirth, C. Innate aminoglycoside resistance of *Achromobacter xylosoxidans* is due to AxyXY-OprZ, an RND-type multidrug efflux pump. *Antimicrob. Agents Chemother.* **2013**, *57*, 603–605. [CrossRef] [PubMed]

35. Nielsen, S.M.; Penstoft, L.N.; Nørskov-Lauritsen, N. Motility, Biofilm Formation and Antimicrobial Efflux of Sessile and Planktonic Cells of *Achromobacter xylosoxidans*. *Pathogens* **2019**, *8*, 14. [CrossRef]

36. Isler, B.; Kidd, T.J.; Stewart, A.G.; Harris, P.; Paterson, D.L. *Achromobacter* Infections and Treatment Options. *Antimicrob. Agents Chemother.* **2020**, *64*, e01025–e01120. [CrossRef] [PubMed]

37. Papalia, M.; Traglia, G.; Ruggiero, M.; Almuzara, M.; Vay, C.; Gutkind, G.; Ramírez, M.S.; Radice, M. Characterisation of OXA-258 enzymes and AxyABM efflux pump in *Achromobacter ruhlandii*. *J. Glob. Antimicrob. Resist.* **2018**, *14*, 233–237. [CrossRef] [PubMed]

38. Schmidt, T.; Schlegel, H.G. Combined nickel-cobalt-cadmium resistance encoded by the *ncc* locus of *Alcaligenes xylosoxidans* 31A. *J. Bacteriol.* **1994**, *176*, 7045–7754. [CrossRef]

39. Hložková, K.; Suman, J.; Strnad, H.; Ruml, T.; Paces, V.; Kotrba, P. Characterization of *pbt* genes conferring increased Pb2+ and Cd2+ tolerance upon *Achromobacter xylosoxidans* A8. *Res. Microbiol.* **2013**, *164*, 1009–1018. [CrossRef] [PubMed]

40. Youenou, B.; Favre-Bonté, S.; Bodilis, J.; Brothier, E.; Dubost, A.; Muller, D.; Nazaret, S. Comparative Genomics of Environmental and Clinical *Stenotrophomonas maltophilia* Strains with Different Antibiotic Resistance Profiles. *Genome Biol. Evol.* **2015**, *7*, 2484–2505. [CrossRef]

41. Crossman, L.C.; Gould, V.C.; Dow, J.M.; Vernikos, G.S.; Okazaki, A.; Sebaihia, M.; Saunders, D.; Arrowsmith, C.; Carver, T.; Peters, N.; et al. The complete genome, comparative and functional analysis of *Stenotrophomonas maltophilia* reveals an organism heavily shielded by drug resistance determinants. *Genome Biol.* **2008**, *9*, R74. [CrossRef]

42. Mathee, K.; Narasimhan, G.; Valdes, C.; Qiu, X.; Matewish, J.M.; Koehrsen, M.; Rokas, A.; Yandava, C.N.; Engels, R.; Zeng, E.; et al. Dynamics of *Pseudomonas aeruginosa* genome evolution. *Proc. Natl. Acad. Sci. USA* **2008**, *105*, 3100–3105. [CrossRef] [PubMed]

43. WHO Publishes List of Bacteria for Which New Antibiotics Are Urgently Needed. Available online: https://www.who.int/news-room/detail/27-02-2017-who-publishes-list-of-bacteria-for-which-new-antibiotics-are-urgently-needed (accessed on 27 May 2021).

44. Pendleton, J.N.; Gorman, S.P.; Gilmore, B.F. Clinical relevance of the ESKAPE pathogens. *Expert Rev. Anti-Infect. Ther.* **2013**, *11*, 297–308. [CrossRef]

45. Cassini, A.; Diaz Högberg, L.; Plachouras, D.; Quattrocchi, A.; Hoxha, A.; Skov Simonsen, G.; Colomb-Cotinat, M.; Kretzschmar, M.E.; Devleesschauwer, B.; Cecchini, M.; et al. Attributable deaths and disability-adjusted life-years caused by infections with antibiotic-resistant bacteria in the EU and the European Economic Area in 2015: A population-level modelling analysis. *Lancet Infect. Dis.* **2019**, *19*, 56–66. [CrossRef]

46. Klockgether, J.; Cramer, N.; Wiehlmann, L.; Davenport, C.F.; Tummler, B. *Pseudomonas aeruginosa* genomic structure and diversity. *Front. Microbiol.* **2011**, *2*, 150. [CrossRef]

47. Elborn, J.S. Cystic fibrosis. *Lancet* **2016**, *388*, 2519–2531. [CrossRef]

48. Maurice, N.M.; Bedi, B.; Sadikot, R.T. *Pseudomonas aeruginosa* Biofilms: Host Response and Clinical Implications in Lung Infections. *Am. J. Respir. Cell Mol. Biol.* **2018**, *58*, 428–439. [CrossRef]

49. Reece, E.; Segurado, R.; Jackson, A.; McClean, S.; Renwick, J.; Greally, P. Co-colonisation with *Aspergillus fumigatus* and *Pseudomonas aeruginosa* is associated with poorer health in cystic fibrosis patients: An Irish registry analysis. *BMC Pulm. Med.* **2017**, *17*, 70. [CrossRef]

50. Riquelme, S.A.; Ahn, D.; Prince, A. *Pseudomonas aeruginosa* and *Klebsiella pneumoniae* Adaptation to Innate Immune Clearance Mechanisms in the Lung. *J. Innate Immun.* **2018**, *10*, 442–454. [CrossRef] [PubMed]

51. Winstanley, C.; O'Brien, S.; Brockhurst, M.A. *Pseudomonas aeruginosa* Evolutionary Adaptation and Diversification in Cystic Fibrosis Chronic Lung Infections. *Trends Microbiol.* **2016**, *24*, 327–337. [CrossRef]

52. Lister, P.D.; Wolter, D.J.; Hanson, N.D. Antibacterial-resistant *Pseudomonas aeruginosa*: Clinical impact and complex regulation of chromosomally encoded resistance mechanisms. *Clin. Microbiol. Rev.* **2009**, *22*, 582–610. [CrossRef]

53. Hancock, R.E.; Speert, D.P. Antibiotic resistance in *Pseudomonas aeruginosa*: Mechanisms and impact on treatment. *Drug Resist. Updates* **2000**, *3*, 247–255. [CrossRef]

54. Tenover, F.C. Mechanisms of Antimicrobial Resistance in Bacteria. *Am. J. Med.* **2006**, *119*, S3–S10. [CrossRef] [PubMed]

55. Drenkard, E. Antimicrobial resistance of *Pseudomonas aeruginosa* biofilms. *Microbes Infect.* **2003**, *5*, 1213–1219. [CrossRef] [PubMed]

56. Breidenstein, E.B.; de la Fuente-Nunez, C.; Hancock, R.E. *Pseudomonas aeruginosa*: All roads lead to resistance. *Trends Microbiol.* **2011**, *19*, 419–426. [CrossRef]

57. Daury, L.; Orange, F.; Taveau, J.C.; Verchere, A.; Monlezun, L.; Gounou, C.; Marreddy, R.K.; Picard, M.; Broutin, I.; Pos, K.M.; et al. Tripartite assembly of RND multidrug efflux pumps. *Nat. Commun.* **2016**, *7*, 10731. [CrossRef] [PubMed]

58. Dreier, J.; Ruggerone, P. Interaction of antibacterial compounds with RND efflux pumps in *Pseudomonas aeruginosa*. *Front. Microbiol.* **2015**, *6*, 660. [CrossRef] [PubMed]

59. Poole, K. Aminoglycoside resistance in *Pseudomonas aeruginosa*. *Antimicrob. Agents Chemother.* **2005**, *49*, 479–487. [CrossRef]

60. Fraud, S.; Campigotto, A.J.; Chen, Z.; Poole, K. MexCD-OprJ multidrug efflux system of *Pseudomonas aeruginosa*: Involvement in chlorhexidine resistance and induction by membrane-damaging agents dependent upon the AlgU stress response sigma factor. *Antimicrob. Agents Chemother.* **2008**, *52*, 4478–4482. [CrossRef]

61. Alcalde-Rico, M.; Olivares-Pacheco, J.; Alvarez-Ortega, C.; Cámara, M.; Martınez, J.L. Role of the multidrug resistance effluxpump MexCD-OprJ in the *Pseudomonas aeruginosa* quorum sensing response. *Front. Microbiol.* **2018**, *9*, 2752. [CrossRef]

62. Li, X.Z.; Plésiat, P.; Nikaido, H. The challenge of efflux-mediated antibiotic resistance in Gram-negative bacteria. *Clin. Microbiol. Rev.* **2015**, *28*, 337–418. [CrossRef]

63. Goli, H.R.; Nahaei, M.R.; Rezaee, M.A.; Hasani, A.; Samadi Kafil, H.; Aghazadeh, M.; Sheikhalizadeh, V. Contribution of *mexAB-oprM* and *mexXY* (*-oprA*) efflux operons in antibiotic resistance of clinical *Pseudomonas aeruginosa* isolates in Tabriz, Iran. *Infect. Genet. Evol.* **2016**, *45*, 75–82. [CrossRef]

64. Köhler, T.; Michea-Hamzehpour, M.; Epp, S.F.; Pechere, J.C. Carbapenem activities against *Pseudomonas aeruginosa*: Respective contributions of OprD and efflux systems. *Antimicrob. Agents Chemother.* **1999**, *43*, 424–427. [CrossRef]

65. Chen, W.; Wang, D.; Zhou, W.; Sang, H.; Liu, X.; Ge, Z.; Zhang, J.; Lan, L.; Yang, C.G.; Chen, H. Novobiocin binding to NalD induces the expression of the MexAB-OprM pump in *Pseudomonas aeruginosa*. *Mol. Microbiol.* **2016**, *100*, 749–758. [CrossRef] [PubMed]

66. Nehme, D.; Li, X.Z.; Elliot, R.; Poole, K. Assembly of the MexAB- OprM multidrug efflux system of *Pseudomonas aeruginosa*: Identification and characterization of mutations in *mexA* compromising MexA multimerization and interaction with MexB. *J. Bacteriol.* **2004**, *186*, 2973–2983. [CrossRef] [PubMed]

67. Andrésen, C.; Jalal, S.; Aili, D.; Wang, Y.; Islam, S.; Jarl, A.; Liedberg, B.; Wretlind, B.; Mårtensson, L.G.; Sunnerhagen, M. Critical biophysical properties in the *Pseudomonas aeruginosa* efflux gene regulator MexR are targeted by mutations conferring multidrug resistance. *Protein Sci.* **2010**, *19*, 680–692. [CrossRef] [PubMed]

68. Evans, K.; Adewoye, L.; Poole, K. MexR repressor of the *mexAB-oprM* multidrug efflux operon of *Pseudomonas aeruginosa*: Identification of MexR binding sites in the *mexA-mexR* intergenic region. *J. Bacteriol.* **2001**, *183*, 807–812. [CrossRef] [PubMed]

69. Choudhury, D.; Ghosh, A.; Dhar Chanda, D.; Das Talukdar, A.; Dutta Choudhury, M.; Paul, D.; Maurya, A.P.; Chakravarty, A.; Bhattacharjee, A. Premature Termination of MexR Leads to Overexpression of MexAB-OprM Efflux Pump in *Pseudomonas aeruginosa* in a Tertiary Referral Hospital in India. *PLoS ONE* **2016**, *11*, e0149156.

70. Pan, Y.P.; Xu, Y.H.; Wang, Z.X.; Fang, Y.P.; Shen, J.L. Overexpression of MexAB-OprM efflux pump in carbapenem-resistant *Pseudomonas aeruginosa*. *Arch. Microbiol.* **2016**, *198*, 565–571. [CrossRef]

71. Wheatley, R.; Diaz Caballero, J.; Kapel, N.; de Winter, F.H.R.; Jangir, P.; Quinn, A.; Del Barrio-Tofiño, E.; López-Causapé, C.; Hedge, J.; Torrens, G.; et al. Rapid evolution and host immunity drive the rise and fall of carbapenem resistance during an acute *Pseudomonas aeruginosa* infection. *Nat. Commun.* **2021**, *12*, 2460. [CrossRef]

72. Mah, T.F.C.; O'Toole, G.A. Mechanisms of biofilm resistance to antimicrobial agents. *Trends Microbiol.* **2001**, *9*, 34–39. [CrossRef]

73. Brauner, A.; Fridman, O.; Gefen, O.; Balaban, N.Q. Distinguishing between resistance, tolerance and persistence to antibiotic treatment. *Nat. Rev. Microbiol.* **2016**, *14*, 320–330. [CrossRef] [PubMed]

74. Masuda, N.; Gotoh, N.; Ohya, S.; Nishino, T. Quantitative correlation between susceptibility and OprJ production in NfxB mutants of *Pseudomonas aeruginosa*. *Antimicrob. Agents Chemother.* **1996**, *40*, 909–913. [CrossRef] [PubMed]

75. Sobel, M.L.; Neshat, S.; Poole, K. Mutations in PA2491 (*mexS*) promote MexT-dependent *mexEF-oprN* expression and multidrug resistance in a clinical strain of *Pseudomonas aeruginosa*. *J. Bacteriol.* **2005**, *187*, 1246–1253. [CrossRef]

76. Maseda, H.; Yoneyama, H.; Nakae, T. Assignment of the substrate- selective subunits of the MexEF-OprN multidrug efflux pump of *Pseudomonas aeruginosa*. *Antimicrob. Agents Chemother.* **2000**, *44*, 658–664. [CrossRef]

77. Ochs, M.M.; McCusker, M.P.; Bains, M.; Hancock, R.E.W. Negative regulation of the *Pseudomonas aeruginosa* outer membrane porin OprD selective for imipenem and basic amino acids. *Antimicrob. Agents Chemother.* **1999**, *43*, 1085–1090. [CrossRef]

78. Köhler, T.; Michéa-Hamzehpour, M.; Henze, U.; Gotoh, N.; Curty, L.K.; Pechère, J.C. Characterization of MexE-MexF-OprN, a positively regulated multidrug efflux system of *Pseudomonas aeruginosa*. *Mol. Microbiol.* **1997**, *23*, 345–354. [CrossRef]

79. Hocquet, D.; Nordmann, P.; El Garch, F.; Cabanne, L.; Plésiat, P. Involvement of the MexXY-OprM efflux system in emergence of cefepime resistance in clinical strains of *Pseudomonas aeruginosa*. *Antimicrob. Agents Chemother.* **2006**, *50*, 1347–1351. [CrossRef] [PubMed]

80. Morita, Y.; Tomida, J.; Kawamura, Y. Primary mechanisms mediating aminoglycoside resistance in the multidrug-resistant *Pseudomonas aeruginosa* clinical isolate PA7. *Microbiology* **2012**, *158*, 1071–1083. [CrossRef]

81. Singh, M.; Sykes, E.M.E.; Li, Y.; Kumar, A. MexXY RND pump of *Pseudomonas aeruginosa* PA7 effluxes bi-anionic β-lactams carbenicillin and sulbenicillin when it partners with the outer membrane factor OprA but not with OprM. *Microbiology* **2020**, *166*, 1095–1106. [CrossRef] [PubMed]

82. Llanes, C.; Hocquet, D.; Vogne, C.; Benali-Baitich, D.; Neuwirth, C.; Plésiat, P. Clinical Strains of *Pseudomonas aeruginosa* Overproducing MexAB-OprM and MexXY Efflux Pumps Simultaneously. *Antimicrob. Agents Chemother.* **2004**, *48*, 1797–1802. [CrossRef]

83. Vogne, C.; Aires, J.R.; Bailly, C.; Hocquet, D.; Plésiat, P. *Pseudomonas aeruginosa* isolates from patients with cystic fibrosis. *Antimicrob. Agents Chemother.* **2004**, *48*, 1676–1680. [CrossRef]

84. Marvig, R.L.; Sommer, L.M.; Molin, S.; Johansen, H.K. Convergent evolution and adaptation of *Pseudomonas aeruginosa* within patients with cystic fibrosis. *Nat. Genet.* **2015**, *47*, 57–64. [CrossRef]

85. Prickett, M.H.; Hauser, A.R.; McColley, S.A.; Cullina, J.; Potter, E.; Powers, C.; Jain, M. Aminoglycoside resistance of *Pseudomonas aeruginosa* in cystic fibrosis results from convergent evolution in the *mexZ* gene. *Thorax* **2017**, *72*, 40–47. [CrossRef]

86. Chuanchuen, R.; Narasaki, C.T.; Schweizer, H.P. The MexJK efflux pump of *Pseudomonas aeruginosa* requires OprM for antibiotic efflux but not for efflux of triclosan. *J. Bacteriol.* **2002**, *184*, 5036–5044. [CrossRef]

87. Li, Y.; Mima, T.; Komori, Y.; Morita, Y.; Kuroda, T.; Mizushima, T.; Tsuchiya, T. A new member of the tripartite multidrug efflux pumps, MexVW-OprM, in *Pseudomonas aeruginosa*. *J. Antimicrob. Chemother.* **2003**, *52*, 572–575. [CrossRef]

88. Lomovskaya, O.; Bostian, K.A. Practical applications and feasibility of efflux pump inhibitors in the clinic—A vision for applied use. *Biochem. Pharmacol.* **2006**, *71*, 910–918. [CrossRef]

89. Mahmood, H.Y.; Jamshidi, S.; Mark Sutton, J.; Rahman, K.M. Current Advances in Developing Inhibitors of Bacterial Multidrug Efflux Pumps. *Curr. Med. Chem.* **2016**, *23*, 1062–1081. [CrossRef]

90. Renau, T.E.; Léger, R.; Filonova, L.; Flamme, E.M.; Wang, M.; Yen, R.; Madsen, D.; Griffith, D.; Chamberland, S.; Dudley, M.N.; et al. Conformationally-restricted analogues of efflux pump inhibitors that potentiate the activity of levofloxacin in *Pseudomonas aeruginosa*. *Bioorg. Med. Chem. Lett.* **2003**, *13*, 2755–2758. [CrossRef]

91. Yoshida, K.; Nakayama, K.; Ohtsuka, M.; Kuru, N.; Yokomizo, Y.; Sakamoto, A.; Takemura, M.; Hoshino, K.; Kanda, H.; Nitanai, H.; et al. MexAB-OprM specific efflux pump inhibitors in *Pseudomonas aeruginosa*. Part 7: Highly soluble and in vivo active quaternary ammonium analogue D13-9001, a potential preclinical candidate. *Bioorg. Med. Chem.* **2007**, *15*, 7087–7097. [CrossRef] [PubMed]

92. Nakashima, R.; Sakurai, K.; Yamasaki, S.; Hayashi, K.; Nagata, C.; Hoshino, K.; Onodera, Y.; Nishino, K.; Yamaguchi, A. Structural basis for the inhibition of bacterial multidrug exporters. *Nature* **2013**, *500*, 102–106. [CrossRef] [PubMed]

93. Ranjitkar, S.; Jones, A.K.; Mostafavi, M.; Zwirko, Z.; Iartchouk, O.; Barnes, S.W.; Walker, J.R.; Willis, T.W.; Lee, P.S.; Dean, C.R. Target (MexB)- and Efflux-Based Mechanisms Decreasing the Effectiveness of the Efflux Pump Inhibitor D13-9001 in *Pseudomonas aeruginosa* PAO1: Uncovering a New Role for MexMN-OprM in Efflux of b-Lactams and a Novel Regulatory Circuit (MmnRS) Controlling MexMN Expression. *Antimicrob. Agents Chemother.* **2019**, *63*, e01718-18.

94. Rathi, E.; Kumar, A.; Kini, S.G. Computational approaches in efflux pump inhibitors: Current status and prospects. *Drug Discov. Today* **2020**. [CrossRef] [PubMed]

95. Fleeman, R.M.; Debevec, G.; Antonen, K.; Adams, J.L.; Santos, R.G.; Welmaker, G.S.; Houghten, R.A.; Giulianotti, M.A.; Shaw, L.N. Identification of a Novel Polyamine Scaffold With Potent Efflux Pump Inhibition Activity Toward Multi-Drug Resistant Bacterial Pathogens. *Front. Microbiol.* **2018**, *9*, 1301. [CrossRef]

96. Lee, M.D.; Galazzo, J.L.; Staley, A.L.; Lee, J.C.; Warren, M.S.; Fuernkranz, H.; Chamberland, S.; Lomovskaya, O.; Miller, G.H. Microbial fermentation-derived inhibitors of efflux-pump-mediated drug resistance. *Farmaco* **2001**, *56*, 81–85. [CrossRef]

97. Chan, B.K.; Sistrom, M.; Wertz, J.E.; Kortright, K.E.; Narayan, D.; Turner, P.E. Phage selection restores antibiotic sensitivity in MDR *Pseudomonas aeruginosa*. *Sci. Rep.* **2016**, *6*, 1–8. [CrossRef]

98. Einarsson, G.G.; Zhao, J.; LiPuma, J.J.; Downey, D.G.; Tunney, M.M.; Elborn, J.S. Community analysis and co-occurrence patterns in airway microbial communities during health and disease. *ERJ Open Res.* **2019**, *5*, 00128–02017. [CrossRef]

99. LiPuma, J.J. The changing microbial epidemiology in cystic fibrosis. *Clin. Microbiol. Rev.* **2010**, *23*, 299–323. [CrossRef]

100. Drevinek, P.; Mahenthiralingam, E. *Burkholderia cenocepacia* in cystic fibrosis: Epidemiology and molecular mechanisms of virulence. *Clin. Microbiol. Infect.* **2010**, *16*, 821–830. [CrossRef]

101. Leitão, J.H.; Sousa, S.A.; Ferreira, A.S.; Ramos, C.G.; Silva, I.N.; Moreira, L.M. Pathogenicity, virulence factors, and strategies to fight against *Burkholderia cepacia* complex pathogens and related species. *Appl. Microbiol. Biotechnol.* **2010**, *87*, 31–40. [CrossRef]

102. Valvano, M.A. Intracellular survival of *Burkholderia cepacia* complex in phagocytic cells. *Can. J. Microbiol.* **2015**, *61*, 607–615. [CrossRef]

103. McClean, S.; Healy, M.E.; Collins, C.; Carberry, S.; O'Shaughnessy, L.; Dennehy, R.; Adams, Á.; Kennelly, H.; Corbett, J.M.; Carty, F.; et al. Linocin and OmpW are involved in attachment of the cystic fibrosis-associated pathogen *Burkholderia cepacia* complex to lung epithelial cells and protect mice against infection. *Infect. Immun.* **2016**, *84*, 1424–1437. [CrossRef] [PubMed]

104. Mesureur, J.; Feliciano, J.R.; Wagner, N.; Gomes, M.C.; Zhang, L.; Blanco-Gonzalez, M.; van der Vaart, M.; O'Callaghan, D.; Meijer, A.H.; Vergunst, A.C. Macrophages, but not neutrophils, are critical for proliferation of *Burkholderia cenocepacia* and ensuing host-damaging inflammation. *PLoS Pathog.* **2017**, *13*, e1006437. [CrossRef] [PubMed]

105. Isles, A.; Maclusky, I.; Corey, M.; Gold, R.; Prober, C.; Fleming, P.; Levison, H. *Pseudomonas cepacia* infection in cystic fibrosis: An emerging problem. *J. Pediatr.* **1984**, *104*, 206–210. [CrossRef]

106. Gibson, R.L.; Burns, J.L.; Ramsey, B.W. Pathophysiology and management of pulmonary infections in cystic fibrosis. *Am. J. Respir. Crit. Care Med.* **2003**, *168*, 918–951. [CrossRef] [PubMed]

107. Snell, G.; Reed, A.; Stern, M.; Hadjiliadis, D. The evolution of lung transplantation for cystic fibrosis: A 2017 update. *J. Cyst. Fibros.* **2017**, *16*, 553–564. [CrossRef]

108. Hatziagorou, E.; Orenti, A.; Drevinek, P.; Kashirskaya, N.; Mei-Zahav, M.; De Boeck, K.; Pfleger, A.; Sciensano, M.T.; Lammertyn, E.; Macek, M.; et al. Changing epidemiology of the respiratory bacteriology of patients with cystic fibrosis–data from the european cystic fibrosis society patient registry. *J. Cyst. Fibros.* **2020**, *19*, 376–383. [CrossRef]

109. Rose, H.; Baldwin, A.; Dowson, C.G.; Mahenthiralingam, E. Biocide susceptibility of the *Burkholderia cepacia* complex. *J. Antimicrob. Chemother.* **2009**, *63*, 502–510. [CrossRef]

110. Peeters, E.; Nelis, H.J.; Coenye, T. In vitro activity of ceftazidime, ciprofloxacin, meropenem, minocycline, tobramycin and trimethoprim/sulfamethoxazole against planktonic and sessile *Burkholderia cepacia* complex bacteria. *J. Antimicrob. Chemother.* **2009**, *64*, 801–809. [CrossRef]

111. Lord, R.; Jones, A.M.; Horsley, A. Antibiotic treatment for *Burkholderia cepacia* complex in people with cystic fibrosis experiencing a pulmonary exacerbation. *Cochrane Database Syst. Rev.* **2020**, *4*, CD009529. [CrossRef]

112. Rhodes, K.A.; Schweizer, H.P. Antibiotic resistance in *Burkholderia* species. *Drug Resist. Updates* **2016**, *28*, 82–90. [CrossRef] [PubMed]

113. Sputael, V.; Van Schandevyl, G.; Hanssens, L. A case report of successful eradication of new isolates of *Burkholderia cenocepacia* in a child with cystic fibrosis. *Acta Clin. Belg.* **2020**, *75*, 421–423. [CrossRef]

114. Khan, S.; Tøndervik, A.; Sletta, H.; Klinkenberg, G.; Emanuel, C.; Onsøyen, E.; Myrvold, R.; Howe, R.A.; Walsh, T.R.; Hill, K.E.; et al. Overcoming drug resistance with alginate oligosaccharides able to potentiate the action of selected antibiotics. *Antimicrob. Agents Chemother.* **2012**, *56*, 5134–5141. [CrossRef] [PubMed]

115. Van den Driessche, F.; Vanhoutte, B.; Brackman, G.; Crabbé, A.; Rigole, P.; Vercruysse, J.; Verstraete, G.; Cappoen, D.; Vervaet, C.; Cos, P.; et al. Evaluation of combination therapy for *Burkholderia cenocepacia* lung infection in different in vitro and in vivo models. *PLoS ONE* **2017**, *12*, e0172723. [CrossRef]

116. Narayanaswamy, V.P.; Duncan, A.P.; LiPuma, J.J.; Wiesmann, W.P.; Baker, S.M.; Townsend, S.M. In vitro activity of a novel glycopolymer against biofilms of *Burkholderia cepacia* complex cystic fibrosis clinical isolates. *Antimicrob. Agents Chemother.* **2019**, *63*, e00498-19. [CrossRef] [PubMed]

117. Silva, E.; Monteiro, R.; Grainha, T.; Alves, D.; Pereira, M.O.; Sousa, A.M. Fostering innovation in the treatment of chronic polymicrobial cystic fibrosis-associated infections exploring aspartic acid and succinic acid as ciprofloxacin adjuvants. *Front. Cell. Infect. Microbiol.* **2020**, *10*, 441. [CrossRef]

118. de la Fuente-Núñez, C.; Reffuveille, F.; Haney, E.F.; Straus, S.K.; Hancock, R.E.W. Broad-spectrum antibiofilm peptide that targets a cellular stress response. *PLoS Pathog.* **2014**, *10*, e1004152. [CrossRef] [PubMed]

119. Vasireddy, L.; Bingle, L.E.H.; Davies, M.S. Antimicrobial activity of essential oils against multidrug-resistant clinical isolates of the *Burkholderia cepacia* complex. *PLoS ONE* **2018**, *13*, e0201835. [CrossRef] [PubMed]

120. Shrestha, C.L.; Zhang, S.; Wisniewski, B.; Häfner, S.; Elie, J.; Meijer, L.; Kopp, B.T. (R)-Roscovitine and CFTR modulators enhance killing of multi-drug resistant *Burkholderia cenocepacia* by cystic fibrosis macrophages. *Sci. Rep.* **2020**, *10*, 21700. [CrossRef]

121. Ganesh, P.S.; Vishnupriya, S.; Vadivelu, J.; Mariappan, V.; Vellasamy, K.M.; Shankar, E.M. Intracellular survival and innate immune evasion of *Burkholderia cepacia*: Improved understanding of quorum sensing-controlled virulence factors, biofilm, and inhibitors. *Microbiol. Immunol.* **2020**, *64*, 87–98. [CrossRef]
122. Scoffone, V.C.; Barbieri, G.; Buroni, S.; Scarselli, M.; Pizza, M.; Rappuoli, R.; Riccardi, G. Vaccines to overcome antibiotic resistance: The challenge of *Burkholderia cenocepacia*. *Trends Microbiol.* **2020**, *28*, 315–326. [CrossRef]
123. Scoffone, V.C.; Chiarelli, L.R.; Trespidi, G.; Mentasti, M.; Riccardi, G.; Buroni, S. *Burkholderia cenocepacia* infections in cystic fibrosis patients: Drug resistance and therapeutic approaches. *Front. Microbiol.* **2017**, *8*, 1592. [CrossRef] [PubMed]
124. Nair, B.M.; Cheung, K.-J.; Griffith, A.; Burns, J.L. Salicylate induces an antibiotic efflux pump in *Burkholderia cepacia* complex genomovar III (*B. cenocepacia*). *J. Clin. Investig.* **2004**, *113*, 464–473. [CrossRef]
125. Guglierame, P.; Pasca, M.R.; De Rossi, E.; Buroni, S.; Arrigo, P.; Manina, G.; Riccardi, G. Efflux pump genes of the resistance-nodulation-division family in *Burkholderia cenocepacia* genome. *BMC Microbiol.* **2006**, *6*, 66. [CrossRef] [PubMed]
126. Holden, M.T.G.; Seth-Smith, H.M.B.; Crossman, L.C.; Sebaihia, M.; Bentley, S.D.; Cerdeño-Tárraga, A.M.; Thomson, N.R.; Bason, N.; Quail, M.A.; Sharp, S.; et al. The genome of *Burkholderia cenocepacia* J2315, an epidemic pathogen of cystic fibrosis patients. *J. Bacteriol.* **2009**, *191*, 261–277. [CrossRef] [PubMed]
127. Buroni, S.; Pasca, M.R.; Flannagan, R.S.; Bazzini, S.; Milano, A.; Bertani, I.; Venturi, V.; Valvano, M.A.; Riccardi, G. Assessment of three resistance-nodulation-cell division drug efflux transporters of *Burkholderia cenocepacia* in intrinsic antibiotic resistance. *BMC Microbiol.* **2009**, *9*, 200. [CrossRef] [PubMed]
128. Bazzini, S.; Udine, C.; Sass, A.; Pasca, M.R.; Longo, F.; Emiliani, G.; Fondi, M.; Perrin, E.; Decorosi, F.; Viti, C.; et al. Deciphering the role of RND efflux transporters in *Burkholderia cenocepacia*. *PLoS ONE* **2011**, *6*, e18902. [CrossRef]
129. Buroni, S.; Matthijs, N.; Spadaro, F.; Van Acker, H.; Scoffone, V.C.; Pasca, M.R.; Riccardi, G.; Coenye, T. Differential roles of RND efflux pumps in antimicrobial drug resistance of sessile and planktonic *Burkholderia cenocepacia* cells. *Antimicrob. Agents Chemother.* **2014**, *58*, 7424–7429. [CrossRef]
130. Scoffone, V.C.; Spadaro, F.; Udine, C.; Makarov, V.; Fondi, M.; Fani, R.; De Rossi, E.; Riccardi, G.; Buroni, S. Mechanism of resistance to an antitubercular 2-thiopyridine derivative that is also active against *Burkholderia cenocepacia*. *Antimicrob. Agents Chemother.* **2014**, *58*, 2415–2417. [CrossRef]
131. Coenye, T.; Van Acker, H.; Peeters, E.; Sass, A.; Buroni, S.; Riccardi, G.; Mahenthiralingam, E. molecular mechanisms of chlorhexidine tolerance in *Burkholderia cenocepacia* biofilms. *Antimicrob. Agents Chemother.* **2011**, *55*, 1912–1919. [CrossRef]
132. Perrin, E.; Maggini, V.; Maida, I.; Gallo, E.; Lombardo, K.; Madarena, M.P.; Buroni, S.; Scoffone, V.C.; Firenzuoli, F.; Mengoni, A.; et al. Antimicrobial activity of six essential oils against *Burkholderia cepacia* complex: Insights into mechanism(s) of action. *Future Microbiol.* **2018**, *13*, 59–67. [CrossRef]
133. Nunvar, J.; Hogan, A.M.; Buroni, S.; Savina, S.; Makarov, V.; Cardona, S.T.; Drevinek, P. The effect of 2-thiocyanatopyridine derivative 11026103 on *Burkholderia cenocepacia*: Resistance mechanisms and systemic impact. *Antibiotics* **2019**, *8*, 159. [CrossRef]
134. Scoffone, V.C.; Ryabova, O.; Makarov, V.; Iadarola, P.; Fumagalli, M.; Fondi, M.; Fani, R.; De Rossi, E.; Riccardi, G.; Buroni, S. Efflux-mediated resistance to a benzothiadiazol derivative effective against *Burkholderia cenocepacia*. *Front. Microbiol.* **2015**, *6*, 815. [CrossRef] [PubMed]
135. Tseng, S.-P.; Tsai, W.-C.; Liang, C.-Y.; Lin, Y.-S.; Huang, J.-W.; Chang, C.-Y.; Tyan, Y.-C.; Lu, P.-L. The contribution of antibiotic resistance mechanisms in clinical *Burkholderia cepacia* complex isolates: An emphasis on efflux pump activity. *PLoS ONE* **2014**, *9*, e104986. [CrossRef]
136. Mira, N.P.; Madeira, A.; Moreira, A.S.; Coutinho, C.P.; Sá-Correia, I. Genomic expression analysis reveals strategies of *Burkholderia cenocepacia* to adapt to cystic fibrosis patients' airways and antimicrobial therapy. *PLoS ONE* **2011**, *6*, e28831. [CrossRef] [PubMed]
137. Dumolin, C.; Peeters, C.; Ehsani, E.; Tahon, G.; De Canck, E.; Cnockaert, M.; Boon, N.; Vandamme, P. *Achromobacter veterisilvae* sp. nov., from a mixed hydrogen-oxidizing bacteria enrichment reactor for microbial protein production. *Int. J. Syst. Evol. Microbiol.* **2020**, *70*, 530–536. [CrossRef]
138. Klinger, J.D.; Thomassen, M.J. Occurrence and antimicrobial susceptibility of gram-negative nonfermentative bacilli in cystic fibrosis patients. *Diagn. Microbiol. Infect. Dis.* **1985**, *3*, 149–158. [CrossRef]
139. European Cystic Fibrosis Society Patient Registry. In Annual Data Report 2018; 2020. Available online: https://www.ecfs.eu/sites/default/files/general-content-files/working-groups/ecfs-patient-registry/ECFSPR_Report_2018_v1.4.pdf (accessed on 27 May 2021).
140. Amoureux, L.; Bador, J.; Siebor, E.; Taillefumier, N.; Fanton, A.; Neuwirth, C. Epidemiology and resistance of *Achromobacter xylosoxidans* from cystic fibrosis patients in Dijon, Burgundy: First French data. *J. Cyst. Fibros.* **2013**, *12*, 170–176. [CrossRef] [PubMed]
141. Trancassini, M.; Iebba, V.; Citerà, N.; Tuccio, V.; Magni, A.; Varesi, P.; De Biase, R.V.; Totino, V.; Santangelo, F.; Gagliardi, A.; et al. Outbreak of *Achromobacter xylosoxidans* in an Italian Cystic fibrosis center: Genome variability, biofilm production, antibiotic resistance, and motility in isolated strains. *Front. Microbiol.* **2014**, *5*, 138. [CrossRef]
142. Firmida, M.C.; Pereira, R.H.; Silva, E.A.; Marques, E.A.; Lopes, A.J. Clinical impact of *Achromobacter xylosoxidans* colonization/infection in patients with cystic fibrosis. *Braz. J. Med. Biol. Res.* **2016**, *49*, e5097. [CrossRef]
143. Papalia, M.; Steffanowski, C.; Traglia, G.; Almuzara, M.; Martina, P.; Galanternik, L.; Vay, C.; Gutkind, G.; Ramírez, M.S.; Radice, M. Diversity of *Achromobacter* species recovered from patients with cystic fibrosis, in Argentina. *Rev. Argent. Microbiol.* **2020**, *52*, 13–18. [CrossRef]

144. Edwards, B.D.; Greysson-Wong, J.; Somayaji, R.; Waddell, B.; Whelan, F.J.; Storey, D.G.; Rabin, H.R.; Surette, M.G.; Parkins, M.D. Prevalence and Outcomes of *Achromobacter* Species Infections in Adults with Cystic Fibrosis: A North American Cohort Study. *J. Clin. Microbiol.* **2017**, *55*, 2074–2085. [CrossRef]

145. Tetart, M.; Wallet, F.; Kyheng, M.; Leroy, S.; Perez, T.; Le Rouzic, O.; Wallaert, B.; Prevotat, A. Impact of *Achromobacter xylosoxidans* isolation on the respiratory function of adult patients with cystic fibrosis. *ERJ Open Res.* **2019**, *5*, 00051–02019. [CrossRef] [PubMed]

146. Jeukens, J.; Freschi, L.; Vincent, A.T.; Emond-Rheault, J.G.; Kukavica-Ibrulj, I.; Charette, S.J.; Levesque, R.C. A Pan-Genomic Approach to Understand the Basis of Host Adaptation in *Achromobacter*. *Genome Biol. Evol.* **2017**, *9*, 1030–1046. [CrossRef]

147. Abbott, I.J.; Peleg, A.Y. *Stenotrophomonas, Achromobacter*, and nonmelioid *Burkholderia* species: Antimicrobial resistance and therapeutic strategies. *Semin. Respir. Crit. Care Med.* **2015**, *36*, 99–110. [PubMed]

148. Pongchaikul, P.; Santanirand, P.; Antonyuk, S.; Winstanley, C.; Darby, A.C. AcGI1, a novel genomic island carrying antibiotic resistance integron In687 in multidrug resistant *Achromobacter xylosoxidans* in a teaching hospital in Thailand. *FEMS Microbiol. Lett.* **2020**, *367*, fnaa109. [CrossRef]

149. Gainey, A.B.; Burch, A.K.; Brownstein, M.J.; Brown, D.E.; Fackler, J.; Horne, B.; Biswas, B.; Bivens, B.N.; Malagon, F.; Daniels, R. Combining bacteriophages with cefiderocol and meropenem/vaborbactam to treat a pan-drug resistant *Achromobacter* species infection in a pediatric cystic fibrosis patient. *Pediatr. Pulmonol.* **2020**, *55*, 2990–2994. [CrossRef]

150. Levesque, R.; Letarte, R.; Pechère, J.C. Comparative study of the beta-lactamase activity found in *Achromobacter*. *Can. J. Microbiol.* **1983**, *29*, 819–826. [CrossRef] [PubMed]

151. Fujii, T.; Sato, K.; Inoue, M.; Mitsuhashi, S. Purification and properties of a beta-lactamase from *Alcaligenes dentrificans* subsp. *xylosoxydans*. *J. Antimicrob. Chemother.* **1985**, *16*, 297–304. [CrossRef]

152. Philippon, A.; Mensah, K.; Fournier, G.; Freney, J. Two resistance phenotypes to beta-lactams of *Alcaligenes denitrificans* subsp. *xylosoxydans* in relation to beta-lactamase types. *J. Antimicrob. Chemother.* **1990**, *25*, 698–700. [CrossRef]

153. Decré, D.; Arlet, G.; Bergogne-Bérézin, E.; Philippon, A. Identification of a carbenicillin-hydrolyzing beta-lactamase in *Alcaligenes denitrificans* subsp. *xylosoxydans*. *Antimicrob. Agents Chemother.* **1995**, *39*, 771–774. [CrossRef]

154. Doi, Y.; Poirel, L.; Paterson, D.L.; Nordmann, P. Characterization of a naturally occurring class D beta-lactamase from *Achromobacter xylosoxidans*. *Antimicrob. Agents Chemother.* **2008**, *52*, 1952–1956. [CrossRef] [PubMed]

155. Li, X.; Hu, Y.; Gong, J.; Zhang, L.; Wang, G. Comparative genome characterization of *Achromobacter* members reveals potential genetic determinants facilitating the adaptation to a pathogenic lifestyle. *Appl. Microbiol. Biotechnol.* **2013**, *97*, 6413–6425. [CrossRef]

156. Nielsen, S.M.; Meyer, R.L.; Nørskov-Lauritsen, N. Differences in Gene Expression Profiles between Early and Late Isolates in Monospecies *Achromobacter* Biofilm. *Pathogens* **2017**, *6*, 20. [CrossRef]

157. Bador, J.; Neuwirth, C.; Liszczynski, P.; Mézier, M.C.; Chrétiennot, M.; Grenot, E.; Chapuis, A.; de Curraize, C.; Amoureux, L. Distribution of innate efflux-mediated aminoglycoside resistance among different *Achromobacter* species. *New Microbes New Infect.* **2015**, *10*, 1–5. [CrossRef]

158. Bador, J.; Neuwirth, C.; Grangier, N.; Muniz, M.; Germé, L.; Bonnet, J.; Pillay, V.G.; Llanes, C.; de Curraize, C.; Amoureux, L. Role of AxyZ Transcriptional Regulator in Overproduction of AxyXY-OprZ Multidrug Efflux System in *Achromobacter* Species Mutants Selected by Tobramycin. *Antimicrob. Agents Chemother.* **2017**, *61*, e00290-17. [CrossRef] [PubMed]

159. Fleurbaaij, F.; Henneman, A.A.; Corver, J.; Knetsch, C.W.; Smits, W.K.; Nauta, S.T.; Giera, M.; Dragan, I.; Kumar, N.; Lawley, T.D.; et al. Proteomic identification of Axc, a novel beta-lactamase with carbapenemase activity in a meropenem-resistant clinical isolate of *Achromobacter xylosoxidans*. *Sci. Rep.* **2018**, *8*, 8181. [CrossRef]

160. Gabrielaite, M.; Nielsen, F.C.; Johansen, H.K.; Marvig, R.L. *Achromobacter* genetic adaptation in cystic fibrosis. *bioRxiv* **2021**, 426490. Available online: https://www.biorxiv.org/content/10.1101/2021.01.13.426490v1.full (accessed on 27 May 2021).

161. Magallon, A.; Roussel, M.; Neuwirthm, C.; Tetu, J.; Cheiakh, A.C.; Boulet, B.; Varin, V.; Urbain, V.; Bador, J.; Amoureux, L. Fluoroquinolone resistance in *Achromobacter* spp.: Substitutions in QRDRs of GyrA, GyrB, ParC and ParE and implication of the RND efflux system AxyEF-OprN. *J. Antimicrob. Chemother.* **2021**, *76*, 297–304. [CrossRef]

162. Morita, Y.; Nakashima, K.; Nishino, K.; Kotani, K.; Tomida, J.; Inoue, M.; Kawamura, Y. Berberine Is a Novel Type Efflux Inhibitor Which Attenuates the MexXY-Mediated Aminoglycoside Resistance in *Pseudomonas aeruginosa*. *Front. Microbiol.* **2016**, *7*, 1223. [CrossRef]

163. Kotani, K.; Matsumura, M.; Morita, Y.; Tomida, J.; Kutsuna, R.; Nishino, K.; Yasuike, S.; Kawamura, Y. 13-(2-Methylbenzyl) Berberine Is a More Potent Inhibitor of MexXY-Dependent Aminoglycoside Resistance than Berberine. *Antibiotics* **2019**, *8*, 212. [CrossRef]

164. Brooke, J.S. *Stenotrophomonas maltophilia*: An emerging global opportunistic pathogen. *Clin. Microbiol. Rev.* **2012**, *25*, 2–41. [CrossRef]

165. Lira, F.; Berg, G.; Martínez, J.L. Double-face meets the bacterial world: The opportunistic pathogen *Stenotrophomonas maltophilia*. *Front. Microbiol.* **2017**, *8*, 1–15. [CrossRef]

166. Biagi, M.; Lamm, D.; Meyer, K.; Vialichka, A.; Jurkovic, M.; Patel, S.; Mendes, R.E.; Bulman, Z.P.; Wenzler, E. Activity of Aztreonam in Combination with Avibactam, Clavulanate, Relebactam, and Vaborbactam against Multidrug-Resistant *Stenotrophomonas maltophilia*. *Antimicrob. Agents Chemother.* **2020**, *64*, e00297-20. [CrossRef] [PubMed]

167. Gould, V.C.; Okazaki, A.; Avison, M.B. Coordinate hyperproduction of SmeZ and SmeJK efflux pumps extends drug resistance in *Stenotrophomonas maltophilia*. *Antimicrob. Agents Chemother.* **2013**, *57*, 655–657. [CrossRef]

168. Alonso, A.; Martinez, J.L. Cloning and characterization of SmeDEF, a novel multidrug efflux pump from *Stenotrophomonas maltophilia*. *Antimicrob. Agents Chemother.* **2000**, *44*, 3079–3086. [CrossRef]
169. Blanco, P.; Corona, F.; Martinez, J.L. Mechanisms and phenotypic consequences of acquisition of tigecycline resistance by *Stenotrophomonas maltophilia*. *J. Antimicrob. Chemother.* **2019**, *74*, 3221–3230. [CrossRef]
170. Chen, C.H.; Huang, C.C.; Chung, T.C.; Hu, R.M.; Huang, Y.W.; Yang, T.C. Contribution of resistance-nodulation-division efflux pump operon *smeU1-V-W-U2-X* to multidrug resistance of *Stenotrophomonas maltophilia*. *Antimicrob. Agents Chemother.* **2011**, *55*, 5826–5833. [CrossRef] [PubMed]
171. Huang, Y.W.; Liou, R.S.; Lin, Y.T.; Huang, H.H.; Yang, T.C. A linkage between SmeIJK efflux pump, cell envelope integrity, and σe-mediated envelope stress response in *Stenotrophomonas maltophilia*. *PLoS ONE* **2014**, *9*, 1–11. [CrossRef] [PubMed]
172. Li, X.Z.; Zhang, L.; Poole, K. SmeC, an outer membrane multidrug efflux protein of *Stenotrophomonas maltophilia*. *Antimicrob. Agents Chemother.* **2002**, *46*, 333–343. [CrossRef]
173. Lin, Y.T.; Huang, Y.W.; Chen, S.J.; Chang, C.W.; Yang, T.C. The SmeYZ efflux pump of *Stenotrophomonas maltophilia* contributes to drug resistance, virulence-related characteristics, and virulence in mice. *Antimicrob. Agents Chemother.* **2015**, *59*, 4067–4073. [CrossRef] [PubMed]
174. Wu, C.J.; Chiu, T.T.; Lin, Y.T.; Huang, Y.W.; Li, L.H.; Yang, T.C. Role of *smeU1VWU2X* operon in alleviation of oxidative stresses and occurrence of sulfamethoxazole-trimethoprim-resistant mutants in *Stenotrophomonas maltophilia*. *Antimicrob. Agents Chemother.* **2018**, *62*, 1–12. [CrossRef]
175. Zhang, L.; Li, X.Z.; Poole, K. SmeDEF multidrug efflux pump contributes to intrinsic multidrug resistance in *Stenotrophomonas maltophilia*. *Antimicrob. Agents Chemother.* **2001**, *45*, 3497–3503. [CrossRef] [PubMed]
176. Wu, C.J.; Huang, Y.W.; Lin, Y.T.; Ning, H.C.; Yang, T.C. Inactivation of SmeSyRy two-component regulatory system inversely regulates the expression of SmeYZ and SmeDEF efflux pumps in *Stenotrophomonas maltophilia*. *PLoS ONE* **2016**, *11*, 1–14. [CrossRef] [PubMed]
177. Hernández, A.; Ruiz, F.M.; Romero, A.; Martínez, J.L. The binding of triclosan to SmeT, the repressor of the multidrug efflux pump SmeDEF, induces antibiotic resistance in *Stenotrophomonas maltophilia*. *PLoS Pathog.* **2011**, *7*, 1–12. [CrossRef]
178. Sanchez, P.; Moreno, E.; Martinez, J.L. The biocide triclosan selects *Stenotrophomonas maltophilia* mutants that overproduce the SmeDEF multidrug efflux pump. *Antimicrob. Agents Chemother.* **2005**, *49*, 781–782. [CrossRef]
179. Kim, H.R.; Lee, D.; Eom, Y.B. Anti-biofilm and anti-virulence efficacy of celastrol against *Stenotrophomonas maltophilia*. *Int. J. Res. Med. Sci.* **2018**, *15*, 617–627. [CrossRef]
180. Lin, C.W.; Huang, Y.W.; Hu, R.M.; Yang, T.C. SmeOP-TolCSm efflux pump contributes to the multidrug resistance of *Stenotrophomonas maltophilia*. *Antimicrob. Agents Chemother.* **2014**, *58*, 2405–2408. [CrossRef] [PubMed]
181. Blanco, P.; Corona, F.; Martínez, J.L. Involvement of the RND efflux pump transporter SmeH in the acquisition of resistance to ceftazidime in *Stenotrophomonas maltophilia*. *Sci. Rep.* **2019**, *9*, 1–14. [CrossRef]
182. Li, L.H.; Zhang, M.S.; Wu, C.J.; Lin, Y.T.; Yang, T.C. Overexpression of SmeGH contributes to the acquired MDR of *Stenotrophomonas maltophilia*. *J. Antimicrob. Chemother.* **2019**, *74*, 2225–2229. [CrossRef] [PubMed]
183. Faure, E.; Kwong, K.; Nguyen, D. *Pseudomonas aeruginosa* in Chronic Lung Infections: How to Adapt Within the Host? *Front. Immunol.* **2018**, *9*, 2416. [CrossRef]
184. Henrichfreise, B.; Wiegand, I.; Pfister, W.; Wiedemann, B. Resistance mechanisms of multiresistant *Pseudomonas aeruginosa* strains from Germany and correlation with hypermutation. *Antimicrob. Agents Chemother.* **2007**, *51*, 4062–4070. [CrossRef]
185. Rees, V.E.; Deveson Lucas, D.S.; López-Causapé, C.; Huang, Y.; Kotsimbos, T.; Bulitta, J.B.; Rees, M.C.; Barugahare, A.; Peleg, A.Y.; Nation, R.L.; et al. Characterization of Hypermutator *Pseudomonas aeruginosa* Isolates from Patients with Cystic Fibrosis in Australia. *Antimicrob. Agents Chemother.* **2019**, *63*, e02538-18. [CrossRef]
186. Díaz-Ríos, C.; Hernández, M.; Abad, D.; Álvarez-Montes, L.; Varsaki, A.; Iturbe, D.; Calvo, J.; Ocampo-Sosa, A.A. New Sequence Type ST3449 in Multidrug-Resistant *Pseudomonas aeruginosa* Isolates from a Cystic Fibrosis Patient. *Antibiotics* **2021**, *10*, 491. [CrossRef]
187. Greipel, L.; Fischer, S.; Klockgether, J.; Dorda, M.; Mielke, S.; Wiehlmann, L.; Cramer, N.; Tümmler, B. Molecular Epidemiology of Mutations in Antimicrobial Resistance Loci of *Pseudomonas aeruginosa* Isolates from Airways of Cystic Fibrosis Patients. *Antimicrob. Agents Chemother.* **2016**, *60*, 6726–6734. [CrossRef]
188. López-Causapé, C.; Sommer, L.M.; Cabot, G.; Rubio, R.; Ocampo-Sosa, A.A.; Johansen, H.K.; Figuerola, J.; Cantón, R.; Kidd, T.J.; Molin, S.; et al. Evolution of the *Pseudomonas aeruginosa* mutational resistome in an international Cystic Fibrosis clone. *Sci. Rep.* **2017**, *7*, 5555. [CrossRef]
189. Costabile, G.; Provenzano, R.; Azzalin, A.; Scoffone, V.C.; Chiarelli, L.R.; Rondelli, V.; Grillo, I.; Zinn, T.; Lepioshkin, A.; Savina, S.; et al. PEGylated mucus-penetrating nanocrystals for lung delivery of a new FtsZ inhibitor against *Burkholderia cenocepacia* infection. *Nanomedicine* **2020**, *23*, 102113. [CrossRef] [PubMed]
190. Chong, S.Y.; Lee, K.; Chung, H.S.; Hong, S.G.; Suh, Y.; Chong, Y. Levofloxacin Efflux and *smeD* in Clinical Isolates of *Stenotrophomonas maltophilia*. *Microb. Drug Resist.* **2017**, *23*, 163–168. [CrossRef] [PubMed]
191. Smith, E.E.; Buckley, D.G.; Wu, Z.; Saenphimmachak, C.; Hoffman, L.R.; D'Argenio, D.A.; Miller, S.I.; Ramsey, B.W.; Speert, D.P.; Moskowitz, S.M. Genetic adaptation by *Pseudomonas aeruginosa* to the airways of cystic fibrosis patients. *Proc. Natl. Acad. Sci. USA* **2006**, *103*, 8487–8492. [CrossRef] [PubMed]

Review

Ever-Adapting RND Efflux Pumps in Gram-Negative Multidrug-Resistant Pathogens: A Race against Time

Martijn Zwama [1,*] and Kunihiko Nishino [1,2,*]

1   SANKEN (The Institute of Scientific and Industrial Research), Osaka University, Ibaraki, Osaka 567-0047, Japan
2   Graduate School of Pharmaceutical Sciences, Osaka University, Suita, Osaka 565-0871, Japan
*   Correspondence: m.zwama@sanken.osaka-u.ac.jp (M.Z.); nishino@sanken.osaka-u.ac.jp (K.N.)

Abstract: The rise in multidrug resistance (MDR) is one of the greatest threats to human health worldwide. MDR in bacterial pathogens is a major challenge in healthcare, as bacterial infections are becoming untreatable by commercially available antibiotics. One of the main causes of MDR is the over-expression of intrinsic and acquired multidrug efflux pumps, belonging to the resistance-nodulation-division (RND) superfamily, which can efflux a wide range of structurally different antibiotics. Besides over-expression, however, recent amino acid substitutions within the pumps themselves—causing an increased drug efflux efficiency—are causing additional worry. In this review, we take a closer look at clinically, environmentally and laboratory-evolved Gram-negative bacterial strains and their decreased drug sensitivity as a result of mutations directly in the RND-type pumps themselves (from *Escherichia coli*, *Salmonella enterica*, *Neisseria gonorrhoeae*, *Pseudomonas aeruginosa*, *Acinetobacter baumannii* and *Legionella pneumophila*). We also focus on the evolution of the efflux pumps by comparing hundreds of efflux pumps to determine where conservation is concentrated and where differences in amino acids can shed light on the broad and even broadening drug recognition. Knowledge of conservation, as well as of novel gain-of-function efflux pump mutations, is essential for the development of novel antibiotics and efflux pump inhibitors.

Keywords: pathogens; multidrug resistance; RND; evolution; efflux pump; adaptation

Citation: Zwama, M.; Nishino, K. Ever-Adapting RND Efflux Pumps in Gram-Negative Multidrug-Resistant Pathogens: A Race against Time. *Antibiotics* **2021**, *10*, 774. https://doi.org/10.3390/antibiotics10070774

Academic Editors: Henrietta Venter, Isabelle Broutin, Attilio V. Vargiu and Gilles Phan

Received: 27 May 2021
Accepted: 14 June 2021
Published: 25 June 2021

Publisher's Note: MDPI stays neutral with regard to jurisdictional claims in published maps and institutional affiliations.

Copyright: © 2021 by the authors. Licensee MDPI, Basel, Switzerland. This article is an open access article distributed under the terms and conditions of the Creative Commons Attribution (CC BY) license (https://creativecommons.org/licenses/by/4.0/).

1. Introduction

Antimicrobial resistance (AMR) undermines our ability to treat infectious diseases, as pathogenic microorganisms become insensitive to our developed antibiotics [1]. Resistance to multiple antibiotics is called multidrug resistance (MDR) and is one of the major concerns in human health worldwide, a trend seen in clinically significant pathogenic organisms [2]. AMR can be caused by alterations in drug targets or the inactivation or alteration of antibiotics [3–5]. Notably, compared to these single factors contributing to the resistance of a single class of antibiotics, MDR can be caused by reduced permeability of bacterial membranes [6] and by the over-expression of multidrug efflux pumps alone, in both Gram-negative and Gram-positive bacterial cells [6–8]. These efflux pumps can be acquired from plasmids and horizontal gene transfer [7,9], and mutations in the regulatory network can significantly increase the expression of both acquired and intrinsic efflux pumps in clinical strains [3]. This over-expression is one of the main reasons for MDR [10]. In Gram-negative bacteria, efflux pumps belonging to the resistance-nodulation-division (RND) superfamily are one of the major contributors to MDR in clinical pathogens today [10, 11]. These efflux pumps can recognize and expel many different classes of antibiotics, including macrolides, β-lactams, aminoglycosides, quinolones, dyes and detergents [12]. It is important to note that these membrane proteins have intrinsic multidrug recognition properties; however, they have been around before the clinical usage of antibiotics, and it has been shown that RND pumps play critical physiological roles in the survival and fitness of bacterial cells [13] and in cell metabolism [14], and that the multidrug recognition ability has been around since ancient transporters [15]. RND pumps form tripartite efflux systems, enabling the export of antibiotics directly to the outside of the cell [16]. The significance of the over-expression caused by increased transcription of the pump operons by mutations in their regulatory genes and proteins (e.g., the AraC family, TetR family and two-component systems (TCS)) [17,18] has been well established for most clinical pathogenic bacteria [3,17–27].

Alarmingly, in recent years, mutations within RND-type exporters themselves have been reported to enhance MDR by an increased efflux activity of the pump proteins. This worrying gain-of-function development adds significantly to the over-expression challenges already set by pathogenic Gram-negative bacteria, as the emergence of amino acid substitutions increases the minimum inhibitory concentrations (MICs) of antibiotics used to treat the pathogenic infections by two-fold or more. This review aims to summarize this recent development in the MDR field for a selection of pathogens: *Escherichia coli*, *Salmonella enterica* subsp. *enterica*, *Neisseria gonorrhoeae*, *Pseudomonas aeruginosa*, *Acinetobacter baumannii* and *Legionella pneumophila*. In addition, we try to summarize phylogenetic connections between efflux pumps in terms of amino acid differences (variation) and conservation within the transmembrane (motor) domain and the periplasmic (drug efflux) domain by analyzing 135 homotrimeric RND multidrug efflux pumps. These insights help us guide the development of novel antibiotics and efflux pump inhibitors.

2. Structure of RND-Type Multidrug Efflux Pumps

Before discussing the RND efflux pumps from different Gram-negative pathogens, we will briefly summarize our current knowledge of arguably the most studied RND pump called AcrB, from *Escherichia coli* (AcrB-Ec). More elaborate and detailed reviews regarding the structure and the mechanism of AcrB-Ec and other multidrug transporters can be read elsewhere [12,28–32]. In short, the first crystal structure of an RND-type multidrug efflux pump (AcrB-Ec) was solved in 2002 [33], paving the way for concise structure–function analysis after previous meticulous biochemical analysis of this efflux pump before this crystal structure was available, e.g., [34,35]. Since then, several research groups have obtained crucial information about AcrB-Ec, and other members of the RND superfamily, by solving crystal structures, analyzing biochemical data, performing molecular dynamics simulations and, more recently, obtaining electron microscope (EM) images of innate conformations of the pumps and even of the entire tripartite complexes. Examples of crystal and EM structures of RND-type multidrug efflux pumps besides AcrB-Ec are MexB from *P. aeruginosa* (MexB-Pa) [36–39], AdeB from *A. baumannii* (AdeB-Ab) [40,41] and MtrD from *N. gonorrhoeae* (MtrD-Ng) [42,43], which we discuss further in this review.

To summarize, RND multidrug efflux pumps are homotrimeric proteins embedded in the inner membrane of Gram-negative bacterial cells and couple with six membrane fusion proteins (MFP). (Among RND multidrug efflux pumps, there are also heteromultimeric pumps [30]; however, this review focuses on the homotrimeric group of pumps). There have been debates on whether the RND pump itself directly, or indirectly through MFPs, couples to the outer membrane protein (OMP) tunnel [44,45], which lies embedded in the outer membrane, and how many proteins of each three of the segments (RND, MFP and OMP) comprise the tripartite complex [46,47] (Figure 1a). However, there has been a growing consensus that one RND efflux pump trimer couples with six MFPs, and that this hexameric MFP tunnel interacts and forms a complex with three OMP monomers by relatively weak tip-to-tip interactions. This consensus is guided mainly by the elucidation of the structures of the entire tripartite complexes of AcrAB–TolC (*E. coli*) and MexAB–OprM (*P. aeruginosa*), obtained by EM imaging [16,38,48–50].

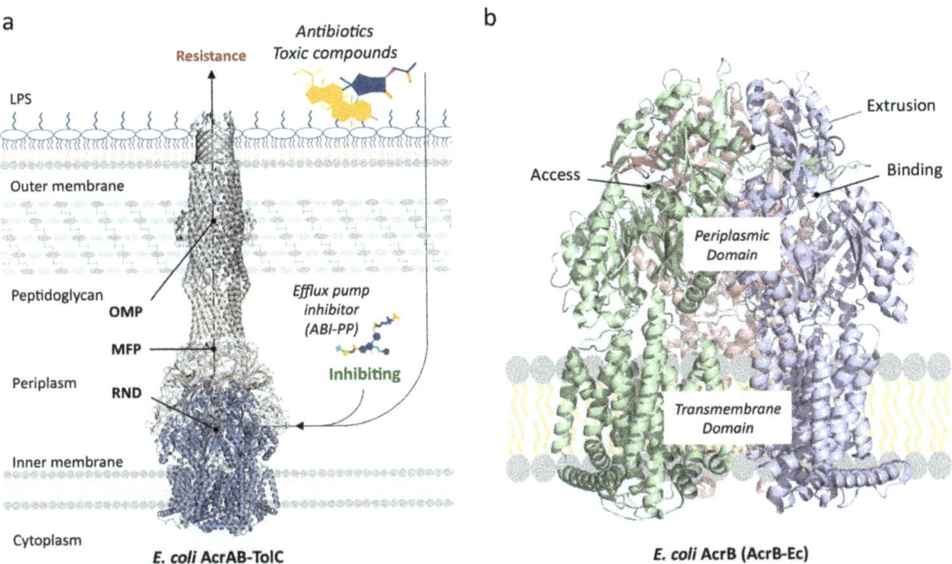

Figure 1. Structure of AcrAB–TolC-Ec and AcrB-Ec. (**a**) The structure of the tripartite complex AcrAB–TolC from *Escherichia coli* (PDB accession code 5O66 [48]). Antibiotics and other toxic compounds enter through the outer membrane and are captured by the RND efflux pump and consequently pumped out of the cells. ABI-PP is an efflux pump inhibitor (EPI), stopping the pump from functioning. (**b**) Structure of AcrB-Ec. Green shows the access monomer, blue the binding monomer and red the extrusion monomer (PDB accession code 3AOD [51]). Abbreviations: OMP, outer membrane protein; MFP, membrane fusion protein; RND, resistance-nodulation-division protein.

The RND efflux pump itself (Figure 1b) consists of three monomers forming a homo-trimer, each showing one of three distinct conformations called access, binding and extru-sion [52] (or alternatively loose (L), tight (T) and open (O) [53]), when actively pumping substrates. The trimer oscillates between these three states, from access to binding to extrusion and back to access, and this movement is called the "functionally rotating mecha-nism" [52]. Throughout this cycle, drugs move through one of the protomers of the pump by a peristaltic motion in the porter domain (Figure 2) [53]. There are two distinct drug-binding pockets within each monomer: a deep distal binding pocket (DBP) and a proximal binding pocket (PBP) [51,54] (Figure 2a,b) separated by the switch-loop [54] (sometimes referred to as the G-loop in the literature [30,55,56]) (Figure 2c). The flexible hoisting-loop enables the significant conformational changes in the porter domain [57]. As shown in Figure 2c, there are several other distinct functional loops within the monomers. Crystal structures of efflux pump inhibitor (EPI) ABI-PP bound to AcrB-Ec and MexB-Pa show the existence of a hydrophobic pit or trap (or inhibitor binding pit) [36], rich in pheny-lalanine residues. Other EPIs (MBX inhibitors) bound to AcrB-Ec (porter domain only) have also been crystallized [58]. It is hypothesized that large drugs, such as erythromycin and rifampicin, bind strongly to the PBP in the access monomer, and smaller drugs, such as minocycline and doxorubicin, bind strongly to the DBP in the binding monomer [51]. However, the large surfactant molecule LMNG (lauryl maltose neopentyl glycol) was recently found to be bound to the DBP of MexB-Pa in the binding monomer [37]. This was also the case for erythromycin, bound in the hydrophobic pit of MtrD-Ng [43], overlapping the ABI-PP binding location in AcrB-Ec and MexB-Pa [36]. Additionally, smaller molecules such as doxorubicin and ethidium have been found to be present in the PBP of AcrB-Ec [54] and AdeB-Ab [41], respectively, besides being found in the DBP. Molecular dynamics simulations have shown that, depending on the molecular properties, pump substrates move within the pockets and have preferred binding sites [55].

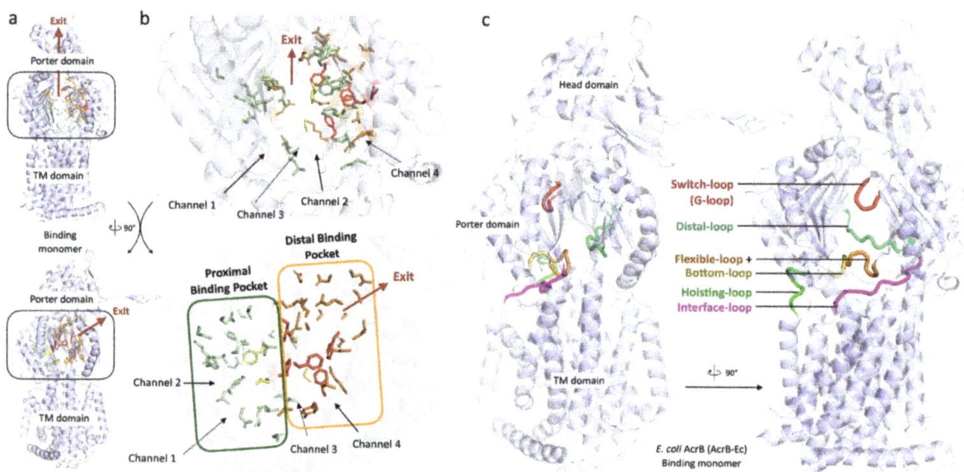

Figure 2. The structure, domains and loops of the RND monomers. (**a**) Side views of the entire protomer of AcrB-Ec. (**b**) The porter domain with the highlighted proximal (PBP) and distal binding pockets (DBP) and their drug-interacting residues. Arrows roughly indicate channels (dashed arrows indicate behind the image). Colors: orange, DBP; green, PBP; yellow, between DBP and PBP; red, Phe residues in the hydrophobic pit. (**c**) Side views of the flexible loops. The cartoon representation is transparent, allowing one to view all loops and residues in their entirety. PDB accession code 3W9H [36].

3. Conservation among RND Efflux Pumps Highlights Important Domains

In this section, we provide an overview of conservation distilled from the comparison of numerous RND multidrug efflux pumps. We previously analyzed about 400 RND genes from Gram-negative gammaproteobacteria [15] (including heavy metal efflux pumps (HME) [59,60], heteromultimeric MdtBC-like pumps [59,61,62] and others such as TriC-like efflux pumps [63,64]). For this review, we specifically selected from that database 133 homotrimeric multidrug efflux pump genes and added the sequences of MtrD-Ng and AdeG-Ab, converted the DNA sequences to amino acid sequences and performed protein multiple sequence alignment [65] on a total of 135 pumps. We also zoomed into 19 better studied and defined efflux pumps from *E. coli* (AcrB-Ec, AcrD-Ec and AcrF-Ec), *Salmonella enterica* (AcrB-Sa, AcrD-Sa and AcrF-Sa), *N. gonorrhoeae* (MtrD-Ng), *P. aeruginosa* (MexB-Pa, MexD-Pa, MexF-Pa, MexI-Pa, MexQ-Pa, MexW-Pa and MexY-Pa), *A. baumannii* (AdeB-Ab, AdeG-Ab and AdeJ-Ab), *L. pneumophila* (LpeB-Lp) and *H. influenzae* (AcrB-Hi). The sequences of all 135 pumps are provided in Supplementary Data S1.

3.1. Conservation Heat Maps Show Distinct Areas of Importance and Adaptation Flexibility

We created two heat maps derived from 135 efflux pumps, each counting, on average, 1043 amino acids which make up one monomer of the trimeric RND pump. The first map is automatically created, based on HMMER homology [66], by using ConSurf [67,68] after first performing multiple sequence alignment by Clustal Omega [65] (Figure 3a). The second map was manually created after using the same Clustal Omega output, based on the following criteria (Figure 3b–e): (A) fully (identically) conserved among all 135 pumps (32 residues, red), (B) conserved among the selected 19 pumps while also highly conserved among all 135 pumps (13 residues, light red), (C) fully conserved among the selected 19 pumps (33 residues, orange) and (D) highly conserved among all 135 pumps (58 residues, yellow). This second map focuses solely on the most conserved residues because the highest conserved residues found by ConSurf are relative and include residues that are between 50% and 100% identically conserved. The residue conservation analysis (including percentages and variability per residue) can be found in Supplementary Table S1. Note that there are a total of 71 highly conserved residues (58 yellow and 13 light red residues) among the total 135 pumps. Additionally, note that there are 78 fully conserved residues among the selected 19 pumps, where these include not only the 33 depicted in orange but those in the red and light red color categories, too, by definition. We included this light red category partly because this includes residues that we know to be crucial for the function of the pump (e.g., AcrB-Ec's D408 and K940, part of the proton relay pathway). These residues are different in only a few (two and three, respectively) of the total 135 pumps (the "K940R" mutation in MexB-Pa resulted in a fully active pump [34]). Nonetheless, we can clearly see distinct areas of conservation and areas where there is basically no conservation (high variability). Conservation suggests specific residues to play an essential role in the functioning of the pumps, may it be for proton relay, remote conformational coupling, stability or flexibility of the pump, stability of the trimeric complex, etc.

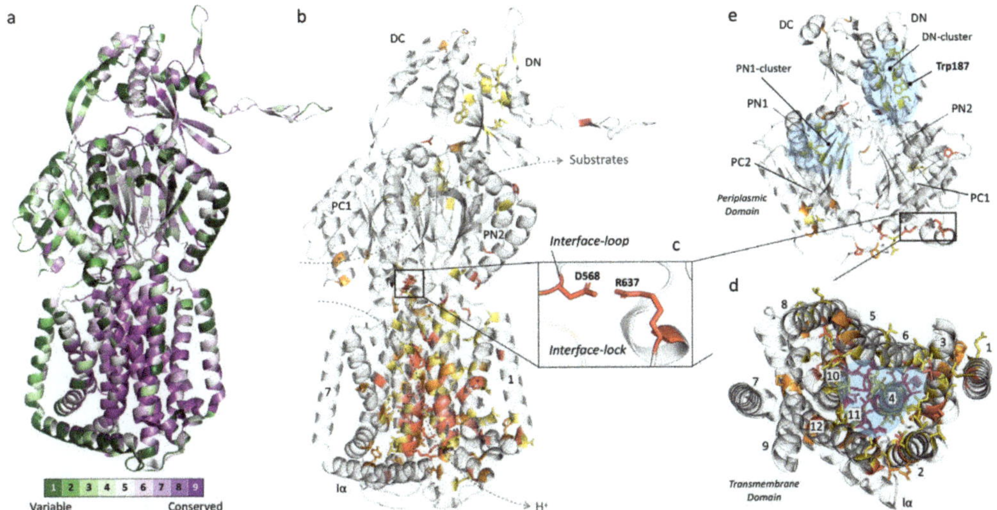

Figure 3. Heat maps of conservation, based on 135 sequences of RND multidrug efflux pumps. (**a**) Side view of AcrB-Ec showing conservation of the monomer, analyzed by using ConSurf [67,68]. Conservation is relative (most conserved "category 9" (dark purple) ranges from 50% to 100% identically conserved depending on the residue; Table 3 and Table S1). (**b**) Manual conservation heat map based on Clustal Omega [65]. Residues can be found in Table 2. (**c**) The "interface-lock" between D568 from the "interface-loop" and R637 from the PC1 subdomain. (**d**) Top-down view of the transmembrane domain. (**e**) Outside-in view of the periplasmic domain. Dashed lines indicate located at the back (for PN1 and PN2). Coordinates from high-resolution AcrB-Ec (using DARPin inhibitors, PDB accession code 4DX5 [54]). Colors: (**a**) Green, variable regions; purple, conserved regions. (**b–e**) Red, fully conserved among all 135 pumps; light red, conserved among the 19 selected pumps while also highly conserved in all 135 pumps; orange, fully conserved in 19 pumps; yellow, highly conserved in all 135 pumps; blue, conserved hydrophobic clusters.

Figure 3a,b show the conservation heat maps for homotrimeric RND multidrug efflux pumps, based on the 135 pump sequences. Table 1 lists the conservation in specific subdomains (based on the AcrB-Ec sequence), Table 2 lists the conserved residues (also based on AcrB-Ec, Figure 3b), and Table 3 lists other highly conserved residues found by ConSurf (Figure 3a). More heat map images can be seen in Supplementary Figures S1 and S2 and ConSurf heat map images in Figure S3. From these overviews, it is visible that the primary conservation is found in the transmembrane domain (TM domain), mainly in two TM helices: TM4 (with the D407 and D408 residues) and TM11 (which contains R971), with complete or high conservation of 63.6% and 39.4%, respectively (Table 1). These residues in these helices (D407, D408 and R971) play crucial roles in the proton transfer and, therefore, energy consumption by the pumps [33,69–71].

Table 1. Conservation and similarity between 135 MDR RND-type efflux pumps.

Domain/Subdomain/Loop		Sequence (Based on AcrB-Ec)	Residue Count	Conserved or Highly (%)
Transmembrane Domain	TM1	FIDRPIFAWVIAIIIMLAGGLAILKL	26	6/26 (23.1%)
	TM2	TPFVKISIHEVVKTLVEAIILVFLVMYLFL	30	8/30 (26.7%)
	TM3	RATLIPTIAVPVVLLGTFAVLAA	23	7/23 (30.4%)
	TM4	NTLTMFGMVLAIGLLVDDAIVVVENVERVMAEE	33	21/33 (63.6%)
	TM5	PKEATRKSMGQIQGALVGIAMVLSAVFVPMAFF	33	9/33 (27.3%)
	TM6	GAIYRQFSITIVSAMALSVLVALILTPALCATML	34	11/34 (32.4%)
	TM7	GRYLVLYLIIVVGMAYLFVRL	21	0/21 (0%)
	TM8	QAPSLYAISLIVVFLCLAALY	21	4/21 (19.0%)
	TM9	IPFSVMLVVPLGVIGALLAATFR	23	3/23 (13.0%)
	TM10	DVYFQVGLLTTIGLSAKNAILIVEFAKDLMDK	32	10/32 (31.3%)
	TM11	GLIEATLDAVRMRLRPILMTSLAFILGVMPLVI	33	13/33 (39.4%)
	TM12	SGAQNAVGTGVMGGMVTATVLAIFFVPVFFVVVRRR	36	5/36 (13.9%)
	Iα	GFFGWFNRMFEKSTHHYTDSVGGILRS	27	1/27 (3.7%)
Periplasmic Domain	PN1	PPAVTISASYPGADAKTVQDTVTQVIEQNMNGIDNLMYMSSNSDSTGTVQI TLTFESGTDADIAQVQVQNKLQLAMPLLPQEVQQQGVSVEK/PRLERYNG	100	7/100 (7%)
	PN2	VVGVINTDGTMTQEDISDYVAANMKDAISRTSGVGDVQLFGSQ/GENYDII AEFNGQPASGLGIKLATGANALDTAAAIRAELAKMEPFFPSGLKIVYPYDT	101	4/101 (4.0%)
	PC1	GVFMTMVQLPAGATQERTQKVLNEVTHYYLTKEKNNVESVFAVNGFGFAG RGQNTGIAFVSLKDWADRPGEENKVEAITMRATRAFSQIKDAMVFAF	97	2/97 (2.1%)
	PC2	GFDFELIDQAGLGHEKLTQARNQLLAEAAKHPDMLTSVRPNGLED/SPRLE RYNGLPSMEILGQAAPGKSTGEAMELMEQLASKLPTGVGYDWTGMSYQ	98	5/98 (5.1%)
	DN	YAMRIWMNPNELNKFQLTPVDVITAIKAQNAQVAAGQLGGTPPVKGQQLN ASIIAQTRLTSTEEFGKILLKVNQDGSRVLLRDVAKIELG	90	8/90 (8.9%)
	DC	TPQFKIDIDQEKAQALGVSINDINTTLGAAWGGSYVNDFIDRGRVKKVYVMS EAKYRMLPDDIGDWYVRAADGQMVPFSAFSSSRWEYG	89	3/89 (3.4%)
Loops	Switch-loop	GFGFAGR	7	1/7 (14.3%)
	Distal-loop	EKSSSSFLM	9	0/9 (0%)
	F + Bottom-loop	PAIVELGTAT	10	0/10 (0%)
	Hoisting-loop	ERLSGNQ	7	0/7 (0%)
	Interface-loop	PSSFLPDEDQG	11	3/11 (27.3%)

Bold underlined, fully conserved between 135 pumps; bold, conserved in 19 selected pumps and highly conserved in all 135. Abbreviations: TM, transmembrane helix; F, flexible.

Table 2. Conserved residues based on 135 RND-type efflux pumps.

Domain		135 Conserved	19 Conserved + 135 Highly	19 Conserved	135 Highly Conserved
	Transmembrane	P9, G23, P373, N391 , L400 , V406 , D407 , I410 , E414 , V452, P455, F470, S481, P490, L888, I943, L944, R971 , R973 , M977 , T978 , P988 , G1010, P1023	A347, V351, I367, S389, D408 * , A409 , V412 , A430, L449, L937, K940 **, A963 , A981	F5, Y327 ***, E346, P368, G378, I402 , G403 , N415 , R418 , P427, G461, G464, L497, Y527, P898, P906, G936, E947, A949, P974 , G985 , A995, G1004, T1015, P1023	R8, A12, L30, T330, S336, I337, V354, L359, V374, L376, L383, L393 , M395 , V399 , L404 , L405 , V411 , V413 , M420 , I438, L442, V443, A451, Y467, I474, A477, A485, L486, L488, F885, L886, Y892, V901, V925, V929, I945, L972 , I975 , S979 , L989 , V1007
Periplasmic	Porter	P36, A52, G171, N298, P318, D568, R637	-	P119, P565, G619, G679, S836, A840	I45, V61, I65, S80, V127, M162, E567, S824, Q865
	Head	G217	-	G740, G796	M184, W187 ****, V203, I207, L251, L262, V265, V771

Analysis after Clustal Omega alignment, based on the categories described in this article (red, light red, orange and yellow). Numbering in accordance with AcrB-Ec. * Exception: D→N in *Marinobacter hydrocarbonoclasticus* and *Alkalilimnicola ehrlichii*; ** exception: K→R in *Idiomarina loihiensis*, *Cellvibrio japonicus* and *Teredinibacter turnerae*; *** exception: 25/135 pumps (mostly Y→F), including Y→I in LpeB-Lp from *L. pneumophila* (the only exception among the 19 selected pumps); **** exception: W→F in "MexI/W" and W→T in LpeB-Lp from *L. pneumophila*. The heat maps of these residues can be seen in Figure 3b–e, Figures S1 and S2 (red to yellow). Underlined, mentioned explicitly in the article text. A green background highlights the conserved residues in TM4, and the blue background highlights the conserved residues in TM11.

Table 3. Additional highly conserved residues.

>90%	80 ≤ 90%	70 ≤ 80%	50 ≤ 80%
G51(NS), Y77(FST), G86(NS), L118(MP), P119(G), G179(AS), R185(N), N211(RS) P223(G), F246(LVY), D264(HNQS), A266(G), V340(GIT), T343(ASV), P358(CL), R363(KN), P368(ATN), G387(DN), E602(Q), Q774(IMRS), R780(DL), A890(GISTV), E893(GN), F948(VY), G994(DS)	I6(TV), F94(AILM), S144(ADFN), V172(AIMT), A279(GST), T394(NS), A401(CSV), V416(ACIM), A457(GSV), T473(ASV), N820(ADLMQS), A889(CSV), V905(AIL), G911(AS), N941(HST), V946(ACFIT), G996(AS)	N109(ADS), V122(AIST), D156(EHNS), R168(AG KQST), Q210(EHNSY), A299(ELPST), L350(ACIV), Q360(AGHR), F396(ALV), M435(ACLSTV), V448(AIT), F453(LY), L492(IMQ), Q469(ES), Q928(DEKLMV), T933(ALMV), A942(G), F982(LMT)	A16(NST), M355(ITV), T365(AIMSV), V372(AI), T431(ASV), S434(AGST), I445(MSTY), I446(FLM(V)), S471(AT(CG)), S375(ACSV(T)), M478(IMTV(A)), T489(S(K)), V884(AI), L891(Q(M)), D924(N(S)), R1000(Q(KL))

Residues were chosen from the relatively highest conserved category after analysis by ConSurf [67,68], and conservation ranges between 50% and 100% (Table S1), excluding the residues mentioned already in Table 2. Residues in brackets (AA) indicate alternative residues among the 135 pumps. Double brackets at "50 ≤ 80%" ((AA)) indicate < 1% occurrence. Numbering and amino acid labeling as in AcrB-Ec. The conservation heat map based on these residues can be seen in Figure 3a and Figure S3 (dark purple). Underlined, mentioned explicitly in this article.

In TM4 (Table 2, green background), which is composed of 33 amino acids, six residues are completely conserved (using AcrB-Ec numbering: N391, L400, V406, D407, I410 and E414), three are conserved in the selected 19 pumps and highly conserved in all 135 (D408, A409 and V412), four are additionally conserved in the selected pumps (I402, G403, N415 and R418) and eight are highly conserved among all pumps. As for TM11 (Table 2, blue background), five residues are fully conserved (R971, R973, M977, T978 and P988), two are conserved in the selected pumps and highly conserved in all (A963 and A981), two additional residues are conserved in the selected pumps (P974 and G985) and four are highly conserved in all pumps. As seen in Table 2 and as mentioned before, D408 (TM4) and K940 (TM10)—which form salt bridges with D407, and provide the energy transduction in the pumps [12]—are not conserved in sequences from two to three organisms (Table 2 and

Table S1). However, we know from experimental data that these residues are crucial to the function of the pump (shown with asterisks (*) in Table 2, more at Discussion (Section 6)). TM7 and Iα (both almost entirely green in Figure 3a) are significantly variable (merely 3.7% and 0% conserved, respectively, Table 1). A multitude of other conserved residues in the TM domain form hydrophobic patches and clusters where these residues come together, especially between TM4, TM5 and TM6 (highlighted in blue in Figure 3d and Figure S1). TM numbering and locations can be seen in Figure 3d and Figure S1.

3.2. Conservation in the Periplasmic Domain

From Figure 3 and Tables 1–3, we can conclude that the TM domain is significantly conserved, while the periplasmic domain (with the porter and head subdomains) is significantly variable (Figure 3e, Figures S2 and S3). As specific included transporters (such as AcrD) have a significantly different drug recognition spectrum, this was partly expected. However, efflux pumps with similar drug spectra also do not show stringent conservation, which we will discuss further in the next section (Section 4). Conserved or highly conserved residues which do exist in the periplasmic domain are not located in the binding pockets but, alternatively, probably provide flexibility to the loops (e.g., P36, P119, G171, P318, P565 (interface-loop), G619 (switch-loop)) and structure (Figure 3e and Table 1; Table 2). Interestingly, it is in the porter domain where the only two interacting fully conserved charged residues in the entire protomer of all 135 pumps are located: D568 and R637 (on the PC1 subdomain), shown in Figure 3b,c,e. D568 is located on a semi-conserved loop (Table 1) we here named the "interface-loop", as it lies in between the TM domain and the porter domain, and we therefore assigned the name "interface-lock" to the residues' interaction. This loop is also the most conserved among all loops (Table 1 and Table S1 and Figure 3). Future experimental data are needed to explain the function of these conserved interacting residues. Another fully conserved residue is N298, for which it has been shown that mutagenesis to a bulky tryptophan (N298W) inhibits the AcrB-Ec efflux ability significantly for all tested compounds [72]. N298 has also been found to be implicated in the binding of carboxylated β-lactams, fusidic acid and DDM in recent crystal structures and biochemical analysis of AcrB-Ec [73]. This same study found that the N298A mutation significantly negatively affected the carboxylated β-lactam MICs (also seen in binding differences in drug-bound crystal structures), but not the erythromycin MICs. On the other hand, our recent mutation N298W significantly negatively affected the MICs of all tested compounds [72]. We discuss the carboxylated β-lactams' conserved drug recognition further in Section 4.6. The three residues (N298, D568 and R637) are even conserved in phylogenetically distanced [15] TriC (data not shown). Two other highly conserved hydrophobic cores (named here the "DN-cluster" and "PN1-cluster" in Figure 3e, highlighted in blue) are present in the porter domain, consisting of hydrophobic residues probably stabilizing the subdomains by holding together the β-sheets (Figure 3e and Figure S2, blue). The PN1-cluster in AcrB-Ec comprises I45, V61, I65 and V127, and the DN-cluster comprises M184, V203, I207, L251, L262, V265 and V771 (Table 2).

An interesting highly conserved tryptophan (W187), partly facing the periplasm, is also observed (Figure 3e and Figure S2, Table 2). According to a cryo-EM structure of AcrAB–TolC, this residue lies between two AcrA-Ec monomers and seems not to be in the direct vicinity of AcrA-Ec (PDB accession code 5O66 [48]). Being close to the arm of the adjacent monomer (pinning through the head domain of the following monomer), it is likely important for the stabilization of the trimer complex, possibly interacting with the neighboring P223 (Table 3, Figure S2) from the other protomer. This P223 residue is also conserved among all pumps, except for the MexI/W-clustered pumps (including LpeB-Lp and AcrB-Hi), where the arm seems to be around six amino acids shorter than the other pumps (data not shown), and, interestingly, MexY-Pa. W187 is present in all analyzed transporters, except for two pumps from *Legionella pneumophila*, being Phe ("MexI/W") or Thr (LpeB-Lp), see Table 2. These two pumps also do not have the P223 residue on their truncated MexI/W-like arms. More images of the heat map of the periplasmic domain can be found in Figures S1–S3. A recent phylogenetic study found that—similar to our recent study on ancient AcrB-Hi (which is close to, or belongs to, the MexI/W cluster) [15]—MexI and MexW form a phylogenetic cluster, in between the Mdt-like cluster (including MdtB and MuxB) and the Acr cluster (including AcrB and MexB) [62]. It would be interesting for future research to study this distinct cluster's members in more detail (see Discussion).

3.3. Partly Conserved Residues in the Binding Pockets

For this review's conservation heat map (Figure 3b–e and Tables 1 and 2), there is a thin line between a residue being classified as fully conserved (red), highly conserved (light red/orange/yellow) or even not listed at all, e.g., when there are only one or two exceptions among the 135 sequences. There may be a few more functionally important relatively conserved residues, which is why we also analyzed the pumps by ConSurf, of which the most conserved residues (corresponding to the darkest purple category in Figure 3a and Figure S3) are listed in Table 3. Other residues of interest can be found in Table S1 (listing all aligned residues, including their conservation percentages and alternative substitutions). Despite no residues in the binding pockets being fully conserved according to the multiple sequence alignment, a few residues are partly conserved among most efflux pumps within the drug-binding regions, we which will discuss further below and in Sections 4 and 6.

Table 4 lists the loop sequences of the 19 selected pumps, and Tables 5 and 6 compare their DBP and PBP residues (including conservation among 135 pumps), based on the AcrB-Ec amino acids and their numbering. Fully conserved residues, as compared to AcrB-Ec, are highlighted with a green background. Two of the most conserved residues in Table 5 (DBP) are AcrB-Ec's F178 and F628 (~70% and ~74% conserved in 135 pumps), located in the hydrophobic pit. F178 is sometimes replaced with Trp (MexY-Pa, MexQ-Pa, LpeB-Lp; red background), Tyr (in AcrD) or Leu (MexW-Pa). F628 is different only in the MexI/W-like proteins, namely MexI/W-Pa, LpeB-Lp and AcrB-Hi, as Gly, Val or Ile, respectively. Another clearly visible relatively conserved residue in the DBP is Y327 (~81% conserved among 135 pumps), which in all 135 pumps is replaced occasionally by Phe (in ~12% of the pumps), by Ile in LpeB-Lp and very rarely by charged residues (see Table S1). This residue has been shown to be part of a postulated entrance channel between the PC2 and PN1 groove, specific for carbonated β-lactams [73], where compounds from the TM1/TM2 interface (fusidic acid, cloxacillin, piperacillin and other carboxylated drugs) translocate via TM2 to this entrance channel [74]. Q176 in the DBP is also party conserved in a selection of pumps (~59% conserved among 136 pumps, including AcrB-Ec, AcrF-Ec, MexB-Pa, MexD-Pa, AdeB-Ab, AcrB-Hi and more, see Table 5). This "Q176D" residue in MexY-Pa contributes to the high number of negatively charged residues in the DBP, possibly explaining aminoglycoside recognition (Section 4.2). A list provided below shows a few occasionally conserved residues for eight well-studied pumps. From this list and Tables 5 and 6, it is visible that compared to AcrB-Ec, MexD-Pa and AcrB-Hi have the least conserved residues of the eight pumps (16 and 17 in both pockets, respectively), and MexB-Pa has the most (31 residues in both pockets). In the PBP, highly conserved residues are L674 and G675 (~72% and 88% conserved, respectively), conserved for most clades of transporters, including AcrD-Ec, while not strictly for the MexI/W-like pumps (MexI/W-Pa, LpeB-Lp and AcrB-Hi, see Table 6). These residues are located at the lower cleft entrance of Channel 2 of the PBP, on the partly conserved flexible-loop (or F-loop, e.g., AcrB-Ec P̲AIVE**LG**T, AcrD-Ec P̲AIS**GLG**S and AdeB-Ab P̲AIDE**LG**T, bold underlined, Table 4). This loop's initial proline (P669) (underlined) also seems to be conserved (~93% conserved among 135 pumps). Mutating the L674 residue to Trp in AcrB-Ec showed decreased drug efflux [51]. On this flexible-loop, another residue (I671 in AcrB-Ec) is partly conserved as a Leu, Ile or Val (~58% conserved among 135 pumps, Table 6), which has been shown to be implicated in drug selectivity of small substrates [75]. A comparison between all the loops (flexible-loop, switch-loop, distal-loop and interface-loop) is provided in Table 4. The switch-loop between the PBP and the DBP consists of seven amino acids, of which one

Gly is fully conserved among 19 pumps (G619 in AcrB-Ec) and ~72% conserved among 135 pumps. Mutating this Gly to Pro in AcrB-Ec inhibits the pump, showing the importance of this loop's flexibility [51]. Despite the observation that the switch-loop is located near erythromycin in the PBP of the access monomer [51], the loop seems unnecessary in the binding of erythromycin, as erythromycin was found in the same location in both Ala-substituted loop mutants and loop-deleted mutants of AcrB-Ec [76]. S824 is also conserved in most of the 19 pumps (Tables 2 and 6, and as seen in the list below), although this residue is far from drug-binding residues in the crystal structures quite deep into the PC2 subdomain, and it is not clear if this residue is important for drug recognition. Interestingly, substitutions of this residue (S824I in LpeB-Lp and S821A in MtrD-Ng) are found in drug-resistant strains (see Section 5). The three highest conserved residues in the PBP are G675 (~88% conserved among 135 pumps), T91 (~70% conserved) and Q569 (~69% conserved). As most residues within the pockets of the pumps are relatively variable while, simultaneously, the TM domain is highly conserved, it underpins just how versatile the adaptations of these pumps may be in regard to substrate recognition, and probably in substrate recognition optimization based on specific physiological functions these pumps play within the natural environments of each bacterial species. Later in this review, we will discuss the differences in substrate recognition and binding pocket residues and other properties to help explain the differences between the pumps in the porter domain.

Table 4. Loop sequences of 19 selected transporters.

Transporters	Flexible-Loop	Distal-Loop	Interface-Loop	Switch-Loop
AcrB-Ec	PAIVELGT	EKSSSSFLM	PSSFL__PDED__QG	GFGFA__GR__
AcrB-Sa	PAIVELGT	EKSSSSFLM	PSSFL__PDED__QG	GFGFA__GR__
AcrF-Sa	PAIVELGT	EKSSSSFLM	PSSFL__PDED__QG	GFSFS__GQ__
AcrF-Ec	PAIVELGT	EKSSSSYLM	PSSFL__PEED__QG	GFSFS__GQ__
MexB-Pa	PSVLELGN	TKAVKNFLM	PEAFV__PAED__LG	GFNFA__GR__
AcrD-Ec	PAISGLGS	RKTGDTNIL	PTSFL__PLED__RG	GSGPG__GN__
AcrD-Sa	PAISGLGS	RKTGDTNIL	PTSFL__PQED__RG	GSGPG__GN__
MexY-Pa	PPLPDLGS	EKAADSIQL	PQAFL__PEED__QG	GFSLY__GD__
MexD-Pa	PPINGLGN	EQTSAGFLL	PEAFV__PAED__LG	GFSFS__GQ__
AdeB-Ab	PAIDELGT	EASSSGFLM	PTAFM__PEED__QG	GWGFS__GA__
AdeJ-Ab	PAMPELGV	TKSGASFLQ	PSSFL__PEED__QG	GFSFT__GV__
MtrD-Ng	PPILELGN	SKARSNFLM	PTSFL__PTED__QG	GFSFS__GS__
MexQ-Pa	PPVPGLGT	QKTSPDILM	PPGFV__PMQD__KY	GLSVN__GF__
MexF-Pa	PPVPGLGT	DKASPDLTM	PTGFV__PQQD__KQ	GLSIN__GF__
AdeG-Ab	PPVMGLGT	LKSSPTLTM	PGGFV__PAQD__KQ	GLSIN__GF__
MexI-Pa	AALPGST–	SSGETTAVA	KRELA__PTED__QA	TWIIN__GT__
LpeB-Lp	PGVDDAG–	QRK–SNGLP	SHETA__PKED__RG	RLTFI__GD__
AcrB-Hi	PEIDTGE–	SSG–GSGIM	SSELT__PNED__KG	GMSIA__GA__
MexW-Pa	PSLPGTG–	EAADASALM	KKELA__PEED__QG	AFQIN__GY__

Bold underlined, fully conserved. A bar (–) indicates a gap in the sequence alignment.

Table 5. Overview of residues of interest in the DBP area of 19 selected RND-type MDR transporters, and conservation among 135 pumps.

Pump \ Position	136	178	610	615	617	628	46	89	128	130	134	135	139	151	176	177	180	273	274	276	277	279	288	290	292	327	571	573	612	620
AcrB-Ec	F	F	F	F	F	F	S	Q	S	E	S	S	V	Q	Q	L	S	E	N	D	I	A	G **	G	K	Y	V	M	V	R
Conservation (%)	49	70	50	60	52	74	38	38	18	27	40	32	50	22	59	44	34	55	31	29	10	40	49	43	33	81	29	24	44	19
AcrB-Sa	F	F	F	F	F	F	S	Q	S	E	S	S	V	Q	Q	L	S	E	N	D	V	A	G ***	G	K	Y	V	M	V	R
AcrF-Sa	F	F	F	F	F	F	S	T	S	E	S	S	V	Q	Q	L	A	E	N	N	V	A	G	G	K	Y	V	L	V	Q
AcrF-Ec	Y	F	F	F	F	F	S	T	S	E	S	S	V	Q	Q	L	A	E	N	N	V	A	G	G	K	Y	V	L	V	Q
MexB-Pa	F	F	F	F	F	F	Q	T	R	T	K	N	V	K	Q	V	S	Q	D	S	I	A	G	A	K	Y	V	F	V	R
AcrD-Ec	N	Y	F	S	P	F	T	S	T	R	D	T	T	K	D	A	S	E	K	D	Y	S	G	G	K	Y	M	T	T	N
AcrD-Sa	N	Y	F	S	P	F	T	T	T	R	D	T	T	K	D	A	S	E	K	D	Y	S	G	G	K	Y	M	T	T	N
MexY-Pa	I	W	Y	F	L	F	S	S	Y	E	D	S	I	A	E	T	A	S	E	G	F	S	G ****	A	K	Y	D	M	V	D
MexD-Pa	F	F	F	F	F	F	T	E	Q	E	A	G	I	T	Q	F	S	E	S	N	I	S	G	A	Q	Y	Y	V	I	Q
AdeB-Ab	F	F	T	W	F	F	S	E	Q	E	S	G	L	E	Q	S	A	Q	A	N	F	I	A	A	Q	Y	W	M	I	A
AdeJ-Ab	F *	F	F	F	F	F	A	S	T	T	A	S	V	D	Q	V	G	D	N	Q	F	S	G *	A	K	Y	V	M	V	V
MtrD-Ng	F	F	I	F	F	F	H	S	T	S	S	N	I	T	R	L	A	E	D	S	S	T	G	A	M	Y	F	M	V	S
MexQ-Pa	I	W	V	L	V	F	T	I	V	Q	P	D	V	P	V	V	A	D	A	A	L	S	A	Q	I	Y	Y	V	F	F
MexF-Pa	L	F	V	L	I	F	R	T	T	D	P	D	V	M	Q	L	L	N	Q	A	L	S	A	P	F	Y	Y	V	F	F
AdeG-Ab	L	F	V	L	I	F	R	T	T	L	P	T	V	M	G	L	S	S	Q	G	L	S	A	P	F	Y	Y	I	F	F
MexI-Pa	A	F	–	W	I	G	T	V	E	S	T	T	Y	I	Q	T	G	A	A	E	T	A	H	G	F	Y	A	L	–	T
LpeB-Lp	G	W	–	L	F	V	S	Q	E	Q	S	N	F	F	E	V	–	D	N	Q	M	V	V	S	N	I	L	G	–	D
AcrB-Hi	G	F	–	M	I	I	S	T	S	S	G	S	Y	S	Q	V	A	E	N	N	S	A	V	A	N	Y	A	I	–	A
MexW-Pa	A	L	–	F	I	G	T	T	Q	E	A	S	Y	N	E	I	N	A	S	D	A	S	Y	G	K	Y	I	F	–	Y

Asterisks: * (F→L) and (G→S) found in AdeJ from *Acinetobacter baumannii*—in experimentally evolved and clinical strains from Australia implicated in meropenem resistance [77,78]; ** (G→C) found in experimentally evolved AcrB from *Escherichia coli*—increases efflux, especially for erythromycin [79]; *** (G→D) found in AcrB from *Salmonella* Typhimurium—in experimentally evolved and clinical strains from the UK, implicated in ciprofloxacin resistance [80,81]; **** (G→A/S) found in MexY from *Pseudomonas aeruginosa*—in clinical strains mainly from the EU and Australia, implicated in tobramycin resistance [82–84]. Colors: green background, conserved residues compared to AcrB-Ec; red background, bulky Trp (potentially) inhibiting inhibitor (EPI) effectiveness. The first six separated columns show the Phe residues (as in AcrB-Ec) in the "hydrophobic pit". Note: *Salmonella* AcrB, AcrD and AcrF alignments based on *Salmonella* Typhi str. CT18 genes STY0519 and STY2719 and *Salmonella* Typhi str. LT2 gene STM3391, respectively. LpeB-Lp alignment based on *Legionella pneumophila* str. Paris gene lpp2880. Conservation based on 135 pumps, and further details (regarding alternative residues and the conservation percentages) can be found in Table S1.

Table 6. Overview of residues in the PBP area of 19 selected RND-type MDR transporters, and conservation among 135 pumps.

Position / Pump	79	91	569	575	577	624	626	662	664	666	667	668	671	673	674	675	676	681	717	719	824	826	828
AcrB-Ec	S	T	Q	M	Q	T	I	M	F	F	N	L	I	E	L	G	T	D	R ***	N	S	E	L
Conservation (%)	55	70	69	26	61	29	25	16	49	56	31	23	58	32	72	88	37	36	49	43	52	43	
AcrB-Sa	S	T	Q	M	Q	T	I	M	F	F	N	L	I	E	L	G	T	D	R	N	S	E	L
AcrF-Sa	S	T	Q	M	Q	S	M	L	F	F	N	M	I	E	L	G	T	D	R	N	S	E	L
AcrF-Ec	S	T	Q	M	Q	A	M	F	I	F	N	M	I	E	L	G	T	D	R	N	S	E	Q
MexB-Pa	S	T	Q	Q	Q	S	M **	M	F	F	A	P	V	E	L	G	N	D	R	N	A	E	L
AcrD-Ec	S	T	R	S	Q	V	R	R	I	S	S	P	I	G	L	G	S	D	R	N	A	E	V
AcrD-Sa	S	T	R	S	Q	V	R	R	F	S	S	P	I	G	L	G	S	D	R	N	A	E	V
MexY-Pa	K *	T	Q	M	M	S	M	T	Y	M	N	S	L	D	L	G	S	D	M	A	S	N	E
MexD-Pa	E	V	L	D	Q	A	L	T	M	V	S	P	I	G	L	G	N	A	M	E	S	R	V
AdeB-Ab	S	T	Q	S	Q	V	V	E	M	V	L	P	I	E	L	G	T	S	W	E	S	S	A
AdeJ-Ab	S	Q	Q	L	Q	A	I	Y	M	L	Q	L	M	E	L	G	V	N	R	E	S	N	Q
MtrD-Ng	S	S	Q	S	Q	M	M	F	I	V	V	P	I	E	L	G	N	S	R ****	G	S	K ****	S
MexQ-Pa	S	T	K	I	Q	A	V	F	G	F	P	P	V	G	L	G	T	K	M	S	S	D	S
MexF-Pa	S	T	K	F	Q	S	I	Y	A	F	P	P	V	G	L	G	T	R	F	S	T	E	N
AdeG-Ab	Q	T	K	F	Q	A	I	Y	A	F	P	P	V	G	L	G	T	K	F	S	S	E	N
MexI-Pa	S	T	Q	A	K	A	F	S	F	F	Q	L	L	G	S	–	–	P	D	D	A	T	Q
LpeB-Lp	T	T	R	Y	P	–	S	W	W	T	G	L	V	D	A	G	–	E	N	D	S *****	T	H
AcrB-Hi	S	T	K	I	N	S	L	S	S	F	N	I	I	T	G	–	–	P	N	D	S	E	S
MexW-Pa	T	S	Q	M	N	A	I	Q	F	F	N	L	L	G	T	G	–	P	D	D	S	I	S

Asterisks: * (K→T) found in MexY from *Pseudomonas aeruginosa*—experimentally evolved strain, implicated in tobramycin resistance [83]. Additionally, K79A in MexY has increased aminoglycoside (paromomycin) resistance [56]; ** (M→V) found in MexB from *Pseudomonas aeruginosa*—clinical strains from Denmark [5]; *** (R→L/Q) found in AcrB from *Salmonella* Typhi and *Salmonella* Paratyphi A—clinical isolates from Bangladesh, Nepal, India and Pakistan, known to cause azithromycin resistance [85–88]; **** (R→C/G/H/L) and (K→D/E/N) found in MtrD from *Neisseria gonorrhoeae*—clinical isolates from India, the USA and the EU, known to cause azithromycin resistance [24,43,89]; ***** (S→I) found in LpeB from *Legionella pneumophila*—spring water isolates from China [90]. Colors: green background, conserved residues compared to AcrB-Ec. A bar (–) indicates a gap after sequence alignment for the specific AcrB-Ec position. Note: *Salmonella* AcrB, AcrD and AcrF alignments based on *Salmonella* Typhi str. CT18 genes STY0519 and STY2719 and *Salmonella* Typhi str. LT2 gene STM3391, respectively. LpeB-Lp alignment based on *Legionella pneumophila* str. Paris gene lpp2880. Conservation based on 135 pumps, and further details (regarding alternative residues and the conservation percentages) can be found in Table S1.

We here list the conserved residues in the DBP compared to AcrB-Ec, among seven additional characterized efflux pumps, in order to provide a quick overview. Bold underlined text highlights conserved in all eight pumps (including AcrB-Ec), and italic underlined text highlights conserved in six out of eight pumps:

- AcrD-Ec   F610, *F628*, S180, E273, D276, *G288*, G290, K292 and **Y327**
- MexY-Pa   F615, *F628*, S46, E130, S135, *G288*, K292, **Y327**, M573 and V612
- MexB-Pa   *Phe-pit*, V139, Q176, S180, I277, A279, *G288*, K292, **Y327**, V571, V612 and R620
- MexD-Pa   *Phe-pit*, E130, Q176, S180, E273, I277, *G288* and **Y327**
- AdeB-Ab   F136, *F178*, F617, F628, S46, E130, S134, Q176, **Y327** and M573
- MtrD-Ng   *Phe-pit* (except F610), S134, L177, E273, *G288*, **Y327**, M573 and V612
- AcrB-Hi   *F178*, S46, S128, S135, Q176, E273, N274, A279 and **Y327**

Additionally, the same is conducted for the residues in the PBP. Bold text means present in six out of eight pumps, and italic underlined text highlights conserved in five out of eight pumps:

- AcrD-Ec   **S79**, *T91*, *Q577*, **I671**, **L674**, **G675**, D681, R717, N719 and E826
- MexY-Pa   T91, Q569, M575, N667, **L674**, **G675**, D681 and **S824**
- MexB-Pa   **S79**, *T91*, Q569, *Q577*, M662, F664, F666, E673, *L694*, G676, D681, R717, N719, E826 and L828
- MexD-Pa   Q577, **I671**, **L674**, **G675** and **S824**
- AdeB-Ab   **S79**, *T91*, Q569, *Q577*, **I671**, E673, **L674**, **G675**, T676 and **S824**
- MtrD-Ng   **S79**, Q569, *Q577*, **I671**, E673, **L674**, **G675**, R717 and **S824**
- AcrB-Hi   **S79**, *T91*, F666, N667, **I671**, **S824** and E826

4. Binding Pocket Differences Help Understand Drug Recognition Spectra

There are several clades of distinct efflux pumps among the homotrimeric RND multidrug efflux pumps with distinct or divergent efflux properties. As mentioned before, we previously analyzed about 400 efflux pump genes and found clades for several groups of pumps, which could be clustered into AcrB/AcrF, AcrD, MexB, MexD/MexY, AdeB, MexF/MexQ and MexI/MexW [15]. Among these pumps, drug recognition can slightly or significantly differ [91]. However, interestingly, phylogenetically distant and ancient AcrB from *H. influenzae* (AcrB-Hi) can export the same compounds as AcrB-Ec (including macrolides, β-lactams and dyes), but it exports bile salts significantly less efficiently [15]. Additionally, ABI-PP could not inhibit AcrB-Hi [15], while it inhibited AcrB-Ec completely [15,36]. Other classes of drugs may also be exported by one pump, but not by another. These include aminoglycosides and monobactams [92]. To further investigate the differences between several transporters, we compared the aforementioned 19 pumps by looking at their amino acids, the hydrophobicity of the pockets and the number of charged and hydrophobic residues, in order to help understand different drug specificities.

4.1. Differences in the Hydrophobic Trap of the Distal Binding Pocket

Table 5 showed the aligned residues within the DBP. These 30 residues in the DBP were selected for comparison based on drug-bound structures and MD simulations, namely: minocycline- and doxorubicin-bound AcrB-Ec [52], erythromycin- and rifampicin-bound AcrB-Ec [51], doxorubicin- and minocycline-bound AcrB-Ec [54], ABI-PP-bound AcrB-Ec and MexB-Pa [36], molecular simulations of multiple drugs to AcrB-Ec [55], a mutation study in AcrB-Ec [75] and ampicillin- and erythromycin-bound MtrD-Ng [43]. The DBP includes the hydrophobic pit, or inhibitor binding pit, which is a phenylalanine-rich pit in, e.g., AcrB-Ec [36,58] and MexB-Pa [36]. For most of the 19 selected pumps (excluding AcrD), these pit residues are hydrophobic residues, except for a Thr in AdeB-Ab; however, this residue has a hydrophobicity between Tyr and Trp, according to the hydrophobicity scale used in this review, based on transmembrane helix insertion [93]. The inhibitor binding pits in MexI-Pa, MexW-Pa, AcrB-Hi and LpeB-Lp ("MexI/W cluster") are significantly different when compared to pumps such as AcrB-Ec, MexB-Pa and MexY-Pa ("Acr cluster"). LpeB-Lp is the least conserved compared to AcrB-Ec when looking at the residues in the DBP and PBP (9 out of the 53 residues, Tables 5 and 6). Interestingly, the differences in AcrB-Hi (16 out of 53 residues conserved) do not contribute to a fundamentally altered drug efflux spectrum; we showed that AcrB-Hi has a similar spectrum to AcrB-Ec (including macrolides, dyes and β-lactams) [15]. Drug-bound structures of AcrB-Ec, AdeB-Ab and MtrD-Ng show the different amino acids mentioned in Table 5, while all have drugs bound at the same location in the DBP, where different subsets of amino acids interact with the drug molecules (Figure 4 and Figure S4). Figure 4a shows ABI-PP bound to AcrB-Ec, tightly bound in the narrow pit [36]. In the same location, we can see much bulkier erythromycin bound in MtrD-Ng (Figure 4b) [43], where the pit seems to be somewhat wider than for ABI-PP-bound AcrB-Ec. Figure 4c shows two ethidium molecules bound to AdeB-Ab's DBP and one in the PBP [41].

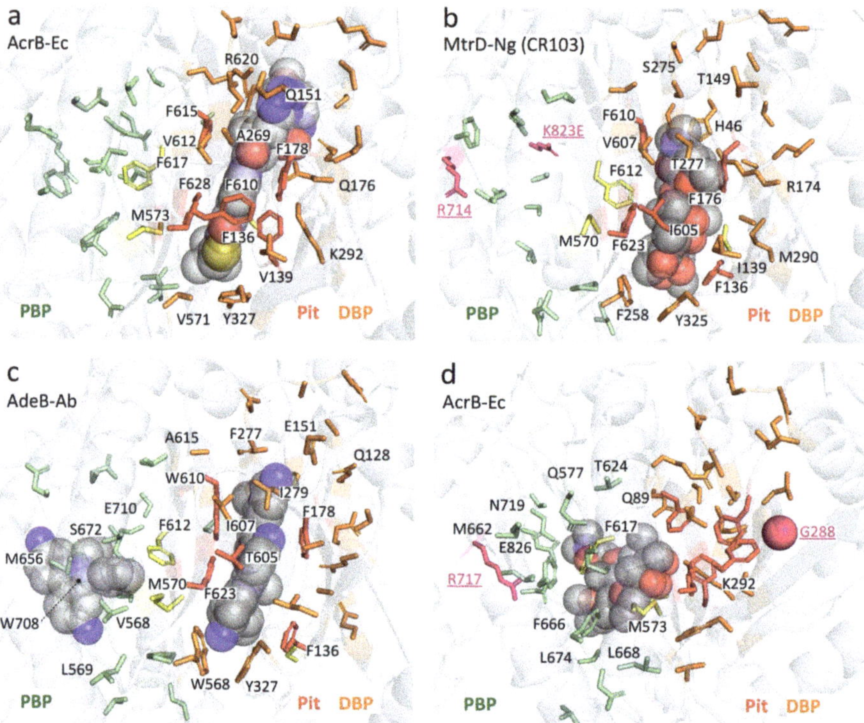

Figure 4. Drug-bound structures of AcrB-Ec, MtrD-Ng and AdeB-Ab. (**a**) ABI-PP bound in the binding monomer of AcrB-Ec (PDB accession code 3W9H [36]). (**b**) Erythromycin bound in the binding monomer of "CR103" MtrD-Ng (PDB accession code 6VKT [43]). (**c**) Ethidium bound in the binding monomer of AdeB-Ab (PDB accession code 7KGG [41]). (**d**) Erythromycin bound to the access monomer of AcrB-Ec (PDB accession code 3AOC [51]). A front view of all four structures can be found in Figure S4. Colors: green sticks show the PBP; orange sticks show the DBP; red sticks show the hydrophobic pit; pink highlights recurring substitution locations in clinical strains.

From the residues in the DBP shown in Table 5 and Figure 4, two conclusions can be drawn: 1) the hydrophobic pit (and DBP) of all transporters contains hydrophobic residues (except for AcrD, which is discussed later on), partly explaining the similar drug recognition spectra and binding structures of the different pumps, and 2) the rest of the DBP is largely not conserved, the only exception being Y327 (see the written list above). As mentioned before, Y327 has been shown to be implicated in carboxylated β-lactam recognition [73]. When looking only at residues as a recognition factor, a few residues are selectively conserved in and near the hydrophobic pit (such as F136, F178, F628 and Y327) and play a role in drug recognition efficiency, and they can also be seen in Figure 4. However, converting AcrB-Ec's Phe residues to Ala did not disrupt the substrate export completely, although the MIC values were usually significantly lower, and the most profound effect was found for F610A [94]. This mutation is believed to alter subtle properties in the pit, resulting in inefficient drug export rather than directly disrupting drug binding [95]. Interestingly, another transporter from *H. influenzae* (AcrB-Hi) only has the F178 residue conserved, and the F136 residue is a Gly, all while this transporter can export the same compounds as AcrB-Ec very efficiently when expressed in *E. coli* cells (when analyzing the protein by homology modeling, the F610 residue may even be a charged Glu) [15]. As seen for AcrB-Hi and other MexI/W proteins (Table 5), alignment causes gaps in areas of interest (binding pockets and the extruded arm too), and actual crystal or EM structures would potentially give interesting new insights into the mechanism of these pumps and actual structural differences between these transporters and other well-defined pumps (such as AcrB-Ec, MexB-Pa, MtrD-Ng and AdeB-Ab). The conserved Y327 interacts with ABI-PP in AcrB-Ec [36] (Figure 4a), erythromycin in MtrD-Ng [43] (Figure 4b) and several substrates in MD simulations [55]. A recent ethidium-bound AdeB-Ab structure shows ethidium interacting with this conserved Tyr, too (Figure 4c). Y327 is located in the lower vicinity of the hydrophobic pit and the DBP (Figure 4), and in AdeB-Ab next to a Trp (W568, which is V571 in AcrB-Ec). This is AdeB-Ab's second Trp in the DBP, together with W610 (which is F615 in AcrB-Ec) on the switch-loop, both interacting with an ethidium molecule [41] (Figure 4c, Table 5). More on Y327 in Section 4.6

4.2. Differences between Distal Binding Pockets Explain Aminoglycoside Selectivity

The biggest outliers in terms of DBP conservation are the AcrDs (AcrD-Ec and AcrD-Sa), where the pit consists of Asn, Ser, Pro, Tyr and two Phes (Table 5). Thus, this pit is significantly more hydrophilic than the pits of the other transporters [96]. Table 7 shows the hydrophilicity (based on the sum of the residues calculated from [93]). AcrD has a DBP hydrophilicity value of around 39–40 kcal mol^{-1} (Table 7, green), while the DBPs of AcrB-Ec, MexY-Pa, AcrB-Hi, MtrD-Ng and AdeB-Ab are significantly more hydrophobic (25.6, 27.0, 17.3, 16.9 and 21.7 kcal mol^{-1}, respectively, Table 7). AcrD's significantly hydrophilic pit (in combination with the many differences in the residues themselves) can explain why AcrD-Ec exports aminoglycosides, while many other drugs (e.g., erythromycin, ciprofloxacin, tetracycline and many more drugs which are substrates of AcrB-Ec) are poorly exported or not exported at all [35,97–99]. AcrD-Ec also has the ability to export, e.g., monobactams, which AcrB-Ec cannot, and this phenotype can be explained by differences in the PBP (Table 6), which will be explained in more detail in Section 4.4.

While the differences in the DBP (both residues and hydrophobicity) explain both aminoglycoside recognition and the inability to export many other drugs by AcrD-Ec, they do not explain a similar phenomenon between MexY-Pa and MexB-Pa [100]. These two pumps are phylogenetically closer to each other than AcrB-Ec and AcrD-Ec [15,100,101], and both show similar hydrophilicity in the DBP of around 26–27 kcal mol^{-1} (Table 7). MexB-Pa and MexY-Pa both have a broad substrate range (especially when compared to AcrD-Ec), including erythromycin, tetracycline, chloramphenicol and more. However, interestingly, MexY-Pa has the ability to also export aminoglycosides [56]. Table 7 shows the number of charged (K, R, D, E) and hydrophobic (I, L, F, V, C, M) residues in both the DBP (from Table 5) and PBP (from Table 6). As seen in Table 7, there is a striking difference in the number of negatively charged residues between MexB-Pa and MexY-Pa in the DBP. MexB-Pa has five positively charged residues (3xK, 2xR, Table 7, orange) and only one negatively charged residue (1xD), while MexY-Pa harbors mainly negatively charged residues (3xE, 3xD, Table 7, green), with only one positively charged residue (1xK). These differences were also observed in computer simulations, where more charged residues are accounted for [102], and a recent study comparing the two pumps in more detail [101].

This significantly negatively charged DBP could explain why MexY-Pa has the ability to export aminoglycosides besides having a broad substrate range (possible by also having a hydrophobic pit), offering a different explanation than for AcrD-Ec.

Table 7. Charged and hydrophobic residues in the binding pockets.

	Proximal Binding Pocket (PBP)					Distal Binding Pocket (DBP)				
Transporters	+	-	Sum	HP	Hydrophilicity (kcal mol^{-1})	+	-	Sum	HP	Hydrophilicity (kcal mol^{-1})
AcrD-Ec	4	2	6	5	27.07	4	4	8	3	39.56
AcrD-Sa	4	2	6	5	27.35	4	4	8	3	39.24
MexY-Pa	1	3	4	7	23.16	1	6	7	8	26.97
MexB-Pa	1	3	4	7	25.75	5	1	6	12	26.19
AcrB-Sa	1	3	4	9	20.54	2	3	5	12	25.93
LpeB-Lp	1	3	4	2	24.35	0	4	4	11	25.87
AcrB-Ec	1	3	4	9	20.54	2	3	5	12	25.64
MexW-Pa	0	2	2	7	20.14	1	3	4	6	23.23
MexD-Pa	1	3	4	9	18.59	0	3	3	11	22.80
AcrF-Ec	1	3	4	8	23.49	1	2	3	11	22.25
AdeB-Ab	0	3	3	7	18.65	0	3	3	9	21.69
AcrF-Sa	1	3	4	9	21.36	1	2	3	12	21.25
AdeJ-Ab	1	2	3	8	23.74	1	2	3	13	18.62
MexF-Pa	2	1	3	6	20.77	1	2	3	15	18.27
AcrB-Hi	1	2	3	5	21.41	0	1	1	7	17.30
MtrD-Ng	2	1	3	8	20.60	1	2	3	12	16.92
MexI-Pa	1	2	3	5	21.57	0	2	2	6	15.47
MexQ-Pa	2	1	3	7	19.95	0	2	2	15	13.59
AdeG-Ab	2	1	3	6	22.82	1	0	1	15	10.16

Colors: green, positive contribution to aminoglycoside recognition; orange, negative contribution or difference explaining aminoglycoside non-recognition; yellow, AcrB-Hi's low charged and hydrophobic residue count (and hydrophilicity, orange), possibly explaining lower substrate export efficiency compared to AcrB-Ec (including the significantly low bile salt MICs). Hydrophobicity based on [93]. Abbreviations and symbols: number of positively charged residues (+), negatively charged residues (-) and hydrophobic residues (HP). Definitions: positively charged, K and R; negatively charged, D and E; hydrophobic, I, L, F, V, C and M residues.

4.3. Bulky Tryptophan in the Inhibitor Binding Pit Prevents Inhibition

Another critical difference (for inhibitor design) between MexY-Pa and MexB-Pa is the presence of a bulky Trp in MexY-Pa, which explains why the inhibitor ABI-PP is not inhibiting this pump [36]. This bulky tryptophan (represented by F178 in AcrB-Ec) is also present in MexQ-Pa and LpeB-Lp (Table 5, red background). Fairly recent studies indicate that LpeB-Lp is an upcoming efflux pump over-expressed in a selection of clinical strains of *L. pneumophila* (the "Paris strain") causing macrolide resistance [103,104]. This fuels the urge for the development of novel universal efflux pump inhibitors which have overcome this bulky Trp hindrance. Two pumps (AdeB-Ab and MexI-Pa) have a Trp in AcrB-Ec's F615 location (located on the switch-loop, Tables 4 and 5); however, these residues possibly do not interfere much with the space of the pit because it is located on the flexible-loop and is located at the "top" of the pit, rather than deeper into the pit itself (Figure 4c, W610 in AdeB-Ab interacting with an ethidium molecule).

4.4. Specific Amino Acids in the Proximal Binding Pocket Explain β-Lactam Selectivity

Table 6 compared the residues in the PBP. As mentioned, AcrD-Ec and MexY-Pa can both export aminoglycosides (while AcrB-Ec and MexB-Pa cannot), explained by the differences in hydrophobicity or the number of negatively charged residues in the DBP. However, AcrD-Ec can also effectively export both monobactams (such as aztreonam) and anionic β-lactams (carbenicillin and sulbenicillin). AcrB-Ec cannot export aztreonam and can only weakly export carbenicillin and sulbenicillin [99]. Also, while MexY-Pa and AcrD-Ec can both export aminoglycosides, MexY-Pa is unable to export carbenicillin and sulbenicillin [105,106]. Three residues of interest are AcrB-Ec's respective Q569, I626 and E673. These residues are charged Arg (R568 and R625) and Gly (G672) in AcrD-Ec and AcrD-Sa. These three residues are implicated in monobactam (aztreonam) and anionic β-lactam (carbenicillin and sulbenicillin) selectivity in AcrD-Ec, and substitution of these residues in AcrB-Ec (Q569R/I626R/E673G) as a triple mutant adds or increases the efflux ability of AcrB-Ec for these three β-lactams [99], providing an explanation to why AcrD-Ec can export aztreonam, carbenicillin and sulbenicillin, while MexY-Pa and AcrB-Ec cannot (or only weakly) [99,100,105]. There are more differences between these efflux pumps, such as the ability of MexB-Pa to efflux imipenem, meropenem, carbenicillin and sulbenicillin, which is not recognized by MexY-Pa [101,105]. Perhaps the differences in charged and hydrophobic residues account for these specificities (Tables 5–7), as MexB-Pa does not have AcrD's Arg and Gly residues in the PBP. Additionally, it is interesting to note that despite the significant substrate specificity differences between AcrB-Ec and AcrD-Ec, both transporters can export certain β-lactams (e.g., nafcillin) and SDS very effectively [99]. Perhaps similarities (10 residues in the list above) in the PBP explain this phenomenon. However, even less conserved MexD-Pa (five conserved residues in the PBP) also has the ability to export nafcillin [105]. A short discussion regarding certain β-lactam (including nafcillin) export abilities by many phylogenetically distinct and distanced efflux pumps is provided in Section 4.6.

4.5. Adaptation through Amino Acids and Hydrophobicity Alterations May Increase Activity

Another interesting PBP difference presented in Table 6 is the presence of a third Trp in AdeB-Ab (W708), which is a charged Arg in AcrB-Ec (R717), located at the entrance of the PBP. This Trp interacted with a third ethidium molecule in a recent cryo-EM structure [41] (Figure S4c). This R717 location is also a hotspot for RND pump mutations in clinical strains (Table 6, and Figure 4b,d in pink) which we will discuss later (Section 5). AdeB-Ab seems to be unique in having two Trp in the DBP and one in the PBP. Just one other transporter in the list of 19 pumps (LpeB-Lp) holds three Trp residues, but in different locations (compared to AcrB-Ec: F178W (same as MexY-Pa, mentioned before) in the DBP, and M662W and F664W at the entrance (Channel 2) of the PBP). We recently found that ethidium efflux is enhanced by double Trp mutations (T37W/A100W) at the Channel 3 entrance in AcrB-Ec [72]. Table 7 shows differences in hydrophobicity of the DBP. AdeG-Ab has the least hydrophilic pocket (ΔG = 10.2 kcal mol^{-1}), compared to AcrB-Ec (25.6 kcal mol^{-1}), MexB-Ec (26.2 kcal mol^{-1}) and AcrD-Ec (39.6 kcal mol^{-1}). Interestingly, there is a significant difference in the number of both charged (K, R, D, E) and hydrophobic (I, L, F, V, C, M) residues between AcrB-Ec and AcrB-Hi (Table 7, yellow). AcrB-Ec has five charged residues, while AcrB-Hi only has one. Additionally, AcrB-Ec has 12 hydrophobic residues, while AcrB-Hi merely has seven. The same goes for the number of hydrophobic residues in the PBP between AcrB-Ec and AcrB-Hi (nine vs. five, respectively). At the same time, as mentioned before, the efflux spectrum of these transporters is almost the same (with the exception of bile salts) [15]. In our previous study, we determined the expression levels of AcrB-Ec and AcrB-Hi in *E. coli* cells to be similar, while AcrB-Hi could export most drugs less effectively with a several-fold lowering of the MICs of certain drugs (such as methicillin and cefcapene pivoxil), a similar MIC for other drugs (such as ethidium or cloxacillin) and, interestingly, a significantly lower ability to export bile salts (including deoxycholic acid) [15]. Perhaps, looking at the presented data in this review, AcrB-Ec (and other evolved transporters) has adapted to have both more charged and hydrophobic residues to increase drug efflux efficiency and accommodate physiologically relevant compounds. For example, AcrB-Ec, AcrD-Ec and MtrABC-Ec have been shown to be involved in enterobactin export [107]. Other differences obtained through evolution may be the Pro ("P223") on the arm (absent in the truncated arms of AcrB-Hi and MexI/W-like pumps) and the Trp residue ("W187") in the DN subdomain, possibly enhancing the stability of the trimer. The differences in hydrophobicity, number of charged residues (specifically positively and negatively charged residues), number of hydrophobic residues and the volume of the pockets can help us understand the substrate recognition differences and the differences in the efficiencies of the export of specific drugs between different pumps.

4.6. Conserved Residues May Partly Explain Conserved Drug Specificities

Comparison of substrate specificities between MexD-Pa, MexY-Pa and MexB-Pa shows that among these pumps, many classes of antibiotics are recognized and exported (including quinolones, macrolides and tetracycline), with distinct differences between them (e.g., for imipenem, carbapenem, carbenicillin, sulbenicillin, ceftazidime, meropenem and more) [105], even though these three pumps differ significantly in pocket residues (Tables 5 and 6). Between MexB-Pa, MexY-Pa and MexD-Pa, the conserved pocket residues are G290, F615, F628 and Y327 in the DBP, and L674 and G675 in the PBP.

Between six pumps, namely, AcrB-Ec, AcrD-Ec, MexB-Pa, MexD-Pa, MexY-Pa and AcrB-Hi, we compared the substrate specificities for a selection of drugs and drug classes listed in Table 8. We found that one class of antibiotic was exported by all six pumps, namely, cloxacillin, oxacillin and/or nafcillin (second-generation narrow-spectrum penicillins) β-lactams. This may also include first-generation penicillins (such as benzylpenicillin) or fourth-generation extended-spectrum β-lactams (such as piperacillin), but these were not tested for all pumps. Therefore, we could at least conclude the second generation to be widely exported (Table 8, green). The six pumps have different hydrophobic properties in the PBP and DBP (Table 7) and, within the binding pockets (DBP and PBP), only have one conserved residue among them: Y327 in the DBP. The overlap in the substrate range may be partly explained by this residue, as well as the aforementioned fully conserved N298 (outside the pockets, near the Channel 3 entrance). A recent study found that a Y327A mutation (postulated to be implicated in a novel substrate entrance Channel 4) caused a decrease in drug resistance against carboxylated β-lactams such as dicloxacillin and oxacillin, and that the N298A mutant of AcrB-Ec decreased drug binding for the specific compounds seen in the crystal structures and is reflected in the MIC data [73]. Another study found that these drugs (in addition to fusidic acid) are translocated via a TM1/TM2 groove [74]. We found one of the implicated TM residues (I337) to be highly conserved within the analyzed 135 pumps (Tables 1 and 2), and a recent study found the mutation I337A to have the largest impact among the tested mutations for the MICs of specific compounds (for fusidic acid, oxacillin, etc.) and hardly for erythromycin [74]. As mentioned above, it would be interesting to determine the structure of phylogenetically distanced AcrB-Hi-like pumps to further understand the recognition determinants and pocket residues. Other differences besides residues in binding pockets (such as differences in the volume of the pockets, movements of loops, distances between loops, interactions with substrates to residues and the importance of water within the pockets and channels) between different pumps are investigated by molecular dynamics to try to explain specific differences between these pumps and their mechanisms (which cannot always be readily understood by only comparing the residues within the pockets) [96,101,102,108,109].

Table 8. Substrate specificities of six different pumps.

Pump	EM	NOV	Tet	Qui	NSP	CAR	SUB	Bile	AG	AZT
AcrB-Ec	✓	✓	✓	✓	✓	✓	✓	✓	X	X
AcrD-Ec	X	✓	X	X	✓	✓	✓	✓	✓	✓
MexB-Pa	✓	✓	✓	✓	✓	✓	✓	n/a	X	✓
MexD-Pa	✓	✓	✓	✓	✓	X	X	n/a	X	X
MexY-Pa	✓	X	✓	✓	✓	X	X	n/a	✓	X
AcrB-Hi	✓	✓	✓	✓	✓	n/a	n/a	✓*	X	X

Comparison between AcrB-Ec, AcrD-Ec, MexB-Pa, MexD-Pa, MexY-Pa and AcrB-Hi for several antibiotics or classes of antibiotics. A check mark (✓) indicates recognition of the substrate by the pump, and a cross (X) indicates no recognition. Green highlights a substrate for all six pumps. Asterisk (*) indicates just slightly exported. Abbreviations: EM, erythromycin; NOV, novobiocin; Tet, tetracyclines; Qui, quinolones (e.g., ciprofloxacin, norfloxacin, enoxacin (fluoroquinolones), nalidixic acid (quinolone)); NSP, (second-generation) narrow-spectrum penicillin β-lactams (e.g., cloxacillin, oxacillin and nafcillin); CAR, carbenicillin; SUB, sulbenicillin; Bile, bile salts (cholic acid, deoxycholic acid); AG, aminoglycosides; AZT, aztreonam (monobactam); n/a, not available. Overview created from references [15,35,56,97,99,101,102,105].

5. Recent Mutations in RND Multidrug Efflux Pumps Cause Enhanced Drug Resistance

As seen in Figure 3 and Tables 1 and 2, the conservation of RND multidrug efflux pumps is mainly present in the transmembrane domain and indicates that the porter domain is flexible to adapt to changes in the environment of the bacterial cells, explaining divergences in drug recognition spectra between different pumps. RND transporters are known to be promiscuous transporters, as they can recognize and transport a large number of structurally different compounds [12]. These substrates are surrounded by a multitude of residues and loops in two voluminous binding pits (Tables 4–7) and enter the pump through a multitude of channels [51,72–75,110]. It is fascinating that these efflux pumps can expel not only a wide range of drugs but also differ significantly in their amino acid composition within the binding pockets, while between these pumps, the substrate recognition spectrum is highly conserved (with the exceptions of some drugs, such as monobactams and aminoglycosides, and divergent efficiencies). Differences and evolved properties in binding pockets described above may have given transporters a more efficient export ability. In this last section, we will describe novel amino acid substitutions in RND multidrug efflux pumps, which have been arising recently in clinically, environmentally and laboratory-evolved strains. Previously displayed Tables 5 and 6 partly identify the location of the mutations in the DBP and PBP, highlighted by asterisks (*). An overview of recent mutations found in different pumps from different organisms can be found in Table 9. These recently spreading mutants significantly further enhance the efflux ability of intrinsically expressed efflux pumps (gain-of-function mutations) and have already proven to be a major problem in treating severe infections. Our over-usage and misusage of antibiotics have been putting extreme selective pressure on bacterial pathogens, causing an uprise of these mutated, highly efficient RND efflux pumps.

5.1. AcrB-Sa Mutants Cause Fluoroquinolone (G288) and Macrolide (R717) Resistance

Blair et al. (2015) reported a mutation in the AcrB-Sa efflux pump found in a post-therapy *Salmonella* Typhimurium clinical isolate, which caused a fatal infection. Table 9 lists recent mutations found in bacterial strains causing increased MDR. The *Salmonella* residue substitution was G288D in AcrB-Sa, a novel mutation causing fluoroquinolone ciprofloxacin resistance (MIC 32- to 64-fold increase pre- vs. post-therapy) [80]. In the same study, concerning the G288D mutation on AcrB-Sa-expressing plasmids, antimicrobial MICs were also increased for other antimicrobials, e.g., chloramphenicol and tetracycline (although doxorubicin export was decreased, also when the mutation was conferred in AcrB-Sa expressed in *E. coli*). Computer simulations in the same study demonstrated that the charged Asp residue protruded through to the hydrophobic pit, altering the hydrophobicity and causing steric clashes with residues in this pit, changing their conformation (especially F178 and Q176), and increased the radius of gyration of the DBP by roughly 10% [80]. Figure 4d shows the location of G288 in AcrB-Ec (shown as a pink ball in the DBP) which is also highlighted in Table 5 and shown in Figure 5. Interestingly, the G288 mutation has also been found in AcrB-Ec, MexY-Pa and AdeJ-Ab (explained in Section 5.3, Section 5.4, Section 5.5). The G288 residue is somewhat conserved, as seen in Table 5 (~50% Gly and ~20% Ala, Table S1). However, interestingly, "G288" is substituted by more bulky residues in the MexI/W-clustered transporters (being Val (~6%), Tyr (~4%) or His (~1.5%)). Additionally, for this reason, and as mentioned before, it would be interesting to study these members in more detail in future research.

In recent years, other AcrB-Sa mutants have been observed, causing untreatable infections in Nepal, Bangladesh, India and Pakistan, in both *Salmonella* Typhi [85–88] and *Salmonella* Paratyphi A [85,88], summarized in Table 9. These clinical isolates are resistant to azithromycin (macrolide) by the mutations R717Q and R717L in AcrB-Sa. Hooda et al. (2019) identified 13 azithromycin-resistant *Salmonella* strains (12 Typhi, 1 Paratyphi A) from around 1000 hospital isolates from Bangladesh, with MIC values between 32 and 64 µg ml^{-1}, with the first strain isolated in 2013 [85]. The 12 *Salmonella* Typhi AcrB-Sa genes had an SNP at R717 substituted with a Glu (R717Q), and the *Salmonella* Paratyphi A AcrB-Sa had an R717L mutation. Both mutations in AcrB-Sa showed a decrease in erythromycin and azithromycin (both macrolides) susceptibility. Similarly, Iqbal et al. (2020) described azithromycin-resistant *Salmonella* Typhi strains from Pakistan. The isolates cause severe problems during treatment, as extensively drug-resistant (XDR) *Salmonella* Typhi has left azithromycin as one of the last treatment options. Here, too, the R717Q mutation in AcrB-Sa was identified as the reason for this resistance [86]. *Salmonella* Typhi isolates from Nepal harboring the R717L mutation have been described by Duy et al. (2020), also responsible for azithromycin resistance. They note that none of the analyzed strains had an acquired AMR gene. Importantly, the authors also described that these mutants had divergently emerged in both Nepal and Bangladesh among the so-called H58 lineage, suggesting that selective pressure caused by treating typhoid fever with azithromycin resulted in these resistant strains independently [111]. Katiyar et al. (2020) analyzed two azithromycin non-susceptible strains (from 133 clinical isolates from patients with typhoid fever) from India, which both had the R717Q mutation in AcrB-Sa [87]. Another recent study by Sajib et al. (2021) predicted that the R717 mutation first occurred somewhere in 2010. They also described a *Salmonella* Typhi isolate from the United Kingdom harboring the AcrB-Sa R717Q mutation. In the same study, the authors analyzed 2519 *Salmonella* Typhi isolates and 506 *Salmonella* Paratyphi A isolates from Bangladesh, of which 104 isolates were azithromycin-non-susceptible. Of these, 32 *Salmonella* Typhi and 6 *Salmonella* Paratyphi A isolates had a significantly high azithromycin MIC (>32 mg ml^{-1}). All of these 32 highly resistant Typhi isolates had the R717 mutation (29 R717Q and 3 R717L), and five Paratyphi A isolates had the R717Q mutation [88]. It is clear that the spontaneous and divergent emergence of the "R717 mutations" in AcrB-Sa should raise great concern for the treatment of typhoid fever by macrolides.

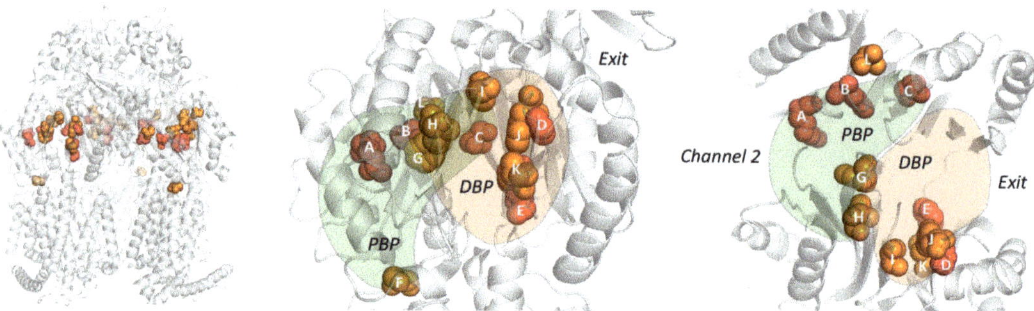

Figure 5. Upcoming gain-of-function mutations in RND-type efflux pumps. Resistant clinically, environmentally and laboratory-evolved strains show an alarming rise in gain-of-function mutations in the binding pockets of homotrimeric RND multidrug efflux pumps. Residues are shown as spheres. From left to right: view of the whole AcrB-Ec trimer, side view and top-down view. Red spheres show the most recurring amino acid substitution (by nonsynonymous mutations in the genes) numbered by letters (A–E), being G228D (AcrB-Sa, "D"), G287A/S (MexY-Pa, "D"), G288C/M/S (AcrB-Ec, "D"), G288S (AdeJ-Ab, "D"), R717Q/L (AcrB-Sa, "A"), R714C/G/H/L (MtrD-Ng, "A"), K823D/E/N (MtrD-Ng, "B"), F136L (AdeJ-Ab, "E") and K79A/T (MexY-Pa, "C"). Orange shows mutations found in resistant strains (letters F–L), potentially increasing drug resistance; however, the direct effects of the specific mutations have not yet been significantly investigated, or the effect is not clear. These include F178S ("J"), A562V ("F") and M626V ("H") (MexB-Pa), S824I ("L") (LpeB-Lp), V139F ("K"), A279T ("I") and F617L ("G") (AcrB-Ec) and S821A ("L") (MtrD-Ng). The mutations P319L and M78I (AcrB-Sa, Table 9) are not shown, as they are both not present in the binding pockets. Colors: green, proximal binding pocket (PBP); light orange, distal binding pocket (DBP). PDB accession code 3W9H [36].

Table 9. RND mutations in recent clinically, environmentally and experimentally evolved strains.

Organism	Pump	Mutations	Country	Resistance	References
Salmonella enterica	AcrB	**G288D** *	UK	Ciprofloxacin (fluoroquinolone)	[80,81]
		P319L *	China	Multiple (fluoroquinolones)	[112]
		P319L/M78I *	China	Multiple (fluoroquinolones)	[112]
		R717Q *	Bangladesh, Pakistan, India	Azithromycin (macrolide)	[85–88]
		R717L *	Bangladesh, Nepal	Azithromycin (macrolide)	[85,88,111]
Salmonella Paratyphi *A*	AcrB	**R717L** *	Bangladesh	Azithromycin (macrolide)	[85]
		R717Q *	Bangladesh	Azithromycin (macrolide)	[88]
Neisseria gonorrhoeae	MtrD	**R714H** *	Europe, Russia	Azithromycin (macrolide)	[89]
		R714L *	USA	Azithromycin (macrolide)	[89]
		R714C *	USA	Azithromycin (macrolide)	[89]
		R714G *	Experimentally	Azithromycin (macrolide)	[43]
		K823N *	Canada	Azithromycin (macrolide)	[89]
		K823E *	USA, India, Canada	Azithromycin (macrolide)	[24,89,113]
		K823E/S821A *	USA	Azithromycin (macrolide)	[24]
		K823D *	USA	(N/D)	[24]
Pseudomonas aeruginosa	MexB	A562V	Denmark	(N/D)	[5]
		M626V	Denmark	(N/D)	[5]
		F178S	Australia	(N/D)	[82]
	MexY	K79T *	Experimentally	Tobramycin (aminoglycoside)	[83]
		K79A *	Experimentally	Paromomycin (aminoglycoside)	[56]
		G287A	Australia, Spain	High tobramycin MIC isolates	[82]
		G287S	Europe (and other), *Experimentally*	Tobramycin (aminoglycoside)	[83,84]
Escherichia coli	AcrB	V139F	Experimentally	(N/D)	[114–117]
		G288S/M/C *	Experimentally	(N/D)	[118]
		G288C *	Experimentally (frequent mutation)	Multiple (especially erythromycin)	[79]
		A279T	Experimentally	(N/D)	[118]
		A279T, F617L	Experimentally (frequent mutation)	(Increased 1-Hexene tolerance)	[119]
Acinetobacter baumannii	AdeJ	F136L *	Australia	Meropenem (carbapenem)	[77,78]
		G288S *	Australia	Meropenem (carbapenem)	[77,78]
Legionella pneumophila	LpeB	S824I (I911L, G1158W, F1124 insert)	China	Azithromycin (macrolide)	[90]

Bold underlined indicates the recurring "**G288**" mutation in AcrB-Ec, AcrB-Sa (*Salmonella* Typhimurium) [80,81], MexY-Pa and AdeJ-Ab. Bold indicates the "**R717**" mutation recurring in AcrD-Sa (*Salmonella* Typhi and Paratyphi A) [85–88,111] and MtrD-Ng [43,89]. Asterisk (*) indicates direct measured increased MICs.

Lastly, recently, P319L and M78I/P319L mutants of AcrB-Sa have been found in *Salmonella* ssp. strains isolated from pork, swine, chicken and duck from Guangdong, Shandong, Hubei, Henan (China), causing increased MICs for multiple substrates, of which the most noticeable are fluoroquinolones (enrofloxacin and norfloxacin), but also for erythromycin and other substrates. These two residues are not located in the binding pockets of AcrB-Sa, but more on the outside of the monomers. The authors argued that the P319L residue might increase the export efficiency by altered interaction with AcrA [112].

5.2. MtrD-Ng Mutations (R714, K823) by Mosaic Patterns Causes Macrolide Resistance

Recently, mutations in the multiple transferable resistance (mtr) efflux pump from *N. gonorrhoeae* (MtrD-Ng), acquired by mosaic-like patterns in the alleles, have become an increasing concern in azithromycin (macrolide) resistance [120] (Table 9). Mosaic patterns arise in *N. gonorrhoeae* acquiring and recombining donor DNA from *Neisseria* spp. (*N. meningitidis* and *N. lactamica*), resulting in multiple mutations in both the repressor (*mtrR*) and efflux pump (*mtrCDE*) genes, and are found worldwide [121]. It has been extensively studied that (as for most other pathogens) increased resistance in clinal strains can be a result of mutations in the regulatory network (e.g., MtrR or MtrA) [122–124]. However, mutations in the efflux pump MtrD-Ng itself (instead of by direct mutations in the 23S rRNA target gene [125]) cause significantly elevated MICs (azithromycin > 2 µg/mL) and are relatively new. Recently, Wadsworth et al. (2018) analyzed 1102 isolates and noticed an increase in mosaic patterns at the *mtrCDE* region, with the highest diversity in the *mtrD* gene [113]. Four residue mutations between isolates were found in MtrD-Ng: I48T (DBP), G59D, K823E (PBP) and F854L. Additionally, in 2018, Rouquette-Loughlin et al. studied eight clinical strains from 2014 and found that mutations directly in the MtrD-Ng protein accounted for an increased azithromycin resistance, which could not be explained only by mutations in the promoter region or in the regulatory network. They identified two mutations, namely, K823E and S821A [24], both on the PC2 subdomain in the PBP of MtrD-Ng. Cryo-EM structures of MtrD-Ng (from transformant "CR103" by [24]) holding the two K823E and S821A mutations were solved recently [43], and the same study identified several other mutations in this pump, including R714G. Both single mutations, K823E and R714G, resulted in an increase in MICs for several substrates (azithromycin, erythromycin and polymixin B) [43]. Additionally, Ma et al. (2020) analyzed 4852 global *N. gonorrhoeae* genomes. Of these, 12 contained the mutation R714H/L/C, and seven contained the mutation K823E/N [89]. They did not observe mutations at positions 74, 669, 821 and 825, as found (and tested) by others [24,43].

Interestingly, the R714H/L/C mutations [43,89] correspond to the R717L/Q mutations discussed before (Section 5.1) present in AcrB-Sa of *Salmonella* clinical isolates from Bangladesh, Pakistan, India and Nepal (Table 9). The location of these mutations (at R714, K823 (and S821)) are all in the PC2 subdomain and face the PBP (Table 6 and Figure 5), which explains the increase in the MICs for macrolides, but, e.g., not for other drugs such as penicillin, ampicillin, ethidium bromide and crystal violet [43]. Although the S821A mutation showed no increase in the MICs of the tested compounds [43], a similar mutation (S824I) was found in the LpeB-Lp efflux pump in *L. pneumophila* clinical isolates from China [90]. However, direct MICs were not determined for this strain, nor was the effect of the mutation determined. It is interesting that these mutations are seen within pumps from different organisms (Table 9), that the Ser residue is highly conserved among the 19 analyzed efflux pumps (represented by an Ala in AcrD-Ec/Sa and MexB-Pa, similar to the S821A mutation in MtrD-Ng) and that the K823E mutation is similar to E826 in wild-type AcrB-Ec/Sa and other pumps (Table 6 and Table S1).

5.3. AdeJ-Ab Mutations (G288, F136) Cause Increased Drug Resistance

Mutations in the previously described G288 location (Section 5.1) have also been found in multiple studies on *A. baumannii* clinical isolates. Hawkey et al. (2018) investigated carbapenem-resistant *A. baumannii* isolates from burn wound sites of a 2013 patient to investigate resistance evolution [77]. They analyzed the collected 20 strains from this patient in addition to strains from three other patients (one before and two after the admission of the main investigated patient). All collected strains were multidrug-resistant (to, e.g., aminoglycosides, fluoroquinolones and more); however, they showed variations in meropenem resistance (MICs ranging from 2 to >32 μg mL^{-1}). All strains which first showed an elevated MIC of $\geq$8 μg mL^{-1} contained the mutation G288S on the RND pump AdeJ (the authors mentioned AdeB, but we believe it to be AdeJ after checking the sequences). A later strain harbors mutation F136L, with an MIC of 8 μg mL^{-1}, and later isolated strains with this F136L mutation also contained a mutation, A515V, in the penicillin-binding protein (PBP3) FtsI, further increasing the MICs ($\geq$32 μg mL^{-1}). After being administered meropenem, the patient was treated with colistin (a last-resort polymyxin treatment) [77]. As explained before, the G288 residue is located near the hydrophobic pit in the DBP, and the G288S mutation possibly altered the drug-binding properties of the pit to meropenem. This can also explain the mutation of F136L, which is also located in the pit, possibly increasing the binding of carbapenems to the pit, although we are not sure of the precise mechanism of these alterations. The F136 location is the least conserved residue of the six Phe residues (with AcrB-Ec as a reference), being a Leu in MexF-Pa and AdeG-Ab, and Ile in MexQ-Pa and MexY-Pa (see Table 5).

Similarly, a recent study by Santos-Lopez et al. (preprint, 2020) investigated the roles of selective pressure by antibiotic treatment of *A. baumannii* laboratory-evolved strains under increased cephalosporin (ceftazidime) and carbapenem (imipenem) concentration conditions. Growth under ceftazidime resulted in mutation in AdeJ, causing resistance to both ceftazidime and imipenem in 16 of the 18 strains (the other two harboring mutations in the *adeIJK* regulatory protein AdeN, or a PBP instead). The mutation in AdeJ found in replicates was G288S. Additionally, other mutations seen in AdeJ were, e.g., F136L, F136S, Q176K, Q176R and A290T, all in the DBP [78]. Q176 in AcrB-Ec interacts with ABI-PP in one of the crystal structures [36], and the G288D mutation in AcrB-Sa alters the Q176 conformation [80]. The recurring F136L and G288S mutations in AdeJ further suggest that these substitutions are significant gain-of-function mutations in this efflux pump.

5.4. Mutations in MexY-Pa (K79, G287) Increase Aminoglycoside Resistance

Greipel et al. (2016) studied 361 isolates of people suffering from cystic fibrosis (CF) by analyzing the genome sequences. The isolates came from multiple EU countries (including Germany, Sweden and the Netherlands) [84]. They found 85 nonsynonymous mutations in the *mexY* gene. In two isolates, the G287S mutation was present [84], which is similar to the G288 mutations in AcrB-Sa/Ec [80] and AdeJ-Ab [77]. They also described a "Q175E" mutation in 327 isolates, similar to the Q176 location in AdeJ-Ab and AcrB-Ec mentioned before; however, in wild-type MexY-Pa, this residue is E175 (Table 5), and thus it may be possible that the other 34 strains have an E175Q mutation in the DBP. However, we cannot confirm this. It is an interesting location as mutations here have also been seen in AdeJ-Ab (Section 5.3), and the Q176 residue is one of the somewhat conserved residues among the 19 selected transporters shown in Table 5 (this residue's conformation is altered in the G288D gain-of-function mutant of AcrB-Sa according to MD simulation (Section 5.1) [80]). Another potentially interesting mutation found in MexY-Pa in the same study was S48N (similar to S48 in AcrB-Ec), located in the DBP, close to the exit. A list of the total 85 nonsynonymous mutations can be found in [84]. Direct MIC measurements looking at the effect of the mutations (G287S, E175Q and S48N) have not been performed. However, the yet again recurring G287S ("G288") mutation is a worrying find. López-Causapé et al. (2017) sequenced and analyzed 28 strains from 18 patients with CF from Spain and Australia, isolated between 1995 and 2012 [82]. They found mutations in more than 100 genes related to AMR. Besides mutations in repressor MexZ, and genes *gyrA* and *fusA1*, they found mutations in the RND-type efflux pumps, including MexY, MexB and MexW. MexY and MexB had the most different numbers of mutations (nine) in many isolates (8 for MexB and 19 for MexY). One of the most recurring mutations was G287A in MexY (similar to G287S mentioned above), seen in three isolates. In MexB, the F178 location was mutated in one strain (F178S). Strains with the MexY (G287A) mutation had a significantly higher MIC for tobramycin compared to other isolates, although this was due to many different other mutations in multiple genes for different isolates, and the effect of the G287A and F178S mutations was not directly observed. However, their analysis by comparing the median MICs for strains with or without a particular mutation suggested an increase in multiple drugs for mutations in MexY (e.g., imipenem, aztreonam, meropenem and tobramycin). For MexB mutations, a similar increase in MICs was observed; however, this was noticeably more significant compared to MexY for aztreonam and meropenem. The complete overview of mutations can be found in [82].

In a recent study, Wardell et al. (2019) showed that 4 out of 13 laboratory-evolved strains, under tobramycin growth conditions, harbor mutations in MexY-Pa, which did not occur in meropenem or ciprofloxacin selected strains. Three of these four had the mutation G287S (the same as the MexY-Pa mutations found by [84]), and one had the mutation K79T [83]. Besides the recurring G287 mutation, K79T catches our attention, as a mutation in the same location (K79A) was found by experimentally evolved MexY-Pa by selective pressure under aminoglycosides. In that study, the K79A mutant caused a significantly higher MIC for aminoglycosides paromomycin, neomycin and spectinomycin [56]. As the K79T mutation was found in a strain with decreased tobramycin susceptibility [82], it is likely that this PBP location mutation increases the substrate recognition of aminoglycosides by MexY-Pa. The same study [83] also found MexY-Pa mutations in 140 out of 558 (25%) clinical isolates and in 15 out of 172 (8.7%) environmental isolates (although specific mutations were not mentioned), again highlighting the significant variability and frequency of mutations in MDR pump genes. Another recent study by Colque et al. (2020) studied 14 clinal isolates from CF patients from Denmark who suffered long-term infections by *P. aeruginosa* between 1991 and 2011. They found two mutations in MexB-Pa (five times in M626V, and once in A562V, both in the PBP) and six in MexY-Pa (although none in a binding pocket) [5]. In MexB-Pa, the M626V mutation is inside the PBP, while A562V is directed to Channel 1 of the monomer. Similarly, multiple mutations in *P. aeruginosa* RND-type pumps (MexY, MexB, MexD, MexK, MexI, MexQ) were found in an MDR clinical isolate ("PA154197") from Hong Kong [126]. In both these studies, the direct effects of the mutations were not determined.

5.5. Experimentally Obtained Mutations in AcrB-Ec (V139, A279, G288)

Cudkowicz et al. (2019) and Langevin et al. (2020) examined the evolution of mutations in *E. coli* and AcrAB–TolC specifically, respectively, under chloramphenicol growth conditions, and both studies observed the V139F mutation in AcrB-Ec [114,115]. This mutation was also seen by Hoeksema et al. (2019) when analyzing the effects of mutations in genes related to AMR, specifically the role of these mutations in the resistance to a second antibiotic after a first antibiotic gave rise to a specific mutation (where V139F was found in strains resistant to tetracycline, which previously acquired resistance to amoxicillin, enrofloxacin or kanamycin) [116]. This Val residue (V139) is located in the hydrophobic pit in the DBP. It is not clear how this mutation exactly enhances the efflux ability of AcrB-Ec, and if the mutation causes increased MICs for multiple drugs and therefore acts as a significant gain-of-function mutation. The recurrence of this mutation, however, makes it a noteworthy one.

Schuster et al. (2014) found a G288S mutation in most of their evolutionarily evolved strains (after in vitro random mutagenesis of the AcrB-Ec gene), along with G288M, G288C and A279T (also in the hydrophobic pit of the DBP). The MIC data for G288S and A279T (single and double) did not indicate a gain-of-function mutation for the tested compounds (even a decrease in MICs for novobiocin and chloramphenicol) [118]. The A279T mutation was also obtained by researchers who optimized AcrB for the export of styrene and alpha-olefins. Out of eight variants, seven contained A279T and five contained F617L [119]. On the other hand, a G288C mutation was found to be the most recurring in another study by Soparkar et al. (2015) when trying to regain the export ability in the loss-of-function mutation F610A in AcrB-Ec (of which the gene was located on a plasmid, transformed into AcrB-deficient cells). They found G288C to be the most effective suppressor alteration, occurring five independent times. The introduction of G288C in AcrB (F610A) increased the MICs for erythromycin, novobiocin, minocycline, nalidixic acid and SDS (when compared to AcrB (F610A)) [79]. The "G288" mutation is the most recurring mutation with the most alternative amino acid substitutions, as seen in Table 9.

6. Discussion

In this review, we provided a conservation analysis of homotrimeric RND-type multidrug efflux pumps, including a more detailed view of 19 selected pumps (which have been better studied). We also looked at the conservation and variation among a selection of pumps to try to summarize, explain and understand the substrate specificities of some pumps, based on specific residues, hydrophobic and hydrophilic residues, in both pockets. The analysis showed that among all efflux pumps, the TM domain was significantly conserved, while the porter domain was largely variable, except for some interesting residues, including the residues D568 from the "interface-loop" and R637 from the PC1 subdomain. Certain residues within the binding pockets were conserved between some pumps, but not all, despite the pumps having a similar efflux spectrum. Interestingly, the least conserved pump (AcrB-Hi) compared with AcrB-Ec can expel the same compounds. We hypothesize that changes in the number of hydrophobic and hydrophilic residues in the pockets may enhance drug efflux and may specifically enhance the efflux of physiologically relevant toxic compounds, such as bile salts. As for the TM domain, the three residues forming salt bridges—playing a crucial role in the proton translocation and therefore energy consumption—are D407, D408 and K940 (numbering based on AcrB-Ec) in most of the 135 analyzed pumps. There were, however, three noticeable outliers for the K940 residue, namely, where the Lys was an Arg residue (for the organisms *I. loihiensis*, *C. japonicus* and *T. turnerae*). This "K940R" residue was also created in MexB-Pa back in 2000 by Guan et al., which resulted in a fully active pump [34], indicating that this region, however critical for the function of the pump, is still slightly flexible by substitutions.

As for the conservation in the porter domain, the conserved residues of interest were the aforementioned "interface-lock" D568 and R637 (both 100% conserved among 135 pumps), N298 (located at the vestibule and close to Channel 3, also 100% conserved), P223 (99% conserved (excluding gaps); present in all analyzed pumps, except for MexI/W members and, interestingly, MexY-Pa (Gly)), W187 (98.5% conserved; possibly stabilizing trimer formation, present in all analyzed pumps, except for two MexI/W-like pumps from *L. pneumophila*) and, to a lesser extent, but still significantly, Y327 (81% conserved; in the hydrophobic pit in the DBP, linked to the recognition of carboxylated drugs), and those partly conserved but noticeable were F178 and F628 (70% and 74% conserved; in the hydrophobic pit), L674 and G675 (72% and 88% conserved; part of the bottom and flexible (F-) loops in the PBP) and the somewhat less conserved Q577 (69% conserved; in the PBP). It is interesting to note that these residues are—as an example—not present in AcrB-Hi (except for F178 and Y327), which has a similar substrate range to AcrB-Ec. We also found it interesting that among the selected transporters, certain carboxylated drugs (including cloxacillin, oxacillin and nafcillin) were exported by all (Table 8), perhaps partly explained by the Y327 residue. A list of all residues with their conservation percentages and alternative substitutions can be found in Table S1. We found that members of the MexI/W cluster (including AcrB-Hi and LpeB-Lp) consistently showed distinct differences when compared to their RND multidrug efflux pump colleagues. These differences are the truncation of (the tip of) the arm, the lack of the conserved P223 on this arm, the gaps after sequence alignment (e.g., in the hydrophobic pit ("F610") and other parts of both the PBP and DBP), the lack of highly conserved F628 in the pit, the lack of conserved L674 on the F-loop in the PBP and, additionally (although only seen in two pumps from *L. pneumophila*, including LpeB-Lp), the conserved W187 substitutions (being a Phe or a Thr). It would be interesting for future research to obtain structures and biochemical data (in addition to AcrB-Hi [17]) of members of this cluster. In particular, the macrolide resistance causing LpeB-Lp pump (gene lpp2880 from clinically relevant *L. pneumophila* str. Paris [103,104]) not only showed the distinct abovementioned MexI/W characteristics but was also an outlier within this cluster, being the only pump lacking both the conserved W187 and Y327 (Table 2) and showing the most gaps after multiple sequence alignment (Table 5; Table 6).

Besides comparing drug recognition and conservation among pumps, we looked into the rise in recent adaptations of the RND pumps occurring in fairly recent multidrug-resistant clinical strains. We found several noticeable recurring amino acid substitutions present in clinically, environmentally and laboratory-evolved strains. Firstly, G288D (AcrB-Sa), G288C/S/M (AcrB-Ec), G288S (AdeJ-Ab) and G287A/S (MexY-Pa), mutations just outside the hydrophobic pit of the DBP, changed—and usually enhanced—MICs for certain drugs (fluoroquinolones in *Salmonella* Typhimurium, aminoglycosides in *P. aeruginosa*, carbapenems in *A. baumannii* and multiple drugs in *E. coli*). In AdeJ-Ab, another noticeable mutation was F136L, decreasing susceptibility to meropenem. Other worrying mutations are R717L/Q (AcrB-Sa in *S.* Typhi and *S.* Paratyphi A), and R714C/G/H/L and K823D/E/N (MtrD-Ng), mutations in the PBP of the RND efflux pump, increasing the MICs for macrolides (such as azithromycin) considerably (possibly also the mutation S821A, recurring as S821I in LpeB-Lp from *L. pneumophila* strains from China). The mutations K79A/T (MexY-Pa) were independently observed by different research groups in laboratory-grown resistant strains. Lastly, V139F (AcrB-Ec), an amino acid located in the hydrophobic pit of the DBP, was found in multiple studies. Other mutations found in clinically, environmentally and laboratory-evolved strains can be found in Table 9. We note that there are likely more mutations and studies regarding mutations in RND multidrug efflux pumps not mentioned in this review.

Additionally, it is possible that the mentioned gain-of-function mutations cause increased MDR, rather than an increased resistance to one class of antibiotics, as many papers reviewed in this review article did not test for multiple classes of drugs but found mutations in the pumps after observing specific resistance in clinical strains (e.g., for carbapenems, fluoroquinolones, macrolides or aminoglycosides). Examples where MICs increased for multiple drugs as a result of a specific mutation are: AcrB-Sa (G288D), increasing the MICs for chloramphenicol, ciprofloxacin (fluoroquinolone) and tetracycline [80]; AcrB-Ec (G288C), increasing the MICs in an F610A background for erythromycin (macrolide), novobiocin, minocycline, nalidixic acid (quinolone) and SDS [79]; MtrD-Ng (R714G), increasing the MICs for azithromycin, erythromycin (macrolides), ethidium and polymyxin B [43]; and MtrD-Ng (K823E), increasing the MICs for azithromycin, erythromycin (macrolides) and polymyxin B [43]. Additionally, the "G288" mutation emerges in different pumps from different organisms, of which strains are resistant to a variety of drugs (including macrolides, fluoroquinolones, aminoglycosides and carbapenems), pointing to an increase in MDR by one gain-of-function mutation. Additionally, a combination of the mutations mentioned in this review may potentially increase MDR, which may result from increased use of alternative antibiotic treatments.

These recent adaptive mutations are worrying, as commonly used antibiotics to treat infections caused by these pathogens are rendered ineffective, and last-resort antibiotics are used (which have more or worse side effects or may not always be an option in underdeveloped regions in the world). An example is the use of colistin for *A. baumannii* infections resistant to carbapenems. Especially worrying are cases where extensively drug-resistant (XDR) pathogens leave a specific class of antibiotics as a last option, after which this XDR pathogen becomes resistant to this antibiotic too, by mutations in the RND pump, noticeably azithromycin resistance in *S.* Typhi strains in India, Nepal, Bangladesh and Pakistan. It is particularly worrying that these mutations—besides being spread through transfer—seem to be appearing independently in different locations, and in different organisms and pumps, further indicating that the misusage and over-usage of antibiotics put extreme selective pressure on these pathogens, giving rise to not only mutations in genes part of expression regulatory pathways but also gain-of-function mutations in the efflux pumps themselves, leaving us with last-resort antibiotics, or worse, when a pump increases (or potentially gains) resistance to the last treatment options.

We hope that this review can help increase our understanding of the mechanisms of drug recognition by RND multidrug efflux pumps and help the development of novel antibiotics and efflux pump inhibitors needed to treat the increasingly spreading and evolving pathogenic bacteria.

Supplementary Materials: The following are available online at https://www.mdpi.com/article/10.3390/antibiotics10070774/s1, Figure S1: Additional images of the TM domain, Figure S2: Additional images of the porter domain, Figure S3: Additional images of the ConSurf output, Figure S4: Front view of the drug-bound structures of AcrB-Ec, MtrD-Ng and AdeB-Ab, Table S1: Overview of residue conservation based on 135 MDR-type RND pumps, Data S1: Sequences of 135 MDR-type RND pumps.

Author Contributions: M.Z. performed the analysis, reviewed the literature and prepared the manuscript and the relevant figures. M.Z. and K.N. edited the manuscript and approved the final version. All authors have read and agreed to the published version of the manuscript.

Funding: This work was supported by the Center of Innovation Program (COI) from the Japan Science and Technology Agency (JST), Grant-in-Aid for Scientific Research (Early-Career Scientists) (Kakenhi 20K16242) and Grant-in-Aid for Scientific Research (Challenging Research (Exploratory)) (Kakenhi 18K19451) from the Japan Society for the Promotion of Science (JSPS), CREST (JPMJCR20H9) and the Dynamic Alliance for Open Innovation Bridging Human, Environment and Materials from the Ministry of Education, Culture, Sports, Science and Technology—Japan (MEXT).

Institutional Review Board Statement: Not applicable.

Informed Consent Statement: Not applicable.

Data Availability Statement: Data is contained within the article or Supplementary Materials.

Conflicts of Interest: The authors declare no conflict of interest.

References

1. World Health Organization. *Global Action Plan on Antimicrobial Resistance*; World Health Organization: Geneva, Switzerland, 2015; ISBN 9789241509763. Available online: https://www.who.int/publications/i/item/9789241509763 (accessed on 16 March 2021).
2. World Health Organization. *Antimicrobial Resistance: Global Report on Surveillance*; World Health Organization: Geneva, Switzerland, 2014; ISBN 9789241564748. Available online: https://apps.who.int/iris/handle/10665/112642 (accessed on 16 March 2021).
3. Blair, J.; Webber, M.A.; Baylay, A.J.; Ogbolu, D.O.; Piddock, L.J.V. Molecular mechanisms of antibiotic resistance. *Nat. Rev. Microbiol.* **2015**, *13*, 42–51. [CrossRef]
4. Walsh, C.T. Where will new antibiotics come from? *Nat. Rev. Genet.* **2003**, *1*, 65–70. [CrossRef]
5. Colque, C.A.; Orio, A.G.A.; Feliziani, S.; Marvig, R.L.; Tobares, A.R.; Johansen, H.K.; Molin, S.; Smania, A.M. Hypermutator *Pseudomonas aeruginosa* exploits multiple genetic pathways to develop multidrug resistance during long-term infections in the Airways of Cystic Fibrosis Patients. *Antimicrob. Agents Chemother.* **2020**, *64*, 119–146. [CrossRef]
6. Nikaido, H. Multidrug resistance in bacteria. *Annu. Rev. Biochem.* **2009**, *78*, 119–146. [CrossRef]
7. Allen, H.K.; Donato, J.; Wang, H.H.; Cloud-Hansen, K.A.; Davies, J.; Handelsman, J. Call of the wild: Antibiotic resistance genes in natural environments. *Nat. Rev. Genet.* **2010**, *8*, 251–259. [CrossRef] [PubMed]
8. Li, X.-Z.; Plésiat, P.; Nikaido, H. The challenge of efflux-mediated antibiotic resistance in Gram-negative bacteria. *Clin. Microbiol. Rev.* **2015**, *28*, 337–418. [CrossRef] [PubMed]
9. Levy, S.B.; Marshall, B. Antibacterial resistance worldwide: Causes, challenges and responses. *Nat. Med.* **2004**, *10*, S122–S129. [CrossRef] [PubMed]
10. Blair, J.; Richmond, E.G.; Piddock, L.J.V. Multidrug efflux pumps in Gram-negative bacteria and their role in antibiotic resistance. *Future Microbiol.* **2014**, *9*, 1165–1177. [CrossRef]
11. Nikaido, H. Multidrug efflux pumps of Gram-negative bacteria. *J. Bacteriol.* **1996**, *178*, 5853–5859. [CrossRef]
12. Zwama, M.; Yamaguchi, A. Molecular mechanisms of AcrB-mediated multidrug export. *Res. Microbiol.* **2018**, *169*, 372–383. [CrossRef]
13. Piddock, L.J.V. Multidrug-resistance efflux pumps? Not just for resistance. *Nat. Rev. Genet.* **2006**, *4*, 629–636. [CrossRef]
14. Wang-Kan, X.; Rodríguez-Blanco, G.; Southam, A.D.; Winder, C.L.; Dunn, W.B.; Ivens, A.; Piddock, L.J.V. Metabolomics reveal potential natural substrates of AcrB in *Escherichia coli* and *Salmonella enterica* Serovar Typhimurium. *mBio* **2021**, *12*, 340. [CrossRef] [PubMed]

15. Zwama, M.; Yamaguchi, A.; Nishino, K. Phylogenetic and functional characterisation of the *Haemophilus influenzae* multidrug efflux pump AcrB. *Commun. Biol.* **2019**, *2*, 1–11. [CrossRef] [PubMed]

16. Du, D.; Wang, Z.; James, N.; Voss, J.E.; Klimont, E.; Ohene-Agyei, T.; Venter, H.; Chiu, W.; Luisi, B.F. Structure of the AcrAB–TolC multidrug efflux pump. *Nat. Cell Biol.* **2014**, *509*, 512–515. [CrossRef] [PubMed]

17. Nishino, K.; Senda, Y.; Yamaguchi, A. The AraC-family regulator GadX enhances multidrug resistance in *Escherichia coli* by activating expression of mdtEF multidrug efflux genes. *J. Infect. Chemother.* **2008**, *14*, 23–29. [CrossRef]

18. Nishino, K.; Nikaido, E.; Yamaguchi, A. Regulation and physiological function of multidrug efflux pumps in *Escherichia coli* and Salmonella. *Biochim. Biophys. Acta (BBA) Proteins Proteom.* **2009**, *1794*, 834–843. [CrossRef]

19. Oethinger, M.; Podglajen, I.; Kern, W.V.; Levy, S.B. Overexpression of the *marA* or *soxS* regulatory gene in clinical topoisomerase mutants of *Escherichia coli*. *Antimicrob. Agents Chemother.* **1998**, *42*, 2089–2094. [CrossRef] [PubMed]

20. Schumacher, M.A.; Miller, M.C.; Grkovic, S.; Brown, M.H.; Skurray, R.A.; Brennan, R.G. Structural mechanisms of QacR induction and multidrug recognition. *Science* **2001**, *294*, 2158–2163. [CrossRef]

21. Poole, K. Efflux-mediated multiresistance in Gram-negative bacteria. *Clin. Microbiol. Infect.* **2004**, *10*, 12–26. [CrossRef]

22. Islam, S.; Jalal, S.; Wretlind, B. Expression of the MexXY efflux pump in amikacin-resistant isolates of *Pseudomonas aeruginosa*. *Clin. Microbiol. Infect.* **2004**, *10*, 877–883. [CrossRef]

23. Webber, M.A.; Talukder, A.; Piddock, L.J.V. Contribution of mutation at amino acid 45 of AcrR to *acrB* expression and ciprofloxacin resistance in clinical and veterinary *Escherichia coli* isolates. *Antimicrob. Agents Chemother.* **2005**, *49*, 4390–4392. [CrossRef]

24. Rouquette-Loughlin, C.E.; Reimche, J.L.; Balthazar, J.T.; Dhulipala, V.; Gernert, K.M.; Kersh, E.N.; Pham, C.D.; Pettus, K.; Abrams, A.J.; Trees, D.L.; et al. Mechanistic basis for decreased antimicrobial susceptibility in a clinical isolate of *Neisseria gonorrhoeae* possessing a mosaic-like *mtr* efflux pump locus. *mBio* **2018**, *9*, e02281-18. [CrossRef]

25. Du, D.; Wang-Kan, X.; Neuberger, A.; Van Veen, H.W.; Pos, K.M.; Piddock, L.J.V.; Luisi, B.F. Multidrug efflux pumps: Structure, function and regulation. *Nat. Rev. Genet.* **2018**, *16*, 523–539. [CrossRef]

26. Yu, K.; Zhang, Y.; Xu, W.; Zhang, X.; Xu, Y.; Sun, Y.; Zhou, T.; Cao, J. Hyper-expression of the efflux pump gene *adeB* was found in *Acinetobacter baumannii* with decreased triclosan susceptibility. *J. Glob. Antimicrob. Resist.* **2020**, *22*, 367–373. [CrossRef]

27. Salehi, B.; Ghalavand, Z.; Yadegar, A.; Eslami, G. Characteristics and diversity of mutations in regulatory genes of resistance-nodulation-cell division efflux pumps in association with drug-resistant clinical isolates of *Acinetobacter baumannii*. *Antimicrob. Resist. Infect. Control.* **2021**, *10*, 1–12. [CrossRef] [PubMed]

28. Yamaguchi, A.; Nakashima, R.; Sakurai, K. Structural basis of RND-type multidrug exporters. *Front. Microbiol.* **2015**, *6*, 327. [CrossRef]

29. Kobylka, J.; Kuth, M.S.; Müller, R.T.; Geertsma, E.R.; Pos, K.M. AcrB: A mean, keen, drug efflux machine. *Ann. N. Y. Acad. Sci.* **2020**, *1459*, 38–68. [CrossRef] [PubMed]

30. Klenotic, P.A.; Moseng, M.A.; Morgan, C.E.; Yu, E.W. Structural and functional diversity of resistance–nodulation–cell division transporters. *Chem. Rev.* **2021**, *121*, 5378–5416. [CrossRef] [PubMed]

31. Alav, I.; Kobylka, J.; Kuth, M.S.; Pos, K.M.; Picard, M.; Blair, J.M.A.; Bavro, V.N. Structure, assembly, and function of tripartite efflux and type 1 secretion systems in Gram-negative bacteria. *Chem. Rev.* **2021**, *121*, 5479–5596. [CrossRef] [PubMed]

32. Zgurskaya, H.I.; Malloci, G.; Chandar, B.; Vargiu, A.V.; Ruggerone, P. Bacterial efflux transporters' polyspecificity—A gift and a curse? *Curr. Opin. Microbiol.* **2021**, *61*, 115–123. [CrossRef] [PubMed]

33. Murakami, S.; Nakashima, R.; Yamashita, E.; Yamaguchi, A. Crystal structure of bacterial multidrug efflux transporter AcrB. *Nature* **2002**, *419*, 587–593. [CrossRef]

34. Guan, L.; Nakae, T. Identification of essential charged residues in transmembrane segments of the multidrug transporter MexB of *Pseudomonas aeruginosa*. *J. Bacteriol.* **2001**, *183*, 1734–1739. [CrossRef] [PubMed]

35. Elkins, C.A.; Nikaido, H. Substrate specificity of the RND-type multidrug efflux pumps AcrB and AcrD of *Escherichia coli* is determined predominately by two large periplasmic loops. *J. Bacteriol.* **2002**, *184*, 6490–6498. [CrossRef] [PubMed]

36. Nakashima, R.; Sakurai, K.; Yamasaki, S.; Hayashi, K.; Nagata, C.; Hoshino, K.; Onodera, Y.; Nishino, K.; Yamaguchi, A. Structural basis for the inhibition of bacterial multidrug exporters. *Nat. Cell Biol.* **2013**, *500*, 102–106. [CrossRef]

37. Sakurai, K.; Yamasaki, S.; Nakao, K.; Nishino, K.; Yamaguchi, A.; Nakashima, R. Crystal structures of multidrug efflux pump MexB bound with high-molecular-mass compounds. *Sci. Rep.* **2019**, *9*, 1–9. [CrossRef] [PubMed]

38. Tsutsumi, K.; Yonehara, R.; Ishizaka-Ikeda, E.; Miyazaki, N.; Maeda, S.; Iwasaki, K.; Nakagawa, A.; Yamashita, E. Structures of the wild-type MexAB–OprM tripartite pump reveal its complex formation and drug efflux mechanism. *Nat. Commun.* **2019**, *10*, 1–10. [CrossRef]

39. Glavier, M.; Puvanendran, D.; Salvador, D.; Decossas, M.; Phan, G.; Garnier, C.; Frezza, E.; Cece, Q.; Schoehn, G.; Picard, M.; et al. Antibiotic export by MexB multidrug efflux transporter is allosterically controlled by a MexA-OprM chaperone-like complex. *Nat. Commun.* **2020**, *11*, 1–11. [CrossRef]

40. Su, C.-C.; Morgan, C.E.; Kambakam, S.; Rajavel, M.; Scott, H.; Huang, W.; Emerson, C.C.; Taylor, D.J.; Stewart, P.L.; Bonomo, R.A.; et al. Cryo-electron microscopy structure of an *Acinetobacter baumannii* multidrug efflux pump. *mBio* **2019**, *10*, e01295-19. [CrossRef]

41. Morgan, C.E.; Glaza, P.; Leus, I.V.; Trinh, A.; Su, C.-C.; Cui, M.; Zgurskaya, H.I.; Yu, E.W. Cryoelectron microscopy structures of AdeB illuminate mechanisms of simultaneous binding and exporting of substrates. *mBio* **2021**, *12*, e03690-20. [CrossRef]

42. Bolla, J.R.; Su, C.-C.; Do, S.V.; Radhakrishnan, A.; Kumar, N.; Long, F.; Chou, T.-H.; Delmar, J.A.; Lei, H.-T.; Rajashankar, K.R.; et al. Crystal structure of the *Neisseria gonorrhoeae* MtrD inner membrane multidrug efflux pump. *PLoS ONE* **2014**, *9*, e97903. [CrossRef]

43. Lyu, M.; Moseng, M.A.; Reimche, J.L.; Holley, C.L.; Dhulipala, V.; Su, C.-C.; Shafer, W.M.; Yu, E.W. Cryo-EM structures of a gonococcal multidrug efflux pump illuminate a mechanism of drug recognition and resistance. *mBio* **2020**, *11*, 697–706. [CrossRef]

44. Touzé, T.; Eswaran, J.; Bokma, E.; Koronakis, E.; Hughes, C.; Koronakis, V. Interactions underlying assembly of *the Escherichia coli* AcrAB-TolC multidrug efflux system. *Mol. Microbiol.* **2004**, *53*, 697–706. [CrossRef]

45. Symmons, M.F.; Bokma, E.; Koronakis, E.; Hughes, C.; Koronakis, V. The assembled structure of a complete tripartite bacterial multidrug efflux pump. *Proc. Natl. Acad. Sci. USA* **2009**, *106*, 7173–7178. [CrossRef]

46. Akama, H.; Matsuura, T.; Kashiwagi, S.; Yoneyama, H.; Narita, S.-I.; Tsukihara, T.; Nakagawa, A.; Nakae, T. Crystal structure of the membrane fusion protein, MexA, of the multidrug transporter in *Pseudomonas aeruginosa*. *J. Biol. Chem.* **2004**, *279*, 25939–25942. [CrossRef]

47. Hayashi, K.; Nakashima, R.; Sakurai, K.; Kitagawa, K.; Yamasaki, S.; Nishino, K.; Yamaguchi, A. AcrB-AcrA fusion proteins that act as multidrug efflux transporters. *J. Bacteriol.* **2016**, *198*, 332–342. [CrossRef]

48. Wang, Z.; Fan, G.; Hryc, C.F.; Blaza, J.N.; Serysheva, I.I.; Schmid, M.F.; Chiu, W.; Luisi, B.F.; Du, D. An allosteric transport mechanism for the AcrAB-TolC multidrug efflux pump. *eLife* **2017**, *6*, e24905. [CrossRef] [PubMed]

49. Kim, J.-S.; Jeong, H.; Song, S.; Kim, H.-Y.; Lee, K.; Hyun, J.; Ha, A.N.-C. Structure of the tripartite multidrug efflux pump AcrAB-TolC suggests an alternative assembly mode. *Mol. Cells* **2015**, *38*, 180–186. [CrossRef] [PubMed]

50. Daury, L.; Orange, F.; Taveau, J.-C.; Verchère, A.; Monlezun, L.; Gounou, C.; Marreddy, R.; Picard, M.; Broutin, I.; Pos, K.M.; et al. Tripartite assembly of RND multidrug efflux pumps. *Nat. Commun.* **2016**, *7*, 10731. [CrossRef] [PubMed]

51. Nakashima, R.; Sakurai, K.; Yamasaki, S.; Nishino, K.; Yamaguchi, A. Structures of the multidrug exporter AcrB reveal a proximal multisite drug-binding pocket. *Nature* **2011**, *480*, 565–569. [CrossRef] [PubMed]

52. Murakami, S.; Nakashima, R.; Yamashita, E.; Matsumoto, T.; Yamaguchi, A. Crystal structures of a multidrug transporter reveal a functionally rotating mechanism. *Nat. Cell Biol.* **2006**, *443*, 173–179. [CrossRef] [PubMed]

53. Seeger, M.A.; Schiefner, A.; Eicher, T.; Verrey, F.; Diederichs, K.; Pos, K.M. Structural asymmetry of AcrB trimer suggests a peristaltic pump mechanism. *Science* **2006**, *313*, 1295–1298. [CrossRef]

54. Eicher, T.; Cha, H.-J.; Seeger, M.A.; Brandstatter, L.; El-Delik, J.; Bohnert, J.A.; Kern, W.V.; Verrey, F.; Grutter, M.G.; Diederichs, K.; et al. Transport of drugs by the multidrug transporter AcrB involves an access and a deep binding pocket that are separated by a switch-loop. *Proc. Natl. Acad. Sci. USA* **2012**, *109*, 5687–5692. [CrossRef] [PubMed]

55. Vargiu, A.V.; Nikaido, H. Multidrug binding properties of the AcrB efflux pump characterized by molecular dynamics simulations. *Proc. Natl. Acad. Sci. USA* **2012**, *109*, 20637–20642. [CrossRef] [PubMed]

56. Lau, C.H.-F.; Hughes, D.; Poole, K. MexY-promoted aminoglycoside resistance in *Pseudomonas aeruginosa*: Involvement of a putative proximal binding pocket in aminoglycoside recognition. *mBio* **2014**, *5*, e01068-14. [CrossRef] [PubMed]

57. Zwama, M.; Hayashi, K.; Sakurai, K.; Nakashima, R.; Kitagawa, K.; Nishino, K.; Yamaguchi, A. Hoisting-loop in bacterial multidrug exporter AcrB is a highly flexible hinge that enables the large motion of the subdomains. *Front. Microbiol.* **2017**, *8*, 2095. [CrossRef] [PubMed]

58. Sjuts, H.; Vargiu, A.V.; Kwasny, S.M.; Nguyen, S.T.; Kim, H.-S.; Ding, X.; Ornik-Cha, A.; Ruggerone, P.; Bowlin, T.L.; Nikaido, H.; et al. Molecular basis for inhibition of AcrB multidrug efflux pump by novel and powerful pyranopyridine derivatives. *Proc. Natl. Acad. Sci. USA* **2016**, *113*, 3509–3514. [CrossRef] [PubMed]

59. Saier, M.H. A functional-phylogenetic classification system for transmembrane solute transporters. *Microbiol. Mol. Biol. Rev.* **2000**, *64*, 354–411. [CrossRef]

60. Nikaido, H. RND transporters in the living world. *Res. Microbiol.* **2018**, *169*, 363–371. [CrossRef]

61. Kim, H.-S.; Nagore, D.; Nikaido, H. Multidrug efflux pump MdtBC of *Escherichia coli* is active only as a B2C heterotrimer. *J. Bacteriol.* **2009**, *192*, 1377–1386. [CrossRef]

62. Górecki, K.; McEvoy, M.M. Phylogenetic analysis reveals an ancient gene duplication as the origin of the MdtABC efflux pump. *PLoS ONE* **2020**, *15*, e0228877. [CrossRef]

63. Mima, T.; Joshi, S.; Gomez-Escalada, M.; Schweizer, H.P. Identification and characterization of TriABC-OpmH, a triclosan efflux pump of *Pseudomonas aeruginosa* requiring two membrane fusion proteins. *J. Bacteriol.* **2007**, *189*, 7600–7609. [CrossRef]

64. Fabre, L.; Ntreh, A.T.; Yazidi, A.; Leus, I.V.; Weeks, J.W.; Bhattacharyya, S.; Ruickoldt, J.; Rouiller, I.; Zgurskaya, H.I.; Sygusch, J. A "drug sweepings State of the TriABC triclosan efflux pump from *Pseudomonas aeruginosa*. *Structure* **2021**, *29*, 261–274. [CrossRef] [PubMed]

65. McWilliam, H.; Li, W.; Uludag, M.; Squizzato, S.; Park, Y.M.; Buso, N.; Cowley, A.; Lopez, R. Analysis tool web services from the EMBL-EBI. *Nucleic Acids Res.* **2013**, *41*, W597–W600. [CrossRef] [PubMed]

66. Eddy, S.R. Accelerated profile HMM searches. *PLoS Comput. Biol.* **2011**, *7*, e1002195. [CrossRef] [PubMed]

67. Berezin, C.; Glaser, F.; Rosenberg, J.; Paz, I.; Pupko, T.; Fariselli, P.; Casadio, R.; Ben-Tal, N. ConSeq: The identification of functionally and structurally important residues in protein sequences. *Bioinformatics* **2004**, *20*, 1322–1324. [CrossRef] [PubMed]

68. Ashkenazy, H.; Abadi, S.; Martz, E.; Chay, O.; Mayrose, I.; Pupko, T.; Ben-Tal, N. ConSurf 2016: An improved methodology to estimate and visualize evolutionary conservation in macromolecules. *Nucleic Acids Res.* **2016**, *44*, W344–W350. [CrossRef]

69. Su, C.-C.; Li, M.; Gu, R.; Takatsuka, Y.; McDermott, G.; Nikaido, H.; Yu, E.W. Conformation of the AcrB multidrug efflux pump in mutants of the putative proton relay pathway. *J. Bacteriol.* **2006**, *188*, 7290–7296. [CrossRef]

70. Seeger, M.A.; von Ballmoos, C.; Verrey, F.; Pos, K.M. Crucial role of Asp408 in the proton translocation pathway of multidrug transporter AcrB: Evidence from site-directed mutagenesis and carbodiimide labeling. *Biochemistry* **2009**, *48*, 5801–5812. [CrossRef] [PubMed]

71. Eicher, T.; Seeger, M.A.; Anselmi, C.; Zhou, W.; Brandstätter, L.; Verrey, F.; Diederichs, K.; Faraldo-Gómez, J.D.; Pos, K.M. Coupling of remote alternating-access transport mechanisms for protons and substrates in the multidrug efflux pump AcrB. *eLife* **2014**, *3*, e03145. [CrossRef]

72. Zwama, M.; Yamasaki, S.; Nakashima, R.; Sakurai, K.; Nishino, K.; Yamaguchi, A. Multiple entry pathways within the efflux transporter AcrB contribute to multidrug recognition. *Nat. Commun.* **2018**, *9*, 1–9. [CrossRef]

73. Tam, H.-K.; Malviya, V.N.; Foong, W.-E.; Herrmann, A.; Malloci, G.; Ruggerone, P.; Vargiu, A.V.; Pos, K.M. Binding and transport of carboxylated drugs by the multidrug transporter AcrB. *J. Mol. Biol.* **2020**, *432*, 861–877. [CrossRef]

74. Oswald, C.; Tam, H.-K.; Pos, K.M. Transport of lipophilic carboxylates is mediated by transmembrane helix 2 in multidrug transporter AcrB. *Nat. Commun.* **2016**, *7*, 13819. [CrossRef]

75. Schuster, S.; Vavra, M.; Kern, W.V. Evidence of a substrate-discriminating entrance channel in the lower porter domain of the multidrug resistance efflux pump AcrB. *Antimicrob. Agents Chemother.* **2016**, *60*, 4315–4323. [CrossRef]

76. Ababou, A.; Koronakis, V. Structures of gate loop variants of the AcrB drug efflux pump bound by erythromycin substrate. *PLoS ONE* **2016**, *11*, e0159154. [CrossRef] [PubMed]

77. Hawkey, J.; Ascher, D.; Judd, L.M.; Wick, R.R.; Kostoulias, X.; Cleland, H.; Spelman, D.W.; Padiglione, A.; Peleg, A.Y.; Holt, K.E. Evolution of carbapenem resistance in *Acinetobacter baumannii* during a prolonged infection. *Microb. Genom.* **2018**, *4*, e000165. [CrossRef] [PubMed]

78. Santos-Lopez, A.; Marshall, C.W.; Welp, A.L.; Turner, C.; Rasero, J.; Cooper, V.S. The roles of history, chance, and natural selection in the evolution of antibiotic resistance. *bioRxiv* **2020**. pre-print. [CrossRef]

79. Soparkar, K.; Kinana, A.D.; Weeks, J.W.; Morrison, K.D.; Nikaido, H.; Misra, R. Reversal of the drug binding pocket defects of the AcrB multidrug efflux pump protein of *Escherichia coli*. *J. Bacteriol.* **2015**, *197*, 3255–3264. [CrossRef] [PubMed]

80. Blair, J.; Bavro, V.N.; Ricci, V.; Modi, N.; Cacciotto, P.; Kleinekathöfer, U.; Ruggerone, P.; Vargiu, A.V.; Baylay, A.J.; Smith, H.E.; et al. AcrB drug-binding pocket substitution confers clinically relevant resistance and altered substrate specificity. *Proc. Natl. Acad. Sci. USA* **2015**, *112*, 3511–3516. [CrossRef]

81. Johnson, R.M.; Fais, C.; Parmar, M.; Cheruvara, H.; Marshall, R.; Hesketh, S.J.; Feasey, M.C.; Ruggerone, P.; Vargiu, A.V.; Postis, V.L.G.; et al. Cryo-EM structure and molecular dynamics analysis of the fluoroquinolone resistant mutant of the AcrB transporter from *Salmonella*. *Microorganisms* **2020**, *8*, 943. [CrossRef]

82. López-Causapé, C.; Sommer, L.M.; Cabot, G.; Rubio, R.; Ocampo-Sosa, A.A.; Johansen, H.K.; Figuerola, J.; Cantón, R.; Kidd, T.J.; Molin, S.; et al. Evolution of the *Pseudomonas aeruginosa* mutational resistome in an international cystic fibrosis clone. *Sci. Rep.* **2017**, *7*, 1–15. [CrossRef]

83. Wardell, S.J.T.; Rehman, A.; Martin, L.W.; Winstanley, C.; Patrick, W.M.; Lamont, I.L. A large-scale whole-genome comparison shows that experimental evolution in response to antibiotics predicts changes in naturally evolved clinical *Pseudomonas aeruginosa*. *Antimicrob. Agents Chemother.* **2019**, *63*, 6726–6734. [CrossRef] [PubMed]

84. Greipel, L.; Fischer, S.; Klockgether, J.; Dorda, M.; Mielke, S.; Wiehlmann, L.; Cramer, N.; Tümmler, B. Molecular epidemiology of mutations in antimicrobial resistance loci of *Pseudomonas aeruginosa* isolates from airways of cystic fibrosis patients. *Antimicrob. Agents Chemother.* **2016**, *60*, 6726–6734. [CrossRef]

85. Hooda, Y.; Sajib, M.S.I.; Rahman, H.; Luby, S.P.; Bondy-Denomy, J.; Santosham, M.; Andrews, J.R.; Saha, S.K.; Saha, S. Molec-ular mechanism of azithromycin resistance among typhoidal *Salmonella* strains in Bangladesh identified through passive pediatric surveillance. *PLoS Negl. Trop. Dis.* **2019**, *13*, e0007868. [CrossRef]

86. Iqbal, J.; Dehraj, I.F.; Carey, M.E.; Dyson, Z.A.; Garrett, D.; Seidman, J.C.; Kabir, F.; Saha, S.; Baker, S.; Qamar, F.N. A race against time: Reduced azithromycin susceptibility in *Salmonella enterica* serovar Typhi in Pakistan. *mSphere* **2020**, *5*, 8299. [CrossRef]

87. Katiyar, A.; Sharma, P.; Dahiya, S.; Singh, H.; Kapil, A.; Kaur, P. Genomic profiling of antimicrobial resistance genes in clinical isolates of *Salmonella* Typhi from patients infected with Typhoid fever in India. *Sci. Rep.* **2020**, *10*, 1–15. [CrossRef]

88. Sajib, M.S.I.; Tanmoy, A.M.; Hooda, Y.; Rahman, H.; Andrews, J.R.; Garrett, D.O.; Endtz, H.P.; Saha, S.K.; Saha, S. Tracking the emergence of azithromycin resistance in multiple genotypes of typhoidal *Salmonella*. *mBio* **2021**, *12*, 109. [CrossRef]

89. Ma, K.C.; Mortimer, T.D.; Grad, Y.H. Efflux pump antibiotic binding site mutations are associated with azithromycin non-susceptibility in clinical *Neisseria gonorrhoeae* isolates. *mBio* **2020**, *11*, 01419-18. [CrossRef]

90. Jia, X.; Ren, H.; Nie, X.; Li, Y.; Li, J.; Qin, T. Antibiotic resistance and azithromycin resistance mechanism of *Legionella pneumophila* serogroup 1 in China. *Antimicrob. Agents Chemother.* **2019**, *63*, 01747-18. [CrossRef] [PubMed]

91. Poole, K. Efflux-mediated antimicrobial resistance. *J. Antimicrob. Chemother.* **2005**, *56*, 20–51. [CrossRef] [PubMed]

92. Nikaido, H.; Pagès, J.-M. Broad-specificity efflux pumps and their role in multidrug resistance of Gram-negative bacteria. *FEMS Microbiol. Rev.* **2012**, *36*, 340–363. [CrossRef] [PubMed]

93. Hessa, T.; Kim, H.; Bihlmaier, K.; Lundin, C.; Boekel, J.; Andersson, H.; Nilsson, I.; White, S.H.; Von Heijne, G. Recognition of transmembrane helices by the endoplasmic reticulum translocon. *Nat. Cell Biol.* **2005**, *433*, 377–381. [CrossRef]

94. Bohnert, J.A.; Schuster, S.; Seeger, M.A.; Fähnrich, E.; Pos, K.M.; Kern, W.V. Site-directed mutagenesis reveals putative substrate binding residues in the *Escherichia coli* RND efflux pump AcrB. *J. Bacteriol.* **2008**, *190*, 8225–8229. [CrossRef] [PubMed]

95. Vargiu, A.V.; Collu, F.; Schulz, R.; Pos, K.M.; Zacharias, M.; Kleinekathöfer, U.; Ruggerone, P. Effect of the F610A mutation on substrate extrusion in the AcrB transporter: Explanation and rationale by molecular dynamics simulations. *J. Am. Chem. Soc.* **2011**, *133*, 10704–10707. [CrossRef] [PubMed]

96. Ramaswamy, V.K.; Vargiu, A.V.; Malloci, G.; Dreier, J.; Ruggerone, P. Molecular rationale behind the differential substrate specificity of bacterial RND multidrug transporters. *Sci. Rep.* **2017**, *7*, 1–18. [CrossRef] [PubMed]

97. Rosenberg, E.Y.; Ma, D.; Nikaido, H. AcrD of *Escherichia coli* is an aminoglycoside efflux pump. *J. Bacteriol.* **2000**, *182*, 1754–1756. [CrossRef]

98. Aires, J.R.; Nikaido, H. Aminoglycosides are captured from both periplasm and cytoplasm by the AcrD multidrug efflux transporter of *Escherichia coli*. *J. Bacteriol.* **2005**, *187*, 1923–1929. [CrossRef]

99. Kobayashi, N.; Tamura, N.; Van Veen, H.W.; Yamaguchi, A.; Murakami, S. β-Lactam selectivity of multidrug transporters AcrB and AcrD resides in the proximal binding pocket. *J. Biol. Chem.* **2014**, *289*, 10680–10690. [CrossRef]

100. Morita, Y.; Tomida, J.; Kawamura, Y. MexXY multidrug efflux system of *Pseudomonas aeruginosa*. *Front. Microbiol.* **2012**, *3*, e408. [CrossRef]

101. Dey, D.; Kavanaugh, L.G.; Conn, G.L. Antibiotic substrate selectivity of *Pseudomonas aeruginosa* MexY and MexB efflux systems is determined by a Goldilocks affinity. *Antimicrob. Agents Chemother.* **2020**, *64*, 1144. [CrossRef]

102. Ramaswamy, V.K.; Vargiu, A.V.; Malloci, G.; Dreier, J.; Ruggerone, P. Molecular determinants of the promiscuity of MexB and MexY multidrug transporters of *Pseudomonas aeruginosa*. *Front. Microbiol.* **2018**, *9*, 1144. [CrossRef]

103. Massip, C.; Descours, G.; Ginevra, C.; Doublet, P.; Jarraud, S.; Gilbert, C. Macrolide resistance in *Legionella pneumophila*: The role of LpeAB efflux pump. *J. Antimicrob. Chemother.* **2017**, *72*, 1327–1333. [CrossRef] [PubMed]

104. Vandewalle-Capo, M.; Massip, C.; Descours, G.; Charavit, J.; Chastang, J.; Billy, P.A.; Boisset, S.; Lina, G.; Gilbert, C.; Maurin, M.; et al. Minimum inhibitory concentration (MIC) distribution among wild-type strains of *Legionella pneumophila* identifies a subpopulation with reduced susceptibility to macrolides owing to efflux pump genes. *Int. J. Antimicrob. Agents* **2017**, *50*, 684–689. [CrossRef] [PubMed]

105. Masuda, N.; Sakagawa, E.; Ohya, S.; Gotoh, N.; Tsujimoto, H.; Nishino, T. Substrate specificities of MexAB-OprM, MexCD-OprJ, and MexXY-OprM efflux pumps in *Pseudomonas aeruginosa*. *Antimicrob. Agents Chemother.* **2000**, *44*, 3322–3327. [CrossRef]

106. Dreier, J.; Ruggerone, P. Interaction of antibacterial compounds with RND efflux pumps in Pseudomonas aeruginosa. *Front. Microbiol.* **2015**, *6*, 660. [CrossRef]

107. Horiyama, T.; Nishino, K. AcrB, AcrD, and MdtABC multidrug efflux systems are involved in enterobactin export in *Esche-richia coli*. *PLoS ONE* **2014**, *9*, e108642. [CrossRef] [PubMed]

108. Vargiu, A.V.; Ramaswamy, V.K.; Malvacio, I.; Malloci, G.; Kleinekathöfer, U.; Ruggerone, P. Water-mediated interactions enable smooth substrate transport in a bacterial efflux pump. *Biochim. Biophys. Acta (BBA) Gen. Subj.* **2018**, *1862*, 836–845. [CrossRef] [PubMed]

109. Atzori, A.; Malviya, V.N.; Malloci, G.; Dreier, J.; Pos, K.M.; Vargiu, A.V.; Ruggerone, P. Identification and characterization of carbapenem binding sites within the RND-transporter AcrB. *Biochim. Biophys. Acta (BBA) Biomembr.* **2019**, *1861*, 62–74. [CrossRef]

110. Husain, F.; Bikhchandani, M.; Nikaido, H. Vestibules are part of the substrate path in the multidrug efflux transporter AcrB of *Escherichia coli*. *J. Bacteriol.* **2011**, *193*, 5847–5849. [CrossRef]

111. Duy, P.T.; Dongol, S.; Giri, A.; To, N.T.N.; Thanh, H.N.D.; Quynh, N.P.N.; Trung, P.D.; Thwaites, G.E.; Basnyat, B.; Baker, S.; et al. The emergence of azithromycin-resistant *Salmonella* Typhi in Nepal. *JAC-Antimicrob. Resist* **2020**, *2*, 01509–01520. [CrossRef]

112. Yang, L.; Shi, H.; Zhang, L.; Lin, X.; Wei, Y.; Jiang, H.; Zeng, Z. Emergence of two AcrB substitutions conferring multidrug resistance to *Salmonella* spp. *Antimicrob. Agents Chemother.* **2021**, *65*, e01589-20. [CrossRef]

113. Wadsworth, C.B.; Arnold, B.J.; Sater, M.R.A.; Grad, Y.H. Azithromycin resistance through interspecific acquisition of an epistasis-dependent efflux pump component and transcriptional regulator in *Neisseria gonorrhoeae*. *mBio* **2018**, *9*, 5555. [CrossRef]

114. Cudkowicz, N.A.; Schuldiner, S. Deletion of the major *Escherichia coli* multidrug transporter AcrB reveals transporter plasticity and redundancy in bacterial cells. *PLoS ONE* **2019**, *14*, e0218828. [CrossRef] [PubMed]

115. Langevin, A.M.; El Meouche, I.; Dunlop, M.J. Mapping the role of AcrAB-TolC efflux pumps in the evolution of antibiotic resistance reveals near-MIC treatments facilitate resistance acquisition. *mSphere* **2020**, *5*, 284. [CrossRef]

116. Hoeksema, M.; Jonker, M.J.; Brul, S.; Ter Kuile, B.H. Effects of a previously selected antibiotic resistance on mutations acquired during development of a second resistance in *Escherichia coli*. *BMC Genom.* **2019**, *20*, 1–14. [CrossRef] [PubMed]

117. Kinana, A.D.; Vargiu, A.V.; Nikaido, H. Some ligands enhance the efflux of other ligands by the *Escherichia coli* multidrug pump AcrB. *Biochemistry* **2013**, *52*, 8342–8351. [CrossRef] [PubMed]

118. Schuster, S.; Kohler, S.; Buck, A.; Dambacher, C.; König, A.; Bohnert, J.A.; Kern, W.V. Random mutagenesis of the multidrug transporter AcrB from *Escherichia coli* for identification of putative target residues of efflux pump inhibitors. *Antimicrob. Agents Chemother.* **2014**, *58*, 6870–6878. [CrossRef] [PubMed]

119. Mingardon, F.; Clement, C.; Hirano, K.; Nhan, M.; Luning, E.G.; Chanal, A.; Mukhopadhyay, A. Improving olefin tolerance and production in *E. coli* using native and evolved AcrB. *Biotechnol. Bioeng.* **2015**, *112*, 879–888. [CrossRef]

120. Shafer, W.M. Mosaic drug efflux gene sequences from commensal *Neisseria* can lead to low-level azithromycin resistance expressed by *Neisseria gonorrhoeae* clinical isolates. *mBio* **2018**, *9*, e01747-18. [CrossRef]

121. Grad, Y.H.; Harris, S.R.; Kirkcaldy, R.D.; Green, A.G.; Marks, D.S.; Bentley, S.D.; Trees, D.; Lipsitch, M. Genomic epidemiology of gonococcal resistance to extended-spectrum cephalosporins, macrolides, and fluoroquinolones in the United States, 2000–2013. *J. Infect. Dis.* **2016**, *214*, 1579–1587. [CrossRef]

122. Pan, W.; Spratt, B.G. Regulation of the permeability of the gonococcal cell envelope by the mtr system. *Mol. Microbiol.* **1994**, *11*, 769–775. [CrossRef]

123. Shafer, W.M.; Balthazar, J.T.; Hagman, K.E.; Morse, S.A. Missense mutations that alter the DNA-binding domain of the MtrR protein occur frequently in rectal isolates of *Neisseria gonorrhoeae* that are resistant to faecal lipids. *Microbiology* **1995**, *141*, 907–911. [CrossRef]

124. Ohneck, E.A.; Zaluck, A.Y.M.; Johnson, P.J.T.; Dhulipala, V.; Golparian, D.; Unemo, M.; Jerse, A.E.; Shafer, W.M. A novel mechanism of high-level, broad-spectrum antibiotic resistance caused by a single base pair change in *Neisseria gonorrhoeae*. *mBio* **2011**, *2*, 1652–1653. [CrossRef] [PubMed]

125. Galarza, P.G.; Abad, R.; Canigia, L.F.; Buscemi, L.; Pagano, I.; Oviedo, C.; Vázquez, J.A. New mutation in 23S rRNA gene associated with high level of azithromycin resistance in *Neisseria gonorrhoeae*. *Antimicrob. Agents Chemother.* **2010**, *54*, 1652–1653. [CrossRef] [PubMed]

126. Cao, H.; Xia, T.; Li, Y.; Xu, Z.; Bougouffa, S.; Lo, Y.K.; Bajic, V.B.; Luo, H.; Woo, P.C.Y.; Yan, A. Uncoupled quorum sensing modulates the interplay of virulence and resistance in a multidrug-resistant clinical *Pseudomonas aeruginosa* isolate belonging to the MLST550 clonal complex. *Antimicrob. Agents Chemother.* **2019**, *63*, e01944-18. [CrossRef]

Review

Clinical Status of Efflux Resistance Mechanisms in Gram-Negative Bacteria

Anne Davin-Regli *, Jean-Marie Pages and Aurélie Ferrand

Membranes et Cibles Thérapeutiques-Faculté de Pharmacie, UMR_MD1, U-1261, Aix-Marseille Université, 27 Boulevard Jean Moulin, 13385 Marseille, France; jean-marie.pages@univ-amu.fr (J.-M.P.); ferrand.aurelie@hotmail.fr (A.F.)
* Correspondence: anne-veronique.regli@univ-amu.fr

Abstract: Antibiotic efflux is a mechanism that is well-documented in the phenotype of multidrug resistance in bacteria. Efflux is considered as an early facilitating mechanism in the bacterial adaptation face to the concentration of antibiotics at the infectious site, which is involved in the acquirement of complementary efficient mechanisms, such as enzymatic resistance or target mutation. Various efflux pumps have been described in the Gram-negative bacteria most often encountered in infectious diseases and, in healthcare-associated infections. Some are more often involved than others and expel virtually all families of antibiotics and antibacterials. Numerous studies report the contribution of these pumps in resistant strains previously identified from their phenotypes. The authors characterize the pumps involved, the facilitating antibiotics and those mainly concerned by the efflux. However, today no study describes a process for the real-time quantification of efflux in resistant clinical strains. It is currently necessary to have at hospital level a reliable and easy method to quantify the efflux in routine and contribute to a rational choice of antibiotics. This review provides a recent overview of the prevalence of the main efflux pumps observed in clinical practice and provides an idea of the prevalence of this mechanism in the multidrug resistant Gram-negative bacteria. The development of a routine diagnostic tool is now an emergency need for the proper application of current recommendations regarding a rational use of antibiotics.

Keywords: RND efflux pumps; prevalence of efflux resistance mechanisms; hospital acquired infections

Citation: Davin-Regli, A.; Pages, J.-M.; Ferrand, A. Clinical Status of Efflux Resistance Mechanisms in Gram-Negative Bacteria. *Antibiotics* **2021**, *10*, 1117. https://doi.org/10.3390/antibiotics10091117

Academic Editors: Isabelle Broutin, Attilio V Vargiu, Henrietta Venter, Gilles Phan and Manuel Simões

Received: 15 July 2021
Accepted: 15 September 2021
Published: 16 September 2021

Publisher's Note: MDPI stays neutral with regard to jurisdictional claims in published maps and institutional affiliations.

Copyright: © 2021 by the authors. Licensee MDPI, Basel, Switzerland. This article is an open access article distributed under the terms and conditions of the Creative Commons Attribution (CC BY) license (https://creativecommons.org/licenses/by/4.0/).

1. Introduction

Membrane-associated antibiotic resistance is a key mechanism in Gram-negative bacteria that efficiently controls the intracellular concentration of various drugs [1]. Two complementary processes, membrane impermeability and the expression of efflux pumps, limit the concentration of deleterious compounds inside bacteria by impairing the entry and expelling the internal molecules [1–4]. Consequently, they play a key role in protecting the bacterial cells against aggressive chemicals, such as antibiotics, disinfectants, conservatives, detergents, etc. In Gram-negative bacteria, the envelope permeability has long been studied, but due to its complex organization and variability, serious gaps in the research remain, regarding the antibiotic translocation and accumulation inside living bacteria. Multidrug efflux pumps active in bacterial cells belong to the ABC, MF, SMR, MATE, PACE and RND superfamilies superfamilies [4,5]. Except ABC transporters using ATP hydrolysis as energy source, the other drug efflux pumps are antiporters, H^+ or Na^+, with the active component inserted in IM (Inner/Cytoplasmic Membrane) [1,4]. This component recognizes and captures the drug substrate from the cytoplasm or IM surface and carries the molecule to an outer membrane channel that releases it to external medium. In Gram-negative bacteria, AcrAB-TolC and MexAB-OprM are the archetypes of the tripartite efflux pump complexes reported in clinical strains (Figure 1). The regulation of the expression of the pump, and its 3D organization, have been extensively studied and reported in numerous reviews with three AcrAB-MexAB in IM for three TolC-OprM partners in OM (Outer

Membrane) [1,4–8]. Interestingly, these transporters exhibit an important versatility and flexibility of substrates, from bile salts to dyes including antibiotics and detergents with a large range of sizes, chemical structures, and charges [1]. The mechanism of efflux as a protection barrier has been genetically and molecularly studied. The functional structures of efflux complexes, e.g., AcrAB-TolC and MexAB-OprM, have been solved at high resolution providing information regarding the energy requirements, dynamic and selectivity of the transport, which includes a recognition, binding and expel step [1,2]. Today, the role of specific amino acids residues during the transport inside the different AcrB-MexB pockets and the subsequent conformational changes of the efflux pump subunits are described [1,4,5].

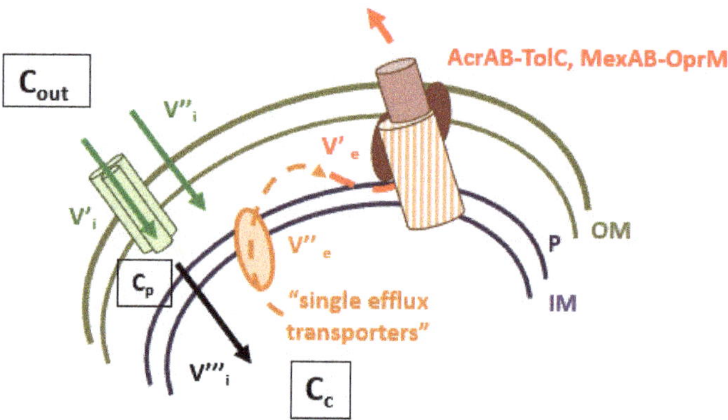

Figure 1. Intracellular concentration of antibiotics and RND efflux pumps. V'_i and V''_i represent the influx rate across the outer membrane, V'''_i represents the diffusion rate across the inner membrane. C_{out}, C_p and C_c represent the external, periplasmic and cytoplasmic concentration of antibiotics, respectively. V''_e and V'_e represent the efflux rate accross the inner and outer membranes. AcrAB-TolC and MexAB-OprM illustrate the archetype of efflux pumps detected in clinical isolates.

Importantly, clinical strains express a basal level of efflux pumps and can overproduce them via mutation in regulators (negative or positive), stimulation by external stresses, or the acquisition of mobile genetic elements coding for additional transport systems [1,5]. A sophisticated regulation cascade can efficiently manage the membrane transporters, porins and efflux components, which are involved in drug transport [5,9].

However, the exact prevalence and contribution of efflux in clinical pathogens are not well defined [1]. This gap is due to a lack of clinical diagnosis assays that impair a rapid and clear identification of active effluxes in hospital microbiology services in contrast to available methods for detecting enzymes or target mutations involved in resistance (inhibitor of β-lactamase, sequence of specific gene, etc.) [1]. Today, the vast majority of clinical studies carried out on efflux resistance use either the ratio of MICs obtained in the absence or presence of efflux inhibitors, the interpretation of the data obtained after the evaluation of the impact of the target mutations or the presence of enzyme barrier, or, more recently, the use of a synthetic substrate of an efflux pump devoid of antibiotic properties (e.g., ethidium bromide or other chemicals). Consequently, it dramatically lacks a real-time approach based on the measurement of the accumulation of an antibiotic correlated with its antibacterial activity to precisely define the role, prevalence, and impact of efflux in clinic.

2. Towards an Detection of Efflux Mechanisms

During the last decade, the respective developments of fluorimetry and mass spectrometry have generated several experimental protocols supporting a serious breakthrough in the precise quantification of the intracellular accumulation of drugs inside bacteria. Consequently, the impact of membrane impermeability and active efflux in resistant isolates have been revealed [3,10], allowing for a clear quantification of the role of these membrane-associated mechanisms in the resistance [1,11,12]. The key point is the requirement of the robust internal controls necessary to validate the drug signals and to allow comparisons between various assays involving various bacterial strains [12]. Unfortunately, it seems difficult today to efficiently transfer these research protocols in simple routine tests for hospital diagnosis. This point could be partly solved by using a first screen in a collection of Multi-Drug Resistant (MDR) isolates and determining the accumulation of selected antibiotics used in clinical settings in a second step. The final analyses will be the determination of the existing correlative index between the accumulation level and susceptibility for molecules that have been classified as a substrate of efflux pumps [3]. This aspect is especially important to define the relevance and contribution of efflux in clinical isolates and the rapid adaptation of bacterial cells under antibiotic pressure.

3. Genetic Regulation

The expression of the various efflux systems is controlled by an interplay of different regulators and regulation cascades, some of which are major players and other substitutes or alternate pathways [9]. The regulation signal seems to be associated with the internal concentration of toxic chemicals that play a role during the trigger of the cascade-inducing efflux expression [9]. In Gram-negative bacteria, the genetic regulation of pumps is mainly managed by two key methods:

1.   Pathway 1: the intervention of a protein-modulating expression of genes by a fixation on their promoters. Regulatory genetic proteins possess α-helix-turn-α-helix (HTH) DNA-binding motifs and can be activators or repressors [13].
2.   Pathway 2: the activation of a two-component systems (TCS) which interferes with gene expressions when environmental stresses require a bacterial adaptation [1].

3.1. HTH Family Regulators

Positive regulation is managed by the AraC-XylS family, which is well-documented in *Enterobacteriaceae* (MarA, RamA, SoxS, Rob, RarA) or H-NS proteins (histone-like structuring nucleoid protein) (SdiA, FIS, CsrA). In this group, representative repressors are TetR/AcrR/RamR/MexR and the OqxR family and are described in *Enterobacteriaceae* and in *Pseudomonadaceae*. They are repressors of the genetic cascade regulation or of the pumps genes operons [9].

3.2. TCS Systems

TCS systems are activated when bacteria sense environmental change such as detrimental hazards to their key physiological activities, and contribute to adaptation/defense. TCS systems (CpxAR, Rcs, BaeSR, PhoPQ, and EnvZ/OmpR) (RocS2-RocA2, ParR-ParS, AmgR-AmgS, CzcR-CzcS and CopR-CopS) can sense the external medium modifications such as pH, osmotic strength, oxidative stress, nutrient starvation, and toxic chemicals, etc. [1].

4. Clinical Impact Situation

The basal efflux present in bacterial cells is essentially the first mechanism with membrane impermeability that faces antibiotics in clinical isolates [14–16]. The basal activity is not clinically detected in a wild strain, although various stimuli are able to quickly initiate their overexpression and synthesis. Although not detected in routine, the efflux paves the way for the most radical mechanisms of resistance [3]. A sub-inhibitory intracellular concentration of the antibiotic then promotes the development or acquisition of more specific mechanisms of resistance-like enzymatic responses or target mutations [15,16]. For example, several mutations are found in genes, *gyrA* and *parC*, coding the gyrase target and they are frequently associated with efflux resulting in high MIC levels [17]. Most of the studied clinical strains exhibit a specific resistance combined to efflux mechanisms (Table 1). The inhibition of efflux mechanisms obtained in the laboratory for these strains never completely restores susceptibility to the antibiotic despite a significative diminution of MIC. Given the lack of automated techniques available to monitor efflux activity, efflux is not usually sought in epidemiological studies involving samples of clinical strains; nevertheless, some studies have been investigated, particularly with *Pseudomonas aeruginosa* [18]. We review the main results and observations of efflux in a clinical context for the various species most concerned.

4.1. Haemophilus influenzae

In *Haemophilus influenzae*, an AcrAB homolog pump system is identified and shares a similar antibiotic efflux profile as AcrAB overexpression generates in *E. coli* [19].

The basal expression of the efflux represents a current mechanism which concerns macrolides and ketolides resistance in most clinical strains [20]. Efflux is frequently found to be associated with mutations in ribosomal proteins resulting in high MICs to these classes of antibiotics [20,21]. Recently, Cherkaoui et al. studied various resistance mechanisms for imipenem heteroresistance in 46 *Haemophilus influenzae* isolates [22]. They concluded that the involvement of the pump AcrAB overexpression caused by a partial deletion in *acrR* led to an overexpression of the AcrAB-TolC efflux complex. For them, efflux was necessary for the development of imipenem heteroresistance [22].

4.2. Helicobacter pylori

In *Helicobacter pylori*, several efflux pumps systems were identified: RND pump complexes named HefABC, HefDEF and HefGHI, ABC and MFS-type efflux pump not yet named [23].

From two series of *H. pylori* clinical isolates resistant to clarithromycin which was the key drug in the eradication therapy for *H. pylori*, two TolC homologs, hp0605 and hp1489, were identified as involved in clarithromycin resistance [24]. Moreover, metronidazole resistance was reported in five MDR clinical isolates exposed to varying concentrations of this antibiotic, and the efflux mechanism was observed in this phenotype [24]. Authors reported that metronidazole could stimulate the expression of efflux pumps genes and expression occurred when higher doses of metronidazole were used [25]. Among 12 clinical strains resistant to clarythromycin, the genetic variants of the three RND pumps of the series of Hef were identified [26]. Authors detected specific single-nucleotide substitutions in resistant strains [26].

4.3. Campylobacter jejuni

Campylobacter jejuni followed by *Campylobacter coli* are the most common bacteria isolated in campylobacteriosis [27]. Macrolides, with erythromycin or fluoroquinolones, are the first line of treatment. The mechanisms of resistance to macrolides and fluoroquinolones are independent and, for macrolides, the resistance mechanisms are also described for other members of this antibiotic class as the ketolide telithromycin [28]. There are two major mechanisms of antibiotic resistance: target mutations and the decrease in the intracellular concentration of antibiotics. The decrease in the intracellular concentration of antibiotics may occur via efflux pump activity, which was described for the first time in 1995 [29].

CmeABC is the major efflux complex that can extrude a large panel of molecules [30]. *CmeB* is the inner membrane transporter (energy-active part), *CmeC* is the outer membrane channel protein, and *CmeA* is the periplasmic fusion protein [30,31]. *cmeR*, located in the promoter region of *cmeABC* operon, encodes *CmeR*, a Tet-like regulator, which controls the efflux complex expression. The *CmeR* acts as a local repressor for *CmeABC* [32].

Interactions with other efflux complexes like *CmeDEF* may result in intrinsic resistance to these antibiotics [33]. A resistance-enhancing variant of *CmeABC* called RE-*CmeABC* is described as a "super efflux pump" which confers to *Campylobacter* at a higher resistance level than *CmeABC*. This super efflux pump can extrude more antibiotics than the *CmeABC* and a mutation is found in the promoter region of RE-*cmeABC*. Furthermore, the study of the structural modeling of *CmeB* points to amino acid changes that suggest a tighter affinity of this variant to the drugs, compared to parental *CmeB* conferring a fitness benefit to *Campylobacter* under selection pressure. The spreading of this pump is characterized by a horizontal transfer and a clonal expansion [34].

There are two levels of macrolide resistance in *Campylobacter* spp.: a low level of resistance (LLR) and a high level of resistance (HLR) [28]. Regarding the macrolide susceptibility, the LLR strains have no target mutations and are susceptible to the PAβN (Phe-Arg-β-naphthylamide) efflux pump inhibitor, while the HLR strains show target mutations that cannot be bypassed with PAβN addition [35]. It has been shown that in *Campylobacter* clinical isolates, concerning fluoroquinolones, strains target mutations and efflux pump activity works synergistically. However, the resistance to fluoroquinolones may occur only with a *gyrA* mutation, with or without an overexpression of the *CmeABC* efflux complexes [36,37]. Conversely, the efflux alone requires the contribution of several pumps to generate fluoroquinolones resistance without target mutations in *Campylobacter* [38,39].

4.4. Enterobacteriaceae

Enterobacteriaceae, which includes over 30 genera and more than 100 species, are common pathogens in the community and in hospital. Some of them are associated with worldwide epidemics, such as *Salmonella*, *Escherichia coli* and *Shigella* sp. Moreover, they are involved in about 40% of hospital acquired infections, half of them associated with *E. coli* and others corresponding to *Klebsiella pneumoniae* and *oxytoca*, *Enterobacter cloacae* and *K. aerogenes*, *Proteus mirabilis* and *Citrobacter flexneri* [9,14,18]. Among them, it must be noted that the *E. coli* AcrAB-TolC represents the archetype of the tripartite RND efflux complex regarding structural organization, genetic regulation, and activity [1–6].

4.4.1. K. pneumoniae

K. pneumoniae comes after *E. coli*, the main enterobacterial species involved in nosocomial infections, with the highest proportion of isolates producing plasmid-mediated extended-spectrum β-lactamases and carbapenemases [40–43]. Two major RND pumps and their regulatory elements are characterized in *K. pneumoniae*, AcrAB and OqxAB [40,44]. The *oqxAB* operon was originally described on plasmid carried by an *E. coli* strain conferring resistance to chloramphenicol and quinolones [44]. In *K. pneumoniae*, it is associated with the resistance of quinoxalines, quinolones, tigecycline, nitrofurantin and chloramphenicol [45]. The major genetic regulators of these two pumps are RamA and RarA, respectively, together belonging to the AraC-type transcriptional regulators in *K. pneumoniae*, [46,47]. AcrAB is involved in the efflux of various antibiotics including the most recent broad-spectrum antibiotics, such as piperatazocillin or ceftolozane-tazobactam, and OqxAB expels fluoroquinolones, tigecyclin and nitrofurantoin [45]. Concerning the detection of efflux mechanisms and their prevalence in the resistance of clinical strains, Kareem et al. showed that 27 of the 43 selected MDR *K. pneumoniae* clinical strains presented an efflux activity by using an EtBr-agar cartwheel screening method [48]. In 36 MDR clinical isolates in Egypt, *acrAB* genes were found in 82% of tested strains [49]. In 2017, in a Taiwan hospital, 17 MDR *K. pneumoniae* responsible for urinary tract infections were found to have an upregulation and an overexpression of *acrB* and/or *oqxB* in 65% of strains [50]. They were associated with mutations or insertions in OmpK36 porin that altered outer membrane permeability. Elgendy et al. found, in Egyptian hospitals, that 12% of *K. pneumoniae* strains exhibited an overexpression of AcrAB and OqxAB associated with tigecycline resistance [51]. In the same aspect, Sekyere et al. observed that, among isolates from Durban hospitals, half of them presented an efflux mechanism associated with fluoroquinolones resistance [52].

In addition to these two major pumps, other transport proteins systems were described and found involved in clinical resistance. By using transcriptome analysis, Majumdar et al. found that the overexpression of the regulator RarA resulted in the overexpression of genes encoding uncharacterized transport proteins (KPN_03141, SdaCB, and LeuE) [47]. Moreover, Srinivasan et al. first characterized in *K. pneumoniae*, the KpnEF efflux pump [53]. Maurya et al. reported the role of new pumps in *K. pneumoniae* in resistance: for the RND family, KexD (for macrolides and tetracycline); for the SMR family pump, KpnEF, and for the MATE family, KdeA (involved in chloramphenicol, norfloxacin, acriflavine, and ethidium bromide (EtBr) resistance) [54]. Interestingly, KpnEF seemed to be involved in the transport of polysaccharides to the outer layer to form the slimy layer in connection with biofilm formation. The KpnEF pump also mediated resistance to antibiotics such as cefepime, ceftriaxone, colistin, erythromycin, rifampin, tetracycline, streptomycin and seemed to have a significative role in the MDR phenotype [53]. Southern hybridization of the genomes of several MDR strains indicated a 70% occurrence for *kpnEF*. Authors concluded that these supplementary pumps did not contribute much to resistance but provided a synergistic/additive effect in the efflux of β-lactam family.

Lv et al. (2021) found in a clinically isolated pan-resistant *Klebsiella pneumoniae* strain, which was resistant to β-lactams, sulfonamides, bacitracin, tetracycline, aminoglycosides, and chloramphenicol, a combinate involvement of enzymatic mechanisms, and efflux pumps co-intervention [55]. Five types of efflux pump families were identified in clinical strains, including RND pumps (AcrAB, AcrD, MdtABC, and KexD), the ABC superfamily (MacAB), SMR (KpnEF), MATE (KdeA), and the MFS superfamily (EmrAB). Moreover, they identified mutations in the deletion mutations of the regulatory genes, *acrR* and *ramR*, which led to the overexpression of the AcrAB efflux pump [55].

4.4.2. *K. aerogenes*

A range of MDR efflux pumps: AcrAB, AcrD, EmrAB, MacAB, MdtABC, OqxAB, RosAB composed the main resistance efflux mechanisms encoded in *K. aerogenes* [46,56]. The EefABC and AcrAB-TolC complexes of *K. aerogenes* were particularly studied at the functional level [57–60].

Regarding the prevalence of efflux and its evolution in clinical strains, rare specific studies have been published: Chevalier et al. studied efflux mechanisms in two groups of *K. aerogenes* collected in two different periods (1995 and 2003) and their level of resistance was studied and compared within the 8-year period. A noticeable increase in efflux mechanism expressions is observed in one decade using PAβN as an efflux reporter [61]. They concluded that approximately 40% of MDR clinical strains exhibited an active efflux in 2003. Tran et al. investigated the occurrence of efflux and other mechanisms in 44 *K. aerogenes* and *Klebsiella pneumoniae* clinical isolates [43]. A phenotypic efflux in the presence of PAβN was detected in *K. aerogenes* isolates, more frequently than in *K. pneumoniae* (100 versus 38% of isolates). Gayet et al. explored two MDR strains resistant to broad spectrum β-lactams antibiotics, fluoroquinolones, chloramphenicol, tetracycline, and kanamycin [62]. In the presence of PaβN, the tested *K. aerogenes* strains resulted in up to a fourfold decrease in the chloramphenicol MIC, suggesting a significative efflux mechanism. The OqxAB pump was also investigated for its involvement in *E. cloacae*, as in *K. aerogenes* strains, and it was shown that it contributed to a decreased susceptibility to fluoroquinolones [63].

4.4.3. *E. coli*

AcrAB is the major MDR efflux pump in *E. coli*, with a constitutive expression in wild-type strains and a significant overexpression during drug exposure [64,65]. Close homologs of AcrAB among the RND pumps family are AcrD, AcrEF, MdtABC and YhiUV (MdtEF), which play a minor role because their in vitro expression is only triggered after AcrB inactivation [65]. Finally, MdfA transporters belong to the MFS family and are known to be related to quinolone resistance [66]. Most of the efflux pumps are conserved across the different *E. coli* phylogroups [64]. Despite the importance of *E. coli* in a clinical setting, limited studies concern the role of efflux in hospital infection situations. In a Japan study, in 2008, authors observed that among 64 *E. coli* isolates obtained from patients with urinary tract infections in posttreatment, 52 isolates (81.3%) presented an overexpression of the *marA* gene that upregulated *acrAB* [66]. Moreover, 26.6% overexpressed *yhiV (mdtF)* and 34.4% *mdfA*. Authors concluded that there was a correlation between the AcrAB and MdfA efflux systems in resistance to fluoroquinolones [66]. In 2018, 200 non-repetitive

E. coli strains isolated from urine samples of patients admitted to the Hospital of Wenzhou in China were collected [67]. Strains presented a high resistance rate to ampicillin, fluoroquinolones, gentamicin and 2.5% were found to be resistant to triclosan. Increased RT-qPCR expressions were noted for most of the efflux pumps encoding genes such as *ydcV, ydcU, ydcS, ydcT, cysP, yihV, acrB, acrD,* and *mdfA.* ABC transport efflux pump genes and *acrB* were the most solicited [67]. Camp et al. recently explored the efflux mechanisms involvement in an international collection of *E. coli* exhibiting an MDR phenotype [68]. By using MICs tests with and without the addition of the AcrAB-TolC efflux inhibitor 1-(1-naphthylmethyl)-piperazine (NMP) and by qRT-PCR, they identified in 50% of strains an efflux mechanism associated with antimicrobial resistance. They showed a significant overexpression of the AcrAB-TolC system in the 17 corresponding strains. Moreover, whole-genome sequencing indicated amino acid substitutions in AcrR, MarR, and SoxR genetic regulators.

4.4.4. E. cloacae

Several teams reported the intervention of efflux pumps belonging to RND and MATE families in resistant *E. cloacae* stains [69]. The involvement of AcrAB, EefABC and OqxAB in the expel of fluoroquinolones, tetracycline and chloramphenicol is well-characterized in some clinical strains [46,69,70]. However, only a few publications have evaluated the efflux prevalence in clinical isolates collections or attempted to follow the evolution of efflux resistance during recent decades. Rosa et al. studied 25 *E. cloacae* carbapenem resistant strains, derived from the same hospital and likely of clonal origin. They found in most strains an overexpression of AcrAB by SDS-PAGE assays associated with carbapenamase KPC expression. They concluded that efflux may be associated with carbapenemase production to confer a high level of resistance to meropenem and imipenem. For them, efflux pumps enhanced carbapenemases action and completed the MDR phenotype [71].

Telke et al. studied two strains of *E. cloacae* and *E. asburiae* which were isolated from stools and presented a colistin hetero-resistant phenotype. They demonstrated that colistin hetero-resistance was due to the over expression of AcrAB upregulated by the *soxRS* regulation [72].

Hang Liu et al. investigated the involvement of efflux mechanisms in 140 *E. cloacae* tigecycline resistant strains isolated from various samples between 2014 and 2017. Efflux pump inhibitory assays with PAβN and the quantification of efflux pump genes expression by qRT-PCR showed that *acrAB* and *oqxAB* were upregulated. Antibiotics MICs with PAβN were reduced in 60% of strains; *acrAB* were upregulated in 78% and *oqxAB* in 28.6% of the strains. The genetic regulator RamA was expressed in 57% of strains [73].

4.5. Burkholderia cepacia, thailandensis and pseudomallei

Several members of the genus *Burkholderia* are dangerous pathogens. Infections caused by these bacteria are difficult to treat because of their significant level of antibiotic resistance [74]. Although efflux pumps are described in several *Burkholderia* species, only specific studies are carried out in *B. cepacia* complex and *B. pseudomallei*, in which they confer resistance to aminoglycosides, chloramphenicol, fluoroquinolones, and tetracyclines as tigecycline [75].

In the *B. cepacia* complex, several efflux pumps (MexA, MexC, MexE, and MexX) are involved in MDR [75,76]. Gautam et al. investigated the relation between the expression of efflux pumps, outer membrane porin OprD, and the β-lactamase AmpC, with antimicrobial susceptibility among 44 clinical isolates of *B. cepacia* complex [77]. They found that the reduced susceptibility to chloramphenicol was correlated with the overexpression of (*mexC, mexE, and mexX*) in the majority (>95%) of the isolates. Increased *mexA* expression showed a significant association with reduced susceptibility to β-lactam and co-trimoxazole. Increased *mexC* and/or *mexX* was associated with a reduced susceptibility to meropenem. Finally, a reduced susceptibility to ceftazidime and levofloxacin was associated with *mexE* and *mexX* expression, respectively [77].

B. thailandensis is closely related to *B. pseudomallei*, but non-pathogenic to humans. In *B. thailandensis*, three efflux complexes, AmrAB-OprA, BpeEF-OprC, and BpeAB-OprB are expressed and provide protection against multiple antibiotics, including polymyxins [78]. Authors show that the inactivation of AmrAB-OprA or BpeAB-OprB potentiates the antibacterial activities of several antibiotics. BpeF expels chloramphenicol, trimethoprim/sulfamethoxazole, and quinolones. Biot et al. demonstrated that, in doxycycline-selected variants, two overexpressed efflux complexes co-exist depending on the selected doxycycline concentration: AmrAB-OprA and BpeEF-OprC [78]. BpeEF-OprC likely took over from AmrAB-OprA at high resistance levels to doxycycline in strains, in which AmrB was downregulated. The expression of the BpeAB-OprB was reduced in doxycycline-resistant variants, whereas the AmrAB-OprA was overexpressed. Interestingly, BpeAB-OprB, could substitute to one of the present complexes when defective. In *B.thailandensis* an efflux pump was present, which belonged to the MSF group, in strains such as EmrB of *E. coli* [79,80]. In *B.thailandensis*, trimethoprim and tetracycline were expelled by this pump and the *emrB* gene was upregulated following the addition of gentamicin, due to the repression of OstR, a member of the MarR family repressor.

B. pseudomallei, the agent of melioidosis, is naturally resistant to many antibiotics [81–83]. Three RND complexes (AmrAB-OprA, BpeAB-OprB and BpeEF-OprC) are characterized in *B. pseudomallei*, two of which confer either intrinsic or acquired resistance to several antibiotic families, using trimethoprim+sulfamethoxazole [81]. It is shown that AmrAB-OprA in *B. pseudomallei* is associated with aminoglycoside and macrolide resistance. The meropenem is the gold standard treatment option. A clinical collection of *B. pseudomallei* isolates, including paired isolates which evolved during treatment, with reduced meropenem susceptibility were studied. Among them, 11 strains developed a decreased susceptibility toward meropenem during treatment. It was identified that, in such strains, multiple mutations affecting RND efflux pump regulators, with the concomitant overexpression of their corresponding pumps, rended strains refractory to treatment. In the expression of the three previously identified RND efflux pumps, as well as four other uncharacterized pumps, clinical isolates were found to be widespread [79]. In 45 of 50 isolates (90%), mRNA was detected for at least one of the seven RND pumps and among them, 41 (82%) expressed multiple pumps, nine expressing all the seven pumps tested. There was no striking correlation between RND efflux pump expression and clinically significant antibiotic resistance, however, RND pumps played important roles in the protection against toxic compounds and resistance.

4.6. Stenotrophomonas maltophilia

Stenotrophomonas maltophilia is a soil-borne bacterium often isolated in nosocomial infections due to its natural resistance to various antibiotics, e.g., carbapenems, and its ability to colonize cancer in immunocompromised or cystic fibrosis patients [84–86]. The species is characterized by a great genetic diversity confirmed by various typing methods and the multilocus sequence typing method [87–90]. Efflux is a major mechanism of adaptation in this species and is involved in intrinsic resistance. The presence of many efflux pumps: 8 RND (SmeABC, SmeDEF, SmeGH, SmeIJK, SmeMN-TolC, SmeOP-TolC, SmeVWX, SmeYZ), and 2 ABC (SmrA, macABC), as well as two MFS transporters encoding genes are described by whole-genome strains [84,91]. Youenov et al. identified 3 MATE, 3 SMR, 4 MFS and 4 ABC pumps in the three clinical strains sequenced [86]. The pumps of Sme series were confirmed as involved in intrinsic as-acquired resistance described in clinical MDR strains [84,92,93]. SmeABC was involved in resistance to β-lactams, aminoglycosides, trimethoprim/sulfamethoxazole, and quinolones; SmeDEF in resistance to chloramphenicol, ceftazidime, tetracycline, fluoroquinolones and triclosan; SmeJK in resistance to aminoglycosides, tetracycline, and ciprofloxacin; SmeOP-TolC in resistance to a low susceptibility of aminoglycosides, nalidixic acid, doxycycline; and, finally, SmeVWX as SmeYZ in resistance to aminoglycosides [84,88,92].

In clinical isolates, the three RND efflux systems, SmeABC, DEF and VWX are the most identified [92–96]. Real-time PCR analysis shows the overexpression of *smeB* in 21 (63.6 %) and *smeF* in 19 (57.5%) of 33 clinical isolates [96]. Fifteen (45.4%) isolates overexpressed both *smeB* and *smeF*. Interestingly, *S. maltophilia* is the only known bacterium in which quinolone-resistance is not associated with mutations in the genes encoding bacterial topoisomerases [88]. SmeDEF and SmeVWX are responsible for a part of the fluoroquinolone's resistance. As illustrated, in 31 clinical *S. maltophilia* isolates presenting an MIC range to ciprofloxacin between 0.5 and >32 μg/mL, there were 11 overexpressed SmeDEF and two overexpressed SmeVWX [95]. The strains overexpressing SmeVWX presented changes at the Gly266 position of SmeRv, the repressor of SmeVWX [95]. In acquired resistance, the overexpression of the SmeDEF and SmeVWX efflux systems in clinical strains are frequently related to mutations in the regulators SmeT and SmeRv, respectively [97]. SmeT appears to play a central role in the adaptive resistance to quinolones and other expelled antibiotics, like tetracycline, chloramphenicol, erythromycin, and aminoglycosides. Resistance from a mutation emerges during a short course of ciprofloxacin, and quinolone monotherapy is not recommended for *S. maltophilia* bacteremia [98].

4.7. Pseudomonas aeruginosa

In *P. aeruginosa*, several multidrug transporters belonging to the RND family, such as MexAB, MexCD, MexEF, MexGHI, MexJK, MexXY and MexVW are reported to expel various antibiotics and biocides [99]. The most prevalent efflux complexes are: (i) MexAB-OprM, that confers resistance against tetracycline, chloramphenicol, quinolones, trimethoprim, and most β-lactams; (ii) MexXY, that contributes to the resistance against aminoglycosides, quinolones, tetracycline, erythromycin, and (iii) MexCD-OprJ for which the upregulation correlates with an increased resistance to ciprofloxacin, cefepime, and chloramphenicol [100,101].

The regulation of the major efflux system MexAB-OprM of *P. aeruginosa* is quite complex. Mutations in at least three different regulatory proteins genes from the TetR family of repressors (*mexR*, *nalC*, and *nalD*) can provide for the increased expression of MexAB-OprM [102]. It must be noted that the pump also plays a role in the production of a few virulence factors [102]. Sobel et al. demonstrated that MexAB-OprM can be overexpressed without mutations in *mexR* or *nalC*, but mutations in *nalD* are reported in some clinical strains [102]. Elsewhere, NalD is an attractive target for developing compounds to dysregulate the major pumps expression. In 2004, a French multicentre study investigated mechanisms of β-lactam resistance in 450 non-redundant strains of *P. aeruginosa* obtained from 15 French university hospitals [101]. The overproduction of the MexAB-OprM efflux system was present in 22.3% of the strains. In 2007, an Algerian study on 199 strains demonstrated an overproduction of MexAB-OprM in 24% of strains [103]. Currently, depending on the studies, overexpression of *mexA* and *mexB* measures between 50 to 88% in clinical *P. aeruginosa* strains [104,105]. A study investigated 147 *P. aeruginosa* isolates from 89 clinical samples from various hospital countries (Australia, USA, and the Netherlands), 20 veterinary and 38 wastewater origins collected from 2012 to 2017 [106]. Among them, 15 isolates showing at least a fourfold reduction in MIC in the presence of PAβN were selected for qRT-PCR. Results indicated an overexpression of the MexA pump in all tested isolates. The highest level of overexpression (sixfold) of MexA was observed in a clinical isolate from a cystic fibrosis patient. MexAB-OprM was investigated for its role in conferring meropenem resistance and the effect of the single-dose exposure of meropenem on the transcription level of *mexA* [107]. Out of 83 meropenem non-susceptible isolates, 38 exhibited efflux pump activity against meropenem; 22 of these overexpressed MexAB-OprM. In most cases, meropenem could increase *oprM* and *oprN* mRNA levels [107].

In a collection of 110 French *P. aeruginosa* strains, Jeannot et al. reported that 3.7% of strains were *nfxB* mutants and exhibited moderate resistance [108]. The alteration of the *nfxB* gene, which coded for the repressor of the *mexCD-oprJ* operon, led to the overproduction of MexCD-OprJ and to fluoroquinolones, macrolides, cefpirome and cefepime resistance [108]. However, some "*nfxB* mutants" are more susceptible than wild-type strains to aminoglycosides, aztreonam, and imipenem. In contrast to MexAB-OprM and MexXY-OprM, the MexCD-OprJ system does not contribute to the natural resistance of *P. aeruginosa*. This phenotype is appreciable, despite high rates of gain-of-efflux mutants (MexAB-OprM, MexXY-OprM) in France. Indeed, very few *nfxB* strains are identified in the clinical setting, probably in connection with the fact that these bacteria present an impaired fitness. Yet, MexCD mutants are quite significant in cystic fibrosis patients and the pump is known to potentiate the effect of mutations in target genes as *gyrA* [108].

Regarding the MexEF-OprN, a study concerning 221 multidrug-resistant strains of *P. aeruginosa* with reduced susceptibility to ciprofloxacin, reported that 43 (19.5%) over-produced the pump [109]. The *mexEF-oprN* operon was controlled by an adjacent gene *mexT*, a LysR-type activator, which itself was regulated by *mexS*, a gene involved in the detoxification of some MexT-activating molecules. Among the 43 strains, only three (13.6%) contained a disrupted *mexS* gene and nine presented a *mexS* mutation, whose inactivation is known to activate the *mexEF-oprN* operon through MexT. Single-point mutations in *mexS* (40.9% of resistant strains) represent a significant cause of MexEF-OprN upregulation. Another study investigated the resistance mechanisms to fluoroquinolones of 85 non-cystic fibrosis *P. aeruginosa* exhibiting a reduced susceptibility to ciprofloxacin [110]. Authors found an upregulation of MexAB-OprM (36% of isolates), MexXY/OprM (46% of isolates) and MexEF-OprN efflux pump (12% of isolates). An analysis of the 10 MexEF-OprN over-producers indicated the presence of various mutations in the *mexT* (two isolates) or *mexS* (five isolates). Importantly, MexEF-OprN represented a key mechanism by which *P. aeruginosa* could acquire higher resistance levels to fluoroquinolones and was underestimated.

MexXY efflux mutants are frequent and described as selected by aminoglycosides, alone or in combination with fluoroquinolones [111]. Moreover, resistance to cefepim and/or ceftazidime is mostly due to the stable overproduction of MexXY that is demonstrated in 32 clinical strains [111]. Thirty three percent of strains resistant to cefepim overexpressed the gene *mexY*. Moreover, the simultaneous overexpression of MexXY and MexAB-OprM in clinical isolates results in a twofold increase in cefepime MIC, compared with single MexXY production [112]. Mutations inside or outside of the regulatory gene, *mexZ* (*agrZ* or *agrW* mutants, respectively), or the TCS system, *armgR/S*, which controls the expression of the operon, *mexXY*, are involved in the pump overexpression [113]. The simultaneous overexpression of MexAB-OprM, MexEF, and MexXY is not exceptional [112]. Additionally, a single clinical strain overexpressed the three RND genes tested in this work. The same authors investigated the resistance mechanisms to β-lactams, aminoglycosides, and the fluoroquinolones of 120 bacteremic strains of *Pseudomonas aeruginosa* [114]. They found that 11 and 36% of the isolates appeared to overproduce the MexAB-OprM and MexXY-OprM efflux systems, respectively. Del Barrio Toffino et al. studied 150 toto-resistant clinical isolates, recovered in 2015 from different sites from nine hospitals in Spain [115]. Via the analysis of the efflux pump gene expression coupled to the sequencing of their regulatory components, they reported that frequent mutations of *mexZ* generated an overexpression of MexXY-OprM. Mutations were detected in most (73%) of the strains analyzed. Moreover, 25% of the strains, were *mexR* (*nalB*), *nalC*, or *nalD* variants and were also overexpressed in the MexAB-OprM pump. *mexT* mutations, associated with MexEF-OprN overexpression and OprD downregulation, were detected in a few strains and two isolates showed MexCD-OprJ overexpression due to *nfxB* mutations, contributing to ciprofloxacin resistance. Additionally, several sequence variations in unique residues were detected in the efflux pump components. Despite this, many of the studied isolates overexpressed several efflux pumps; the major regulator, *mexZ*, controlling MexXY expression was very often mutated (70.5%), highlighting the relevance of MexXY overexpression as reported in other studies [116]. Among 57 unrelated strains from non-cystic fibrosis patients, 44 (77.2%), named *agrZ* mutants presented mutations inactivating the local repressor gene, *mexZ* [117]. These mutations, were located, in the dimerization domain, the DNA-binding domain or the affected amino acid positions of the TetR-like regulators. Five strains

(8.8%), harbored single amino acid variations in ParRS, a two-component system known to positively control *mexXY* expression. Some studies demonstrated the involvement of other genetic alterations in the upregulation of MexXY in clinical strains [118]. For example, mutations in the elongation factor G (EF-G1A) potentiate aminoglycoside resistance in MexXY pump mutants [117]. In conclusion, clinical strains of *P. aeruginosa* exploit three distinct regulatory pathways, mutations in the local repressor, MexZ, in the MexZ antirepressor, ArmZ, and/or in the two-component regulatory system, ParRS, that contribute to overproduce the MexXY-OprM. The use or combination of these ways explain the high prevalence of MexXY-OprM mutants in the clinical samples [115]. Very recently, an Italian study investigated the prevalence of aminoglycoside resistance in 147 *P. aeruginosa* strains isolated from respiratory samples from Cystic Fibrosis patients. Of these, 78 (53%) were resistant to at least one aminoglycoside and overexpressed the MexXY-OprM system [119].

From 122 isolates obtained from three hospitals in Iran, most of strains expressed *mexB* (69%), *mexC* (28.7%), *mexE* (43.4%), and *mexY* (74.6%), suggesting that *mexB* and *mexY* were highly expressed in ICU wards [120].

Finally, several studies demonstrated the involvement of two or three different pumps in the resistance of clinical strains: 33 clinical strains of *P. aeruginosa* from French hospitals, resistant to ciprofloxacin and 30 non-clinical strains originating from the hospital waterborne environment, were collected during a 5-month period and included in the study [121]. The overexpression of *mexB*, *mexF* and *mexY* was detected in 27, 12, and 45% of the clinical strains, respectively. In the 30 non-clinical strains, no overexpression could be found for all genes studied. In both clinical and environmental strains, a positive correlation was found between the expressions of *oprD*, *mexB* and *mexF*. However, in clinical strains, no statistically significant link could be found between the extent of reduction in ciprofloxacin MICs in the presence of PAβN and the expression of any of the three efflux genes studied [121]. In Iran, 154 *P. aeruginosa* strains recovered from the clinical burn wound and resistant to ciprofloxacin, were studied [122]. The *mexA*, *mexC* and *mexE* genes were recovered in 95.4% isolates simultaneously, but according to the phenotypic investigation of the efflux pump, with MICs studies with CCCP, half of the genes showed an overexpression of multidrug efflux pumps. Furthermore, 59 *P. aeruginosa* clinical isolates were obtained from 57 patients and the efflux pump inhibitor PAβN was used to determine the contribution of the RND pumps in the resistance to carbapenems and ciprofloxacin [123]. Authors found that 17 strains showed a decrease in ciprofloxacin associated with efflux mechanisms. A total of 634 *P. aeruginosa* clinical isolates were collected from various clinical specimens from patients treated by carbapenems [124]. Most of the isolates displayed the gene overexpression of *mexCD-oprJ* (75%) followed by *mexXY-oprM* (62%), while *mexAB-oprM* and *mexEF-oprN* genes were overexpressed in 21.8% and 18.7% of the isolates, respectively. A correlation was found between the MexXY and MexCDJ efflux pump expression and meropenem, or imipenem resistance in more than 60% of resistant strains. Three of the isolates showed an overexpression of the four tested efflux pumps [124]. A Polish study, investigated the role of the efflux pump in 73 cefepime and/or ceftazidime-resistant strains isolated between 2004 and 2011 by using the inhibitor, PAβN. The restoration of susceptibility to cefepime and/or ceftazidime in the majority of ESβL-positive *P. aeruginosa* strains with low and moderate levels of resistance to cefepime indicated that RND efflux pumps had a significative impact on susceptibility to β-lactams [14].

4.8. Acinetobacter baumannii

Acinetobacter baumannii is a major nosocomial pathogen associated with a high mortality in immunocompromised patients, and the emergence of multidrug-resistant strains has increasingly been reported [125,126]. Efflux mechanisms, in the species, are suspected to contribute to intrinsic resistance to broad spectrum antibiotics since several years [127]. Three RND efflux systems, AdeABC, AdeFGH, AdeIJK and AdeXYZ are characterized in *Acinetobacter* species [126–128]. AdeABC, the most prevalent efflux system, is associated with resistance to β-lactams, such as: ticarcillin, cephalosporins, aztreonam, carbapenem, tetracyclines, tigecycline, aminosides, fluoroquinolones, lincosamides, rifampin, chloramphenicol, cotrimoxazole, novobiocin, and fusidic acid [127,129]. The over-expression of AdeABC is controlled by the AdeRS two-component system, and mutations in the *adeRS* genes efficiently contribute to multiresistance [126]. *AdeFGH*-overexpressing mutants are resistant to fluoroquinolones, chloramphenicol, trimethoprim, and clindamycin and have a decreased susceptibility to tetracyclines, tigecycline, and sulfamethoxazole without affecting β-lactams and aminoglycosides phenotypes [130]. AdeIJK is responsible for the natural resistance of the species and associated with resistance to cyclins, third generation cephalosporins, sulfamides, fluoroquinolones, and cyclines. However, its overexpression above a certain threshold is toxic for the host and its contribution to the acquired resistance is limited [126,131,132]. The expression of AdeFGH and AdeIJK is controlled by AdeL, a LysR-type transcriptional regulator and AdeN, a TetR-like transcriptional regulator, respectively [133]. Non-RND efflux systems, such as MFS pumps, CraA, AmvA, TetA, TetB, CmlA, FloR, or MATE pumps, similar to AbeM, have narrow-spectrum efflux profiles and are mainly encoded by mobile genetic elements [125,126]. AmvA extrudes mainly dyes, disinfectants, and detergents [134]. Erythromycin is the only antibiotic for which the activity is significantly increased, with a fourfold decrease in the MIC when the structural gene is inactivated. CraA is homologous to the MdfA efflux pump of *E. coli*, which extrudes only chloramphenicol [135]. The inactivation of CraA in *A. baumannii* results in a 128-fold decrease in chloramphenicol resistance. The system is found in all 82 *A. baumannii* strains tested and contributes to the intrinsic resistance to chloramphenicol. Several Tet efflux pumps conferring tetracycline resistance are acquired by *A. baumannii* clinical isolates [136]. TetA and TetB are the most prevalent, with TetA conferring resistance to tetracycline only and TetB also expelling minocycline [137]. CmlA and FloR efflux systems confer resistance to phenicols and are found in the AbaR1 resistance island [138]. AbeM, belonging to the MATE family, extrudes aminoglycosides, fluoroquinolones, chloramphenicol, trimethoprim, ethidium bromide, and dyes [139]. Regarding the SMR family, AbeS, a chromosomally efflux pump displaying homology with the EmrE system of *E. coli*, has been recently characterized in an *A. baumannii* MDR clinical isolate and is responsible for resistance to erythromycin, fluoroquinolones, and chloramphenicol [136,140,141]. Several studies have identified the involvement of efflux in MDR phenotypes of collection of clinical strains. Lari et al. observed that 14 of the 16 efflux-positive isolates with full resistance to ciprofloxacin, overexpressed *adeB* [142]. In 95 *A. baumannii* clinical strains, phenotypic assays with PAβN demonstrated that 40% of the isolates expressed a sensitive PAβN-efflux mechanism, mainly against tigecycline [143]. The *adeA* gene was detected in more than 73% of clinical strains, and the expression of pump genes, *adeB*, *adeG* and *adeJ* was observed in some of these by qRT-PCR [144]. Bratu et al. reported that in the absence of cephalosporinase activity, the AdeABC efflux system was responsible for the reduced susceptibility to cefepime [145]. Sixty-eight ciprofloxacin-resistant clinical isolates were investigated for their efflux pumps genes profile [146]. They all had an overexpression of *adeB* and, in 73% of them, mutations were characterized in the regulatory system of the pump AdeRS. In particular, mutations within the two-component AdeRS correlated with the high tigecycline MICs in southern European countries. Several substitutions in *adeS* were found to be responsible for *adeABC* overexpression [145]. Yoon et al. screened 14 MDR non-redundant clinical isolates for the presence and overexpression of the three Ade efflux systems and analyzed the sequences of the regulators, AdeRS and AdeL [147]. The gene, *adeB*, was detected in 13 of 14 isolates; *adeG* and the intrinsic *adeJ* gene were detected in all strains. The significant overexpression of *adeB* was observed in 10 strains, whereas only seven had moderately increased levels of expression of AdeFGH; however, nonoverexpressed AdeIJK pumps. Thirteen strains displayed a reduced susceptibility to tigecycline, but there was no correlation between tigecycline MICs and the levels of AdeABC expression. No mutations were found in the AdeL regulator of the nine strains expressing AdeFGH. In contrast, functional mutations were found in the conserved domains of AdeRS in all the strains that overexpressed AdeABC, with two mutational hot spots:

one located in AdeS and the other in the DNA-binding domain of AdeR compatible with horizontal gene transfer. The authors confirmed the high incidence of AdeABC efflux pump overexpression in MDR strains because of a panel of single mutations in the corresponding two-component regulatory system. Among 47 *A. baumannii* strains from the blood of septicemic neonates from an Indian hospital, efflux-based fluoroquinolones resistance was found in 65% of strains with at least two different active pumps; AdeAB and AdeIJ, or AdeFG in 38% of strains [148]. Amino acid changes in the regulators (AdeRS/AdeN/AdeL), either as single or multiple substitutions, were responsible for the overexpression of the pumps and detected simultaneously among 64% of resistant strains. The AdeFG efflux pump was specifically associated with fluoroquinolones resistance in 200 clinical strains from Iran [131]. It was found in 90% of clinical strains and was responsible for high levels of resistance to chloramphenicol, clindamycin, fluoroquinolones, and trimethoprim [131]. From a total 313 isolates, 113 tigecycline-resistant isolates were analyzed [149]. The most frequent mechanism associated with tigecycline resistance was the disruption of the repressor, AdeN, to the AdeIJK system, by IS elements or nucleotide deletions causing premature stop codons. The overexpression of *adeB, adeJ, adeM and adeG* genes was detected among 70%, 53%, 30% and 23% of 30 *A. baumannii* strains, respectively, which showed at least a fourfold decrease in MICs for FQs in presence of efflux inhibitors [150]. The relative expression level of *adeB* was highest (2.2- to 34-fold) among all the pumps tested. This suggested that, although the involvement of *adeB* was most described in FQ resistance, the overexpression of other pumps cannot be overlooked in the future. Since the AdeIJK pump was intrinsic to *A. baumannii*, it was detected among large numbers of strains, but, in comparison to *AdeB*, the relative fold change of *AdeJ* production was low, supporting the fact that the overexpression of AdeIJK was phenotypically toxic to *A. baumannii* [132]. In five isolates, the overexpression of AdeABC was not detected either in AdeRS or mutations within AdeRS. In these isolates, an association of overexpression of other pumps (*adeJ* or *adeG* or *abeM*) with elevated FQ MICs was observed. In this study, at least two different pumps were overexpressed simultaneously among 38% of fluoroquinolone-resistant strains [132]. One hundred *A. baumannii* clinical isolates were assayed for the inhibitory effect of reserpine and CCCP on the antimicrobial susceptibility and expression of 4 RND-type multidrug efflux systems, including AdeABC, AdeDE, AdeIJK, and AdeFGH, using RT-PCR [151]. Ten *A. baumannii* isolates expressing AdeABC, AdeIJK, or AdeFGH were randomly selected for the determination of the transcription level and regulatory mutations. The reserpine and CCCP experiment showed that the mulidrug resistance phenotype in most *A. baumannii* isolates was associated with the presence of active efflux pumps under these conditions. Most isolates expressed at least one of the RND-type efflux pumps tested (97%). AdeIJK expression was most common (97%) as none of the isolates produced AdeDE; 52% of isolates simultaneously produced up to three RND-type efflux systems (AdeABC, AdeFGH, and AdeIJK). Importantly, authors couldn't correlate the expression of RND-type efflux pumps and the MDR phenotype [151].

Table 1. The various pumps identified in Gram-negative clinical strains as involved in antibiotic resistance, N.D; (Not Determined).

Species	Efflux Pump Family	Characterized Efflux System Involved in Resistance of Clinical Strains	Identified Regulator(s)	Refs
Haemophilus influenzae	RND	AcrAB homolog	AcrR homolog	[19,22]
Helicobacter pylori	RND	HefABC, HefDEF, HefGHI	N.D.	[23,26]
Campylobacter jejuni	RND	CmeABC, CmeDEF, RE-CmeABC	CmeR	[30,33,34]

Table 1. *Cont.*

Species	Efflux Pump Family	Characterized Efflux System Involved in Resistance of Clinical Strains	Identified Regulator(s)	Refs
Klebsiella pneumoniae	RND	AcrAB, OqxAB, KexD, AcrCD, MdtABC	MarA, RamA, SoxS, Rob, RarA, AcrR, H-NS, CpxAR, SdiA, RamR	[7,40,44,54,55]
	SMR	KpnEF	N.D.	[23,53–55]
	MATE	KdeA	N.D.	[54,55]
	ABC	MacAB	N.D.	[55]
	MFS	EmrAB	N.D.	[53]
Klebsiella aerogenes	RND	AcrAB, AcrCD, MdtABC, OqxAB, EefABC, RosAB	MarA, RamA, SoxS, Rob, RarA, AcrR, H-NS, CpxAR, SdiA, RamR	[7,46,56–60]
	ABC	MacAB	N.D.	[46,56]
	MFS	EmrAB	N.D.	[46,56]
Escherichia coli	RND	AcrAB, AcrCD, AcrEF, MdtABC, MdtEF, YihV	MarA, SoxS, Rob, RarA, AcrR, H-NS, CpxAR, SdiA, BaeSR, MarR, SoxR	[7,64,65]
	MFS	MdfA	N.D.	[66]
	ABC	YdcV, YdcU, YdcS, YdcT, CysP	N.D.	
Enterobacter cloacae	RND	AcrAB, EefABC, OqxAB	MarA, RamA, SoxS, Rob, RarA, AcrR, H-NS, CpxAR, SdiA, SoxRS, RamR	[7,46,69,70]
Burkholderia cepacia	RND	MexAB, MexCD, MexEF, MexXY	LysR family, Tet-R type regulator, MerR-type regulator	[75,76,152]
Burkholderia thailandensis	RND	AmrAB, BpeEF, BpeAB,	OstR	[78,80]
	MFS	EmrAB	N.D.	[79]
Burkholderia pseudomallei	RND	AmrAB, BpeEF, BpeAB	N.D.	[79,81]
Stenotrophomonas maltophilia	RND	SmeABC, SmeDEF, SmeGH, SmeIJK, SmeMN, SmeOP, SmeVWX, SmeYZ	SmeSR, SmeRySy, Tet-R type regulator, SmeT, SmeRv	[84,92,153]
	ABC	SmrA, macABC	N.D.	[84,86,91]
Pseudomonas aeruginosa	RND	MexAB, MexCD, MexXY, MexEF, MexGHI, MexJK, MexVW	MexR, NfxB, MexT, MexZ, MexL, NalD, NalC, ParRS, ArmZ, ArmgR/S, MexS	[99–102,153]
Acinetobacter baumannii	RND	AdeABC, AdeFGH, AdeIJK, AdeXYZ, AdeDE	AdeRS, AdeL, AdeN	[126–128,133,151]
	MFS	CraA, AmvA, TetA, TetB, CmlA, FloR	N.D.	[125,126]
	MATE	AbeM	N.D.	[125,126,139]
	SMR	AbeS	N.D.	[136,140,141]

5. Conclusions, Therapeutic Solutions to Efflux

Clinically used antibiotics are well-recognized and actively expelled by efflux pumps which are expressed in clinical resistant isolates [1,2,4]. The appropriate methods are now available (e.g., fluorimetry, mass spectrometry) to monitor the intra-bacterial accumulation of antibiotics and these methods can be used to measure the effectiveness of countermeasure developed to block efflux activity, and to compare these measurements with the expected increase in bacterial susceptibility [1]. However, this approach is not widely used and only the direct effect of the combination of antibiotic + adjuvant (by fractional inhibitory concentration, FIC, or other calculations [154]) on bacterial susceptibility is currently used, generating some uncertainties in the characterization of the mode of action for a putative adjuvant. Regarding the possible ways to circumvent this efflux mechanism and restore the antibiotic action of resistant bacteria, numerous publications described the collection of molecules that collapse the efflux in Gram-negative bacteria, and synthetic or natural compounds [6,11,155,156].

Impairing the efflux activity can be achieved by using different approaches: (i) decreasing the affinity of efflux system for the antibiotic, (ii) increasing the antibiotic penetration rate, (iii) saturating the efflux capability by competitors that mimic the substrate, (iv) dissipating the energy source of expel, (v) blocking the exit channel, and (vi) downregulating or destabilizing the efflux system.

5.1. Working on Molecule Profile

An interesting strategy regarding the molecule itself was recently opened with the study reporting the susceptibility of antibiotics belonging to the fluoroquinolone (FQ) family to efflux pump [3]. In this paper, the authors have used the Structure Concentration Intracellular Activity Relationship (SICAR) indexes [3,12] to illustrate the penetration and the accumulation rate of various fluoroquinolones in isogenic stains expressing the various levels of the AcrAB pump. This ranking using the SICAR$_{EFF}$ allows for the identity of some chemical chains that are noticeably involved in expelling efficacy and recognized by AcrB sites [3]. Importantly, this study paves the way for future rational pharmacomodulations of the antibiotic class to decrease its sensitivity to efflux and to consequently restore the internal concentration necessary to kill the bacteria.

Alternatively, it has been reported that when the antibiotic concentration in the medium increases, the total efficacy of efflux decreases [10]. This suggest that the level of resistance conferred by efflux depends on the antibiotic permeation rate across the bacterial membrane and its intrabacterial accumulation [1,7]. This "saturation" of efflux capability can be also reached by using appropriate lures that mimic the antibiotic structure and induce a competition for the efflux sites inside the pump.

5.2. Working on Pump Activity-Structure

Since RND efflux pumps require the inner membrane energy (proton motive force) as the driving force during drug transport, an attractive idea was to dry up the energy source and collapse the energy–transport coupling [2]. Unfortunately, carbonyl cyanide *m*-chlorophenylhydrazone is more efficient and widely used as an energy dissipator, and it is also a strong poison which cannot be used in clinic.

Regarding alternative methods, several molecules are described as restoring antibiotic susceptibility by destabilizing the membrane or inducing some hindrances for the access of AcrB affinity pockets that recognize and capture the antibiotics [4,156,157]. The feasibility of blocking the TolC channel, carried out in vitro, could be an attractive way to restore the antibiotic concentration inside the bacterial cells with many advantages (specificity, target located at the outer membrane surface), but the proof of this idea remains to be demonstrated on intact resistant bacteria [152]. Finally, the downregulation of efflux pump genes, by blocking the role of activators, such as MarA or RamA, seeFms to be difficult to exploit due to the redundancy of regulators and the presence of genetic overlaps controlling the expression of efflux systems.

5.3. Assays and Clinic Aspects of Efflux Inhibitors

At this moment, several β-lactamase inhibitors are routinely used in combination with β-lactams to combat and eradicate β-lactamase producers. In contrast, no efflux blockers passed the different steps to be validated and accepted for clinical use. Different aspects are involved in the lack of development of efflux blockers and are discussed in recent publications: chemical structures different from usual antibiotics, combination aspects. and rules in regulatory affairs (patents and licensing, etc.) [155,158,159].

Today, the clear identification and characterization of the efflux role in resistant clinical isolates in the research laboratory using recently developed methods (mass spectrometry, fluorimetry, and rate-killing assays, etc.), and its contribution to the spreading of the resistance phenotype in hospital wards clearly demonstrate the need for a robust diagnostic of efflux strains and theie associations with other membrane-associated mechanisms of resistance that contribute to MDR emergence. In addition, the molecular dissection of expel transport, and the dialogue between the antibiotic and bacterial transporter (pump and channel) are essential to circumvent the efflux capability and restore the antibiotic concentration on the target inside the bacterial cell.

Author Contributions: A.D.-R. and A.F. performed the analysis, reviewed the literature, and prepared the manuscript and the table; J.-M.P. wrote the discussion section, edited the manuscript, and approved the final version. All authors have read and agreed to the published version of the manuscript.

Funding: This work was supported by Aix-Marseille Univ. INSERM, and Service de Santé des Armées.

Data Availability Statement: Not applicable.

Acknowledgments: In Memoriam to Leonard Amaral, who died in 2020, a key member of COST European program ATENS which focused on antibiotic efflux, who has strongly contributed to the efflux studies in several European laboratories and stimulated lot of research in the domain during recent decades.

Conflicts of Interest: The authors declare no conflict of interest.

References

1. Masi, M.; Réfregiers, M.; Pos, K.M.; Pagès, J.M. Mechanisms of envelope permeability and antibiotic influx and efflux in Gram-negative bacteria. *Nat. Microbiol.* **2017**, *2*, 17001. [CrossRef]
2. Nikaido, H.; Pagès, J.M. Broad-specificity efflux pumps and their role in multidrug resistance of Gram-negative bacteria. *FEMS Microbiol. Rev.* **2012**, *36*, 340–363. [CrossRef]
3. Vergalli, J.; Atzori, A.; Pajovic, J.; Dumont, E.; Malloci, G.; Masi, M.; Vargiu, A.V.; Winterhalter, M.; Réfrégiers, M.; Ruggerone, P.; et al. The challenge of intracellular antibiotic accumulation, a function of fluoroquinolone influx versus bacterial efflux. *Commun. Biol.* **2020**, *3*, 198. [CrossRef]
4. Rybenkov, V.V.; Zgurskaya, H.I.; Ganguly, C.; Leus, I.V.; Zhang, Z.; Moniruzzaman, M. The Whole Is Bigger than the Sum of Its Parts: Drug Transport in the Context of Two Membranes with Active Efflux. *Chem. Rev.* **2021**, *121*, 5597–5631. [CrossRef]
5. Du, D.; Wang-Kan, X.; Neuberger, A.; van Veen, H.W.; Pos, K.M.; Piddock, L.J.V.; Luisi, B.F. Multidrug efflux pumps: Structure, function and regulation. *Nat. Rev. Microbiol.* **2018**, *16*, 523–539. [CrossRef]
6. Zgurskaya, H.I.; Malloci, G.; Chandar, B.; Vargiu, A.V.; Ruggerone, P. Bacterial efflux transporters' polyspecificity—A gift and a curse? *Curr. Opin. Microbiol.* **2021**, *61*, 115–123. [CrossRef]
7. Ferrand, A.; Vergalli, J.; Pagès, J.M.; Davin-Regli, A. An Intertwined Network of Regulation Controls Membrane Permeability Including Drug Influx and Efflux in *Enterobacteriaceae*. *Microorganisms* **2020**, *8*, 833. [CrossRef]
8. Tsutsumii, K.; Yonehara, R.; Ishizaka-Ikeda, E.; Miyazaki, N.; Maeda, S.; Iwasaki, K.; Nakagawa, A.; Yamashita, E. Structures of the wild-type MexAB-OprM tripartite pump reveal its complex for-mation and drug efflux mechanism. *Nat. Commun.* **2019**, *10*, 1520. [CrossRef]
9. Dumont, E.; Vergalli, J.; Conraux, L.; Taillier, C.; Vassort, A.; Pajovic, J.; Réfrégiers, M.; Mourez, M.; Pagès, J.M. Antibiotics and efflux: Combined spectrofluorimetry and mass spectrometry to evaluate the involvement of concentration and efflux activity in antibiotic intracellular accumulation. *J. Antimicrob. Chemother.* **2019**, *74*, 58–65. [CrossRef]
10. Glavier, M.; Puvanendran, D.; Salvador, D.; Decossas, M.; Phan, G.; Garnier, C.; Frezza, E.; Cece, Q.; Schoehn, G.; Picard, M.; et al. Antibiotic export by MexB multidrug efflux transporter is allosterically controlled by a MexA-OprM chaperone-like complex. *Nat. Commun.* **2020**, *11*, 4948. [CrossRef]

11. Farhat, N.; Ali, A.; Bonomo, R.A.; Khan, A.U. Efflux pumps as interventions to control infection caused by drug-resistance bacteria. *Drug Discov. Today* **2020**, *25*, 2307–2316. [CrossRef]

12. Vergalli, J.; Dumont, E.; Pajović, J.; Cinquin, B.; Maigre, L.; Masi, M.; Réfrégiers, M.; Pagès, J.M. Spectrofluorimetric quantification of antibiotic drug concentration in bacterial cells for the characterization of translocation across bacterial membranes. *Nat. Protoc.* **2018**, *13*, 1348–1361. [CrossRef]

13. Holden, E.R.; Webber, M.A. MarA, RamA, and SoxS as Mediators of the Stress Response: Survival at a Cost. *Front Microbiol.* **2020**, *11*, 828. [CrossRef]

14. Laudy, A.E.; Osińska, P.; Namysłowska, A.; Zając, O.; Tyski, S. Modification of the Susceptibility of Gram-Negative Rods Producing ESβLS to β-Lactams by the Efflux Phenomenon. *PLoS ONE* **2015**, *10*, e0119997. [CrossRef]

15. Langevin, A.M.; El Meouche, I.; Dunlop, M.J. Mapping the Role of AcrAB-TolC Efflux Pumps in the Evolution of Antibiotic Resistance Reveals Near-MIC Treatments Facilitate Resistance Acquisition. *mSphere* **2020**, *5*, e01056-20. [CrossRef]

16. Ebbensgaard, A.E.; Løbner-Olesen, A.; Frimodt-Møller, J. The Role of Efflux Pumps in the Transition from Low-Level to Clinical Antibiotic Resistance. *Antibiotics* **2020**, *9*, 855. [CrossRef]

17. Piddock, L.J.V. The 2019 Garrod Lecture: MDR efflux in Gram-negative bacteria-how understanding resistance led to a new tool for drug discovery. *J. Antimicrob. Chemother.* **2019**, *74*, 3128–3134. [CrossRef]

18. Li, X.-Z.; Plésiat, P.; Nikaido, H. The Challenge of Efflux-Mediated Antibiotic Resistance in Gram-Negative Bacteria. *Clin. Microbiol. Rev.* **2015**, *28*, 337–418. [CrossRef]

19. Zwama, M.; Yamaguchi, A.; Nishino, K. Phylogenetic and functional characterisation of the *Haemophilus influenzae* multidrug efflux pump AcrB. *Commun. Biol.* **2019**, *2*, 340. [CrossRef] [PubMed]

20. Bogdanovich, T.; Bozdogan, B.; Appelbaum, P.C. Effect of Efflux on Telithromycin and Macrolide Susceptibility in *Haemophilus influenzae*. *Antimicrob. Agents Chemother.* **2006**, *50*, 893–898. [CrossRef]

21. Peric, M.; Bozdogan, B.; Jacobs, M.R.; Appelbaum, P.C. Effects of an Efflux Mechanism and Ribosomal Mutations on Macrolide Susceptibility of *Haemophilus influenzae* Clinical Isolates. *Antimicrob. Agents Chemother.* **2003**, *47*, 1017–1022. [CrossRef]

22. Cherkaoui, A.; Gaïa, N.; Baud, D.; Leo, S.; Fischer, A.; Ruppe, E.; Schrenzel, F.P. Molecular characterization of fluoroquinolones, macrolides, and imipenem resistance in *Haemophilus influenzae*: Analysis of the mutations in QRDRs and assessment of the extent of the AcrAB-TolC-mediated resistance. *J. Eur. J. Clin. Microbiol. Infect. Dis.* **2018**, *37*, 2201–2210. [CrossRef]

23. Raj, D.S.; Kesavan, D.K.; Muthusamy, N.; Umamaheswari, S. Efflux pumps potential drug targets to circumvent drug Resistance Multi drug efflux pumps of *Helicobacter pylori*. *Mater.Today Proc.* **2021**, *45*, 2976–2981. [CrossRef]

24. Cai, Y.; Wang, C.; Chen, Z.; Xu, Z.; Li, H.; Li, W.; Sun, Y. Transporters HP0939, HP0497, and HP0471 participate in intrinsic multidrug resistance and biofilm formation in *Helicobacter pylori* by enhancing drug efflux. *Helicobacter* **2020**, *25*, e12715. [CrossRef]

25. Mehrabadi, J.F.; Sirous, M.; Daryani, M.E.; Eshraghi, S.; Akbari, B.; Shirazi, M.H. Assessing the role of the RND efflux pump in metronidazole resistance of *Helicobacter pylori* by RT-PCR assay. *J. Infect. Dev. Ctries* **2011**, *5*, 88–93. [CrossRef] [PubMed]

26. Iwamoto, A.; Tanahashi, T.; Okada, R.; Yoshida, Y.; Kikuchi, K.; Keida, Y.; Murakami, Y.; Yang, L.; Yamamoto, K.; Nishiumi, S.; et al. Whole-genome sequencing of clarithromycin resistant *Helicobacter pylori* characterizes unidentified variants of multidrug resistant efflux pump genes. *Gut Pathog.* **2014**, *6*, 27. [CrossRef]

27. Moore, J.E.; Corcoran, D.; Dooley, J.S.G.; Fanning, S.; Lucey, B.; Matsuda, M.; McDowell, D.A.; Mégraud, F.; Cherie Millar, B.; O'Mahony, R.; et al. Campylobacter. *Vet. Res.* **2005**, *36*, 351–382. [CrossRef] [PubMed]

28. Mamelli, L.; Prouzet-Mauléon, V.; Pagès, J.-M.; Mégraud, F.; Bolla, J.-M. Molecular basis of macrolide resistance in *Campylobacter*: Role of efflux pumps and target mutations. *J. Antimicrob. Chemother.* **2005**, *56*, 491–497. [CrossRef] [PubMed]

29. Charvalos, E.; Tselentis, Y.; Hamzehpour, M.M.; Kohler, T.; Pechere, J.C. Evidence for an efflux pump in multidrug-resistant *Campylobacter jejuni*. *Antimicrob. Agents Chemother.* **1995**, *39*, 2019–2022. [CrossRef]

30. Lin, J.; Michel, L.O.; Zhang, Q. CmeABC Functions as a Multidrug Efflux System in *Campylobacter jejuni*. *Antimicrob. Agents Chemother.* **2002**, *46*, 2124–2131. [CrossRef]

31. Pumbwe, L.; Piddock, L.J.V. Identification and molecular characterisation of CmeB, a *Campylobacter jejuni* multidrug efflux pump. *FEMS Microbiol. Lett.* **2002**, *206*, 185–189. [CrossRef]

32. Quinn, T.; Bolla, J.-M.; Pagès, J.-M.; Fanning, S. Antibiotic-resistant *Campylobacter*: Could efflux pump inhibitors control infection? *J. Antimicrob. Chemother.* **2007**, *59*, 1230–1236. [CrossRef]

33. Akiba, M.; Lin, J.; Barton, Y.-W.; Zhang, Q. Interaction of CmeABC and CmeDEF in conferring antimicrobial resistance and maintaining cell viability in *Campylobacter jejuni*. *J. Antimicrob. Chemother.* **2006**, *57*, 52–60. [CrossRef] [PubMed]

34. Yao, H.; Shen, Z.; Wang, Y.; Deng, F.; Liu, D.; Naren, G.; Dai, L.; Su, C.-C.; Wang, B.; Wang, S.; et al. Emergence of a Potent Multidrug Efflux Pump Variant That Enhances *Campylobacter* Resistance to Multiple Antibiotics. *mBio* **2016**, *7*, e01543-16. [CrossRef] [PubMed]

35. Mamelli, L. A phenylalanine–arginine β-naphthylamide sensitive multidrug efflux pump involved in intrinsic and acquired resistance of *Campylobacter* to macrolides. *Int. J. Antimicrob. Agents* **2003**, *22*, 237–241. [CrossRef]

36. Luo, N.; Sahin, O.; Lin, J.; Michel, L.O.; Zhang, Q. In Vivo Selection of Campylobacter Isolates with High Levels of Fluoroquinolone Resistance Associated with gyrA Mutations and the Function of the CmeABC Efflux Pump. *Antimicrob. Agents Chemother.* **2003**, *47*, 390–394. [CrossRef]

37. Ge, B.; McDermott, P.F.; White, D.G.; Meng, J. Role of Efflux Pumps and Topoisomerase Mutations in Fluoroquinolone Resistance in *Campylobacter jejuni* and *Campylobacter coli*. *Antimicrob. Agents Chemother.* **2005**, *49*, 3347–3354. [CrossRef]

38. Pumbwe, L.; Randall, L.P.; Woodward, M.J.; Piddock, L.J.V. Evidence for Multiple-Antibiotic Resistance in *Campylobacter jejuni* Not Mediated by CmeB or CmeF. *Antimicrob. Agents Chemother.* **2005**, *49*, 1289–1293. [CrossRef]

39. Hakanen, A.J. Multidrug resistance in *Campylobacter jejuni* strains collected from Finnish patients during 1995–2000. *J. Antimicrob. Chemother.* **2003**, *52*, 1035–1039. [CrossRef] [PubMed]

40. Bialek-Davenet, S.; Lavigne, J.P.; Guyot, K.; Mayer, N.; Tournebize, R.; Brisse, S.; Leflon-Guibout, V.; Nicolas-Chanoine, M.H. Differential contribution of AcrAB and OqxAB efflux pumps to multidrug resistance and virulence in *Klebsiella pneumoniae. J. Antimicrob. Chemother.* **2015**, *70*, 81–88. [CrossRef]

41. Nicolas-Chanoine, M.H.; Mayer, N.; Guyot, K.; Dumont, E.; Pagès, J.M. Interplay Between Membrane Permeability and Enzymatic Barrier Leads to Antibiotic-Dependent Resistance in *Klebsiella Pneumoniae. Front. Microbiol.* **2018**, *29*, 1422. [CrossRef] [PubMed]

42. Pagès, J.M.; Lavigne, J.P.; Leflon-Guibout, V.; Marcon, E.; Bert, F.; Noussair, L.; Nicolas-Chanoine, M.H. Efflux pump, the masked side of beta-lactam resistance in *Klebsiella pneumoniae* clinical isolates. *PLoS ONE* **2009**, *4*, e4817. [CrossRef]

43. Tran, Q.T.; Dupont, M.; Lavigne, J.P.; Chevalier, J.; Pagès, J.M.; Sotto, A.; Davin-Regli, A. Occurrence of efflux mechanism and cephalosporinase variant in a population of *Enterobacter aerogenes* and *Klebsiella pneumoniae* isolates producing extended-spectrum beta-lactamases. *Antimicrob. Agents Chemother.* **2009**, *53*, 1652–1656. [CrossRef]

44. Davin-Regli, A.; Masi, M.; Bialek, S.; Nicolas-Chanoine, M.H.; Pagès, J.-M. Antimicrobial resistance and drug efflux pumps in *Enterobacter* and *Klebsiella*. In *Efflux-Mediated Drug Resistance in Bacteria: Mechanisms, Regulation and Clinical Implications*; Li, X.-Z., Elkins, C.A., Zgurskaya, H.I., Eds.; Springer: Berlin/Heidelberg, Germany, 2016.

45. Li, J.; Xu, Q.; Ogurek, S.; Li, Z.; Wang, P.; Xie, Q.; Sheng, Z.; Wang, M. Efflux Pump AcrAB Confers Decreased Susceptibility to Piperacillin–Tazobactam and Ceftolozane–Tazobactam in Tigecycline-Non-Susceptible *Klebsiella pneumoniae. Infect. Drug Resist.* **2020**, *13*, 4309–4319. [CrossRef]

46. Davin-Regli, A.; Lavigne, J.-P.; Pagès, J.-M. *Enterobacter* spp.: Update on Taxonomy, Clinical Aspects, and Emerging Antimicrobial Resistance. *Clin. Microbiol. Rev.* **2019**, *32*, e00002-19. [CrossRef]

47. De Majumdar, S.; Veleba, M.; Finn, S.; Fanning, S.; Schneiders, T. Elucidating the regulon of multidrug resistance regulator RarA in *Klebsiella pneumoniae. Antimicrob. Agents Chemother.* **2013**, *57*, 1603–1609. [CrossRef]

48. Kareem, S.M.; Al-kadmy, I.M.S.; Kazaal, S.S.; Mohammed Ali, A.N.; Aziz, S.N.; Makharita, R.R.; Algammal, A.M.; Al-Rejaie, S.; Behl, T.; Batiha, G.E.S.; et al. Detection of *gyrA* and *parC* Mutations and Prevalence of Plasmid-Mediated Quinolone Resistance Genes in *Klebsiella pneumoniae. Infect. Drug Resist.* **2021**, *14*, 555–563.

49. Wasfi, R.; Elkhatib, W.F.; Ashour, H.M. Molecular typing and virulence analysis of multidrug resistant *Klebsiella pneumoniae* clinical isolates recovered from Egyptian hospitals. *Sci. Rep.* **2016**, *6*, 38929. [CrossRef]

50. Lee, Y.J.; Huang, C.H.; Ilsan, N.A.; Lee, I.H.; Huang, T.W. Molecular Epidemiology and Characterization of Carbapenem-Resistant *Klebsiella pneumoniae* Isolated from Urine at a Teaching Hospital in Taiwan. *Microorganisms* **2021**, *9*, 271. [CrossRef] [PubMed]

51. Elgendy, S.G.; Abdel Hameed, M.R.; El-Mokhtar, M.A. Tigecycline resistance among *Klebsiella pneumoniae* isolated from febrile neutropenic patients. *J. Med. Microbiol.* **2018**, *67*, 972–975. [CrossRef]

52. Sekyere, J.O.; Amoako, D.G. Genomic and phenotypic characterisation of fluoroquinolone resistance mechanisms in *Enterobacteriaceae* in Durban, South Africa. *PLoS ONE* **2017**, *12*, e0178888. [CrossRef] [PubMed]

53. Srinivasan, V.B.; Rajamohan, G. KpnEF, a New Member of the *Klebsiella pneumoniae* Cell Envelope Stress Response Regulon, Is an SMR-Type Efflux Pump Involved in Broad-Spectrum Antimicrobial Resistance. *Antimicrob. Agents Chemother.* **2013**, *57*, 4449–4462. [CrossRef]

54. Maurya, N.; Jangra, M.; Tambat, R.; Nandanwar, H. Alliance of Efflux Pumps with beta-Lactamases in Multidrug-Resistant *Klebsiella pneumoniae* Isolates. *Microb. Drug Resist.* **2019**, *25*, 1155–1163. [CrossRef]

55. Lv, F.; Cai, J.; He, Q.; Wang, W.Q.; Luo, Y.; Wang, X.; Mi, N.; Zhao, Z.; Li, G.; Luo, W. Overexpression of Efflux Pumps Mediate Pan Resistance of *Klebsiella pneumoniae* Sequence Type 11. *Microb. Drug Resist.* **2021**. Online ahead of print. [CrossRef] [PubMed]

56. Passarelli-Araujo, H.; Palmeiro, J.K.; Moharana, K.C.; Pedrosa-Silva, F.; Dalla-Costa, L.M.; Venancio, T.M. Genomic analysis unveils important aspects of population structure, virulence, and antimicrobial resistance in *Klebsiella aerogenes. FEBS J.* **2019**, *286*, 3797–3810. [CrossRef] [PubMed]

57. Malléa, M.; Chevalier, J.; Bornet, C.; Eyraud, A.; Davin-Regli, A.; Bollet, C.; Pagès, J.M. Porin alteration and active efflux: Two in vivo drug resistance strategies used by *Enterobacter aerogenes. Microbiology* **1998**, *144*, 3003–3009. [CrossRef]

58. Masi, M.; Pagès, J.M.; Villard, C.; Pradel, E. The eefABC multidrug efflux pump operon is repressed by H-NS in *Enterobacter aerogenes. J. Bacteriol.* **2005**, *187*, 3894–3897. [CrossRef]

59. Masi, M.; Pagès, J.M.; Pradel, E. Production of the cryptic EefABC efflux pump in *Enterobacter aerogenes* chloramphenicol-resistant mutants. *J. Antimicrob. Chemother.* **2006**, *57*, 1223–1226. [CrossRef] [PubMed]

60. Martins, M.; Couto, I.; Viveiros, M.; Amaral, L. Identification of efflux-mediated multi-drug resistance in bacterial clinical isolates by two simple methods. *Methods Mol. Biol.* **2010**, *642*, 143–157.

61. Chevalier, J.; Mulfinger, C.; Garnotel, E.; Nicolas, P.; Davin-Régli, A.; Pagès, J.-M. Identification and Evolution of Drug Efflux Pump in Clinical *Enterobacter aerogenes* Strains Isolated in 1995 and 2003. *PLoS ONE* **2008**, *3*, e3203. [CrossRef]

62. Gayet, S.; Chollet, R.; Molle, G.; Pagès, J.-M.; Chevalier, J. Modification of Outer Membrane Protein Profile and Evidence Suggesting an Active Drug Pump in *Enterobacter aerogenes* Clinical Strains. *Antimicrob. Agents Chemother.* **2003**, *47*, 1555–1559. [CrossRef]

63. Lavigne, J.P.; Sotto, A.; Nicolas-Chanoine, M.H.; Bouziges, N.; Bourg, G.; Davin-Regli, A.; Pagès, J.M. Membrane permeability, a pivotal function involved in antibiotic resistance and virulence in *Enterobacter aerogenes* clinical isolates. *Clin. Microbiol. Infect.* **2012**, *18*, 539–545. [CrossRef]

64. Teelucksingh, T.; Thompson, L.K.; Cox, G. The Evolutionary Conservation of *Escherichia coli* Drug Efflux Pumps Supports Physiological Functions. *J. Bacteriol.* **2020**, *202*, e00367-20. [CrossRef] [PubMed]

65. Anes, J.; McCusker, M.P.; Fanning, S.; Martins, M. The ins and outs of RND efflux pumps in *Escherichia coli*. *Front Microbiol.* **2015**, *6*, 587. [CrossRef]

66. Yasufuku, T.; Shigemura, K.; Shirakawa, T.; Matsumoto, M.; Nakano, Y.; Tanaka, K.; Arakawa, S.; Kinoshita, S.; Kawabata, M.; Fujisawa, M. Correlation of Overexpression of Efflux Pump Genes with Antibiotic Resistance in *Escherichia coli* Strains Clinically Isolated from Urinary Tract Infection Patients. *J. Clin. Microbiol.* **2011**, *49*, 189–194. [CrossRef]

67. Zeng, W.; Xu, W.; Xu, Y.; Liao, W.; Zhao, Y.; Zheng, X.; Xu, C.; Zhou, T.; Cao, J. The prevalence and mechanism of triclosan resistance in *Escherichia coli* isolated from urine samples in Wenzhou, China. *Antimicrob. Resist. Infect. Control.* **2020**, *9*, 161. [CrossRef] [PubMed]

68. Camp, J.; Schuster, S.; Vavra, M.; Schweigger, T.; Rossen, J.W.A.; Reuter, S.; Kern, W.V. Limited Multidrug Resistance Efflux Pump Overexpression among Multidrug-Resistant *Escherichia coli* Strains of ST131. *Antimicrob. Agents Chemother.* **2021**, *65*, e01735-20. [CrossRef] [PubMed]

69. Pérez, A.; Poza, M.; Fernández, A.; Fernández, M.; Mallo, S.; Merino, M.; Rumbo-Feal, S.; Cabral, M.P.; Bou, G. Involvement of the AcrAB-TolC Efflux Pump in the Resistance, Fitness, and Virulence of *Enterobacter cloacae*. *Antimicrob. Agents Chemother.* **2012**, *56*, 2084–2090. [CrossRef]

70. Davin-Regli, A.; Pagès, J.-M. *Enterobacter aerogenes* and *Enterobacter cloacae*; versatile bacterial pathogens confronting antibiotic treatment. *Front. Microbiol.* **2015**, *6*, 392. [CrossRef]

71. Rosa, J.F.; Rizek, C.; Marchi, A.P.; Guimaraes, T.; Miranda, L.; Carrilho, C.; Levin, A.S.; Costa, S.F. Clonality, outer-membrane proteins profile and efflux pump in KPC- producing *Enterobacter* sp. in Brazil. *BMC Microbiol.* **2017**, *17*, 69. [CrossRef]

72. Telke, A.A.; Olaitan, A.O.; Morand, S.; Rolain, J.M. soxRS induces colistin hetero-resistance in *Enterobacter asburiae* and *Enterobacter cloacae* by regulating the AcrAB-TolC efflux pump. *J. Antimicrob. Chemother.* **2017**, *72*, 2715–2721. [CrossRef]

73. Liu, H.; Jia, X.; Zou, H.; Sun, S.; Li, S.; Wang, Y.; Xia, Y. Detection and characterization of tigecycline heteroresistance in *E. cloacae*: Clinical and microbiological findings. *Emerg. Microbes Infect.* **2019**, *8*, 564–574. [CrossRef] [PubMed]

74. Podnecky, N.L.; Rhodes, K.A.; Schweizer, H.P. Efflux pump-mediated drug resistance in *Burkholderia*. *Front. Microbiol.* **2015**, *6*, 305. [CrossRef] [PubMed]

75. Rajendran, R.; Quinn, R.F.; Murray, C.; McCulloch, E.; Williams, C.; Ramage, G. Efflux pumps may play a role in tigecycline resistance in *Burkholderia* species. *Int. J. Antimicrob. Agents* **2010**, *36*, 151–154. [CrossRef]

76. Biot, F.V.; Valade, E.; Garnotel, E.; Chevalier, J.; Villard, C.; Thibault, F.M.; Vidal, D.R.; Pagès, J.M. Involvement of the efflux pumps in chloramphenicol selected strains of *Burkholderia thailandensis*: Proteomic and mechanistic evidence. *PLoS ONE* **2011**, *6*, e16892. [CrossRef]

77. Gautam, V.; Kumar, S.; Patil, P.P.; Meletiadis, J.; Patil, P.B.; Mouton, J.W.; Sharma, M.; Daswal, A.; Singhal, L.; Ray, P.; et al. Exploring the Interplay of Resistance Nodulation Division Efflux Pumps, AmpC and OprD in Antimicrobial Resistance of *Burkholderia cepacia* Complex in Clinical Isolates. *Microb. Drug Resist.* **2020**, *26*, 1144–1152. [CrossRef]

78. Biot, F.V.; Lopez, M.M.; Poyot, T.; Neulat-Ripoll, F.; Lignon, S.; Caclard, A.; Thibault, F.M.; Peinnequin, A.; Pagès, J.-M.; Valade, E. Interplay between Three RND Efflux Pumps in Doxycycline-Selected Strains of *Burkholderia thailandensis*. *PLoS ONE* **2013**, *8*, e84068. [CrossRef]

79. Sabrin, A.; Gioe, B.W.; Gupta, A.; Grove, A. *An EmrB multidrug efflux pump in Burkholderia thailandensis* with unexpected roles in antibiotic resistance. *J. Biol. Chem.* **2019**, *294*, 1891–1903. [CrossRef]

80. Krishnamoorthy, G.; Weeks, J.W.; Zhang, Z.; Chandler, C.E.; Xue, H.; Schweizer, H.P.; Ernst, R.K.; Zgurskaya, H.I. Efflux Pumps of *Burkholderia thailandensis* Control the Permeability Barrier of the Outer Membrane. *Antimicrob. Agents Chemother.* **2019**, *63*, e00956-19. [CrossRef]

81. Kumar, A.; Mayo, M.; Trunck, L.A.; Cheng, A.C.; Currie, B.J.; Schweizer, H.P. Expression of resistance-nodulation-cell-division efflux pumps in commonly used *Burkholderia pseudomallei* strains and clinical isolates from northern. Australia. *Trans. R. Soc. Trop. Med. Hyg.* **2008**, *102* (Suppl. 1), S145–S151. [CrossRef]

82. Sarovich, D.S.; Webb, J.R.; Pitman, M.C.; Viberg, L.T.; Mayo, M.; Baird, R.W.; Robson, J.M.; Currie, B.J.; Price, E.P. Raising the Stakes: Loss of Efflux Pump Regulation Decreases Meropenem Susceptibility in *Burkholderia pseudomallei*. *Clin. Infect. Dis.* **2018**, *67*, 243–250. [CrossRef]

83. Webb, J.R.; Price, E.P.; Somprasong, N.; Schweizer, H.P.; Baird, R.W.; Currie, B.J.; Sarovich, D.S. Development and validation of a triplex quantitative real-time PCR assay to detect efflux pump-mediated antibiotic resistance in *Burkholderia pseudomallei*. *Future Microbiol.* **2018**, *13*, 1403–1418. [CrossRef]

84. Crossman, L.C.; Gould, V.C.; Dow, J.W.; Vernikos, G.S.; Okazaki, A.; Sebaihia, M.; Saunders, D.; Arrowsmith, C.; Carver, T.; Peters, N.; et al. The complete genome, comparative and functional analysis of *Stenotrophomonas maltophilia* reveals an organism heavily shielded by drug resistance determinants. *Genome Biol.* **2008**, *9*, R74. [CrossRef]

85. Sánchez, M.B. Antibiotic resistance in the opportunistic pathogen *Stenotrophomonas maltophilia*. *Front. Microbiol.* **2015**, *30*, 658. [CrossRef]

86. Youenou, B.; Favre-Bonté, S.; Bodilis, J.; Brothier, E.; Dubost, A.; Muller, D.; Nazaret, S. Comparative Genomics of Environmental and Clinical *Stenotrophomonas maltophilia* Strains with Different Antibiotic Resistance Profiles. *Genome Biol. Evol.* **2015**, *7*, 2484–2505. [CrossRef]

87. Davin-Regli, A.; Bollet, C.; Auffray, J.P.; Saux, P.; De Micco, P. Use of random amplified polymorphic DNA for epidemiological typing of *Stenotrophomonas maltophilia*. *J. Hosp. Infect.* **1996**, *32*, 39–50. [CrossRef]

88. Cho, H.H.; Sung, J.Y.; Kwon, K.C.; Koo, S.H. Expression of Sme efflux pumps and multilocus sequence typing in clinical isolates of *Stenotrophomonas maltophilia*. *Ann. Lab. Med.* **2012**, *32*, 38–43. [CrossRef]

89. Rizek, C.F.; Jonas, D.; Garcia Paez, J.I.; Rosa, J.F.; Perdigão Neto, L.V.; Martins, R.R.; Moreno, L.Z.; Rossi Junior, A.; Levin, A.S.; Costa, S.F. Multidrug-resistant *Stenotrophomonas maltophilia*: Description of new MLST profiles and resistance and virulence genes using whole-genome sequencing. *J. Glob. Antimicrob. Resist.* **2018**, *15*, 212–214. [CrossRef]

90. Bostanghadiri, N.; Ghalavand, Z.; Fallah, F.; Yadegar, A.; Ardebili, A.; Tarashi, S.; Pournajaf, A.; Mardaneh, J.; Shams, S.; Hashemi, A. Characterization of Phenotypic and Genotypic Diversity of *Stenotrophomonas maltophilia* Strains Isolated From Selected Hospitals in Iran. *Front. Microbiol.* **2019**, *10*, 1191. [CrossRef] [PubMed]

91. Sánchez, M.B.; Martínez, J.L. Overexpression of the Efflux Pumps SmeVWX and SmeDEF Is a Major Cause of Resistance to Cotrimoxazole in *Stenotrophomonas maltophilia*. *Antimicrob. Agents Chemother.* **2018**, *62*, e00301-18. [CrossRef]

92. Lin, Y.T.; Huang, Y.W.; Chen, S.J.; Chang, C.W.; Yang, T.C. The SmeYZ efflux pump of *Stenotrophomonas maltophilia* contributes to drug resistance, virulence-related characteristics, and virulence in mice. *Antimicrob. Agents Chemother.* **2015**, *59*, 4067–4073. [CrossRef]

93. Chong, S.Y.; Lee, K.; Chung, H.S.; Hong, S.G.; Suh, Y.; Chong, Y. Levofloxacin Efflux and smeD in Clinical Isolates of *Stenotrophomonas maltophilia*. *Microb. Drug Resist.* **2017**, *23*, 163–168. [CrossRef] [PubMed]

94. Zhang, L.; Li, X.-Z.; Poole, K. Multiple Antibiotic Resistance in *Stenotrophomonas maltophilia*: Involvement of a Multidrug Efflux System. *Antimicrob. Agents Chemother.* **2000**, *44*, 287–293. [CrossRef]

95. García-León, G.; Ruiz de Alegría Puig, C.; García de la Fuente, C.; Martínez-Martínez, L.; Martínez, J.L.; Sánchez, M.B. High-level quinolone resistance is associated with the overexpression of smeVWX in *Stenotrophomonas maltophilia* clinical isolates. *Clin. Microbiol. Infect.* **2015**, *21*, 464–467. [CrossRef] [PubMed]

96. Liaw, S.J.; Lee, Y.L.; Hsueh, P.R. Multidrug resistance in clinical isolates of Stenotrophomonas maltophilia: Roles of integrons, efflux pumps, phosphoglucomutase (SpgM), and melanin and biofilm formation. *Int. J. Antimicrob. Agents.* **2010**, *35*, 126–130. [CrossRef]

97. Wu, C.-J.; Huang, Y.-W.; Lin, Y.-T.; Ning, H.-C.; Yang, T.-C. Inactivation of SmeSyRy Two-Component Regulatory System Inversely Regulates the Expression of SmeYZ and SmeDEF Efflux Pumps in *Stenotrophomonas maltophilia*. *PLoS ONE* **2016**, *11*, e0160943. [CrossRef]

98. Pak, T.R.; Altman, D.R.; Attie, O.; Sebra, R.; Hamula, C.L.; Lewis, M.; Deikus, G.; Newman, L.C.; Fang, G.; Hand, J.; et al. Whole-genome sequencing identifies emergence of a quinolone resistance mutation in a case of *Stenotrophomonas maltophilia* bacteremia. *Antimicrob. Agents Chemother.* **2015**, *59*, 7117–7120. [CrossRef]

99. Dreier, J.; Ruggerone, P. Interaction of antibacterial compounds with RND efflux pumps in *Pseudomonas aeruginosa*. *Front. Microbiol.* **2015**, *8*, 660. [CrossRef] [PubMed]

100. Vettoretti, L.; Plésiat, P.; Muller, C.; El Garch, F.; Phan, G.; Attrée, I.; Ducruix, A.; Llanes, C. Efflux unbalance in *Pseudomonas aeruginosa* isolates from cystic fibrosis patients. *Antimicrob. Agents Chemother.* **2009**, *53*, 1987–1997. [CrossRef]

101. Cavallo, J.D.; Hocquet, D.; Plesiat, P.; Fabre, R.; Roussel-Delvallez, M. GERPA (Groupe d'Etude de la Résistance de *Pseudomonas aeruginosa* aux antibiotiques) Susceptibility of *Pseudomonas aeruginosa* to antimicrobials: A 2004 French multicentre hospital study. *J. Antimicrob. Chemother.* **2007**, *59*, 1021–1024. [CrossRef]

102. Sobel, M.L.; Hocquet, D.; Cao, L.; Plesiat, P.; Poole, K. Mutations in PA3574 (*nalD*) Lead to Increased MexAB-OprM Expression and Multidrug Resistance in Laboratory and Clinical Isolates of *Pseudomonas aeruginosa*. *Antimicrob. Agents Chemother.* **2005**, *49*, 1782–1786. [CrossRef] [PubMed]

103. Drissi, M.; Baba Ahmed, Z.; Dehecq, B.; Bakour, R.; Plésiat, P.; Hocquet, D. Antibiotic susceptibility and mechanisms of beta-lactam resistance among clinical strains of *Pseudomonas aeruginosa*: First report in Algeria. *Med. Mal. Infect.* **2008**, *38*, 187–191. [CrossRef]

104. Pourakbari, B.; Yaslianifard, S.; Yaslianifard, S.; Mahmoudi, S.; Keshavarz-Valian, S.; Mamishi, S. Evaluation of efflux pumps gene expression in resistant *Pseudomonas aeruginosa* isolates in an Iranian referral hospital. *Iran J. Microbiol.* **2016**, *8*, 249–256.

105. Kishk, R.M.; Abdalla, M.O.; Hashish, A.A.; Nemr, N.A.; El Nahhas, N.; Alkahtani, S.; Abdel-Daim, M.M.; Kishk, S.M. Efflux *MexAB*-Mediated Resistance in *P. aeruginosa* Isolated from Patients with Healthcare Associated Infections. *Pathogens* **2020**, *9*, 471. [CrossRef] [PubMed]

106. Amsalu, A.; Sapula, S.A.; De Barros Lopes, M.; Hart, B.J.; Nguyen, A.H.; Drigo, B.; Turnidge, J.; Leong, L.E.; Vente, H. Efflux Pump-Driven Antibiotic and Biocide Cross-Resistance in *Pseudomonas aeruginosa* Isolated from Different Ecological Niches: A Case Study in the Development of Multidrug Resistance in Environmental Hotspots. *Microorganisms* **2020**, *8*, 1647. [CrossRef]

107. Choudhury, D.; Paul, D.; Ghosh, A.S.; Talukdar, A.D.; Choudhury, M.D.; Maurya, A.P.; Chanda, D.D.; Chakravarty, A.; Bhattacharjee, A. Effect of single-dose carbapenem exposure on transcriptional expression of bla $_{NDM-1}$ and *mexA* in *Pseudomonas aeruginosa*. *J. Glob. Antimicrob. Resist.* **2016**, *7*, 72–77. [CrossRef] [PubMed]

108. Jeannot, K.; Elsen, S.; Köhler, T.; Attree, I.; van Delden, C.; Plésiat, P. Resistance and Virulence of *Pseudomonas aeruginosa* Clinical Strains Overproducing the MexCD-OprJ Efflux Pump. *Antimicrob. Agents Chemother.* **2008**, *52*, 2455–2462. [CrossRef]

109. Richardot, C.; Juarez, P.; Jeannot, K.; Patry, I.; Plésiat, P.; Llanes, C. Amino Acid Substitutions Account for Most MexS Alterations in Clinical *nfxC* Mutants of *Pseudomonas aeruginosa*. *Antimicrob. Agents Chemother.* **2016**, *60*, 2302–2310. [CrossRef]
110. Lanes, C.; Köhler, T.; Patry, I.; Dehecq, B.; van Delden, C.; Plésiat, P. Role of the MexEF-OprN Efflux System in Low-Level Resistance of Pseudomonas aeruginosa to Ciprofloxacin. *Antimicrob Agents Chemother.* **2011**, *55*, 5676–5684. [CrossRef]
111. Hocquet, D.; Nordmann, P.; El Garch, F.; Cabanne, L.; Plésiat, P. Involvement of the MexXY-OprM Efflux System in Emergence of Cefepime Resistance in Clinical Strains of *Pseudomonas aeruginosa*. *Antimicrob. Agents Chemother.* **2006**, *50*, 1347–1351. [CrossRef]
112. Hocquet, D.; Roussel-Delvallez, M.; Cavallo, J.D.; Plésiat, P. MexAB-OprM- and MexXY-overproducing mutants are very prevalent among clinical strains of *Pseudomonas aeruginosa* with reduced susceptibility to ticarcillin. *Antimicrob. Agents Chemother.* **2007**, *51*, 1582–1583. [CrossRef] [PubMed]
113. Hocquet, D.; Muller, A.; Blanc, K.; Plésiat, P.; Talon, D.; Monnet, D.L.; Bertrand, X. Relationship between antibiotic use and incidence of MexXY-OprM overproducers among clinical isolates of *Pseudomonas aeruginosa*. *Antimicrob. Agents Chemother.* **2008**, *52*, 1173–1175. [CrossRef] [PubMed]
114. Hocquet, D.; Berthelot, P.; Roussel-Delvallez, M.; Favre, R.; Jeannot, K.; Bajolet, O.; Marty, N.; Grattard, F.; Mariani-Kurkdjian, P.; Bingen, E.; et al. *Pseudomonas aeruginosa* may accumulate drug resistance mechanisms without losing its ability to cause bloodstream infections. *Antimicrob. Agents Chemother.* **2007**, *51*, 3531–3536. [CrossRef]
115. Del Barrio-Tofiño, E.; López-Causapé, C.; Cabot, G.; Rivera, A.; Benito, N.; Segura, C.; Montero, M.M.; Sorlí, L.; Tubau, F.; Gómez-Zorrilla, S.; et al. Genomics and Susceptibility Profiles of Extensively Drug-Resistant *Pseudomonas aeruginosa* Isolates from Spain. *Antimicrob. Agents Chemother.* **2017**, *61*, e01589-17. [CrossRef] [PubMed]
116. Vettoretti, L.; Floret, N.; Hocquet, D.; Dehecq, B.; Plésiat, P.; Talon, D.; Bertrand, X. Emergence of extensive-drug-resistant *Pseudomonas aeruginosa* in a French university hospital. *Eur. J. Clin. Microbiol. Infect. Dis.* **2009**, *28*, 1217–1222. [CrossRef] [PubMed]
117. Bolard, A.; Plésiat, P.; Jeannot, K. Mutations in Gene *fusA1* as a Novel Mechanism of Aminoglycoside Resistance in Clinical Strains of *Pseudomonas aeruginosa*. *Antimicrob. Agents Chemother.* **2018**, *62*, e01835-17. [CrossRef]
118. Guénard, S.; Muller, C.; Monlezun, L.; Benas, P.; Broutin, I.; Jeannot, K.; Plésiat, P. Multiple mutations lead to MexXY-OprM-dependent aminoglycoside resistance in clinical strains of *Pseudomonas aeruginosa*. *Antimicrob. Agents Chemother.* **2014**, *58*, 221–228. [CrossRef]
119. Mangiaterra, G.; Cedraro, N.; Citterio, B.; Simoni, S.; Vignaroli, C.; Biavasco, F. Diffusion and Characterization of *Pseudomonas aeruginosa* Aminoglycoside Resistance in an Italian Regional Cystic Fibrosis Centre. *Adv. Exp. Med. Biol.* **2021**, *1323*, 71–80.
120. Zahedi Bialvaei, A.; Rahbar, M.; Hamidi-Farahani, R.; Asgari, A.; Esmailkhani, A.; Mardani Dashti, Y.; Soleiman-Meigooni, S. Expression of RND efflux pumps mediated antibiotic resistance in *Pseudomonas aeruginosa* clinical strains. *Microb. Pathog.* **2021**, *153*, 104789. [CrossRef]
121. Serra, C.; Bouharkat, B.; Touil-Meddah, A.T.; Guénin, S.; Mullié, C. MexXY Multidrug Efflux System Is More Frequently Overexpressed in Ciprofloxacin Resistant French Clinical Isolates Compared to Hospital Environment Ones. *Front. Microbiol.* **2019**, *10*, 366. [CrossRef]
122. Adabi, M.; Talebi-Taher, M.; Arbabi, L.; Afshar, M.; Fathizadeh, S.; Minaeian, S.; Moghadam-Maragheh, N.; Majidpour, A. Spread of Efflux Pump Overexpressing-Mediated Fluoroquinolone Resistance and Multidrug Resistance in Pseudomonas aeruginosa by using an Efflux Pump Inhibitor. *Infect. Chemother.* **2015**, *47*, 98–104. [CrossRef] [PubMed]
123. López-García, A.; Rocha-Gracia, R.D.C.; Bello-López, E.; Juárez-Zelocualtecalt, C.; Sáenz, Y.; Castañeda-Lucio, M.; López-Pliego, L.; González-Vázquez, M.C.; Torres, C.; Ayala-Nuñez, T.; et al. Characterization of antimicrobial resistance mechanisms in carbapenem-resistant *Pseudomonas aeruginosa* carrying IMP variants recovered from a Mexican Hospital. *Infect. Drug Resist.* **2018**, *11*, 1523–1536. [CrossRef] [PubMed]
124. Hassuna, N.A.; Darwish, M.K.; Sayed, M.; Ibrahem, R.A. Molecular Epidemiology and Mechanisms of High-Level Resistance to Meropenem and Imipenem in *Pseudomonas aeruginosa*. *Infect. Drug Resist.* **2020**, *13*, 285–293. [CrossRef] [PubMed]
125. Mugnier, P.D.; Poirel, L.; Naas, T.; Nordmann, P. Worldwide dissemination of the bla$_{OXA-23}$ carbapenemase gene of *Acinetobacter baumannii*. *Emerg. Infect. Dis.* **2010**, *16*, 35–40. [CrossRef]
126. Rumbo, C.; Gato, E.; López, M.; Ruiz de Alegría, C.; Fernández-Cuenca, L.; Martínez-Martínez, J.; Vila, J.; Pachón, J.M.; Cisneros, J.; Rodríguez-Baño, A.; et al. Contribution of Efflux Pumps, Porins, and β-Lactamases to Multidrug Resistance in Clinical Isolates of *Acinetobacter baumannii*. *Antimicrob. Agents Chemother.* **2013**, *57*, 5247–5257. [CrossRef]
127. Hu, W.S.; Yao, S.-M.; Fung, C.-P.; Hsieh, Y.-P.; Liu, C.-P.; Lin, J.-F. An OXA-66/OXA-51-Like Carbapenemase and Possibly an Efflux Pump Are Associated with Resistance to Imipenem in *Acinetobacter baumannii*. *Antimicrob. Agents Chemother.* **2007**, *51*, 3844–3852. [CrossRef]
128. Coyne, S.; Courvalin, P.; Périchon, B. Efflux-mediated antibiotic resistance in *Acinetobacter* spp. *Antimicrob. Agents Chemother.* **2011**, *55*, 947–953. [CrossRef]
129. Lin, L.; Ling, B.D.; Li, X.Z. Distribution of the multidrug efflux pump genes, *adeABC*, *adeDE* and *adeIJK*, and class 1 integron genes in multiple-antimicrobial-resistant clinical isolates of *Acinetobacter baumannii–Acinetobacter calcoaceticus* complex. *Intern. J. Antimicrob. Agents* **2009**, *33*, 27–32. [CrossRef]
130. Kaviani, R.; Pouladi, I.; Niakan, M.; Mirnejad, R. Molecular Detection of *Adefg* Efflux Pump Genes and their Contribution to Antibiotic Resistance in *Acinetobacter baumannii* Clinical Isolates. *Rep. Biochem. Mol. Biol.* **2020**, *8*, 413–418.

131. Lee, Y.-T.; Chen, H.-Y.; Yang, Y.-S.; Chou, Y.-C.; Chang, T.-Y.; Hsu, W.-J.; Lin, I.-C.; ACTION Study Group; Sun, J.-R. AdeABC Efflux Pump Controlled by AdeRS Two Component System Conferring Resistance to Tigecycline, Omadacycline and Eravacycline in Clinical Carbapenem Resistant *Acinetobacter nosocomialis*. *Front. Microbiol.* **2020**, *11*, 584789. [CrossRef]
132. Damier-Piolle, L.; Magnet, S.; Bremont, S.; Lambert, T.; Courvalin, P. *AdeIJK*, a resistance-nodulation-cell division pump effluxing multiple antibiotics in *Acinetobacter baumannii*. *Antimicrob. Agents Chemother.* **2008**, *52*, 557–562. [CrossRef]
133. Xu, C.; Bilya, S.R.; Xu, W. AdeABC efflux gene in *Acinetobacter baumannii*. *New Microbes New Infect.* **2019**, *30*, 100549. [CrossRef]
134. Rajamohan, G.; Srinivasan, V.B.; Gebreyes, W.A. Novel role of *Acinetobacter baumannii* RND efflux transporters in mediating decreased susceptibility to biocides. *J. Antimicrob. Chemother.* **2010**, *65*, 228–232. [CrossRef] [PubMed]
135. Roca, I.; Marti, S.; Espinal, P.; Martínez, P.; Gibert, I.; Vila, J. CraA: An MFS efflux pump associated with chloramphenicol resistance in *Acinetobacter baumannii*. *Antimicrob. Agents Chemother.* **2009**, *53*, 4013–4014.
136. Vila, J.; Marti, S.; Sanchez-Cespedes, J. Porins, efflux pumps and multidrug resistance in *Acinetobacter baumannii*. *J. Antimicrob. Chemother.* **2007**, *59*, 1210–1215. [CrossRef]
137. Beheshti, M.; Ardebili, A.; Beheshti, F.; Lari, A.R.; Siyadatpanah, A.; Pournajaf, A.; Gautam, D.; Dolma, K.G.; Nissapatorn, V. Tetracycline resistance mediated by *tet* efflux pumps in clinical isolates of *Acinetobacter baumannii*. *Rev. Inst. Med. Trop. Sao Paulo* **2020**, *62*, e88. [CrossRef] [PubMed]
138. Fournier, P.E.; Vallenet, D.; Barbe, V.; Audic, S.; Ogata, H.; Poirel, L.; Richet, H.; Robert, C.; Mangenot, S.; Abergel, C.; et al. Comparative genomics of multidrug resistance in *Acinetobacter baumannii*. *PLoS Genet.* **2006**, *2*, e7. [CrossRef]
139. Pérez-Varela, M.; Corral, J.; Aranda, J.; Barbé, J. Roles of Efflux Pumps from Different Superfamilies in the Surface-Associated Motility and Virulence of *Acinetobacter baumannii* ATCC 17978. *Antimicrob. Agents Chemother.* **2019**, *63*, e02190-18. [CrossRef]
140. Srinivasan, V.B.; Rajamohan, G.; Gebreyes, W.A. The role of AbeS, a novel efflux pump member of the SMR family of transporters, in resistance to antimicrobial agents in *Acinetobacter baumannii*. *Antimicrob. Agents Chemother.* **2009**, *53*, 5312–5316.
141. Singkham-in, U.; Higgins, P.-G.; Wannigama, D.L.; Hongsing, P.; Chatsuwan, T. Rescued chlorhexidine activity by resveratrol against carbapenem-resistant *Acinetobacter baumannii* via down-regulation of AdeB efflux pump. *PLoS ONE* **2020**, *15*, e0243082.
142. Lari, A.R.; Ardebili, A.; Hashemi, A. AdeR-AdeS mutations & overexpression of the AdeABC efflux system in ciprofloxacin-resistant *Acinetobacter baumannii* clinical isolates. *Indian J. Med. Res.* **2018**, *147*, 413–421.
143. Salehi, B.; Ghalavand, Z.; Yadegar, A.; Eslami, G. Characteristics and diversity of mutations in regulatory genes of resistance-nodulation-cell division efflux pumps in association with drug-resistant clinical isolates of *Acinetobacter baumannii*. *Antimicrob. Resist. Infect. Control.* **2021**, *10*, 53. [CrossRef]
144. Basatian-Tashkan, B.; Niakan, M.; Khaledi, M.; Afkhami, H.; Sameni, F.; Bakhti, S.; Mirnejad, R. Antibiotic resistance assessment of *Acinetobacter baumannii* isolates from Tehran hospitals due to the presence of efflux pumps encoding genes (*adeA* and *adeS* genes) by molecular method. *BMC Res. Notes.* **2020**, *13*, 543. [CrossRef]
145. Bratu, S.; Landman, D.; Martin, D.A.; Georgescu, C.; Quale, J. Correlation of antimicrobial resistance with beta-lactamases, the OmpA-like porin, and efflux pumps in clinical isolates of *Acinetobacter baumannii* endemic to New York City. *Antimicrob. Agents Chemother.* **2008**, *52*, 2999–3005. [CrossRef]
146. Ardebili, A.; Rastegar Lari, A.; Talebi, M. Correlation of Ciprofloxacin Resistance with the AdeABC Efflux System in *Acinetobacter baumannii* clinical isolates. *Ann. Lab. Med.* **2014**, *34*, 433–438. [CrossRef]
147. Yoon, E.J.; Courvalin, P.; Grillot-Courvalin, C. RND-type efflux pumps in multidrug-resistant clinical isolates of *Acinetobacter baumannii*: Major role for AdeABC overexpression and AdeRS mutations. *Antimicrob. Agents Chemother.* **2013**, *57*, 2989–2995. [CrossRef]
148. Roy, S.; Chatterjee, S.; Bhattacharjee, A.; Chattopadhyay, P.; Saha, B.; Dutta, S.; Basu, S. Overexpression of Efflux Pumps, Mutations in the Pumps' Regulators, Chromosomal Mutations, and AAC(6')-Ib-cr Are Associated With Fluoroquinolone Resistance in Diverse Sequence Types of Neonatal Septicaemic *Acinetobacter baumannii*: A 7-Year Single Center Study. *Front. Microbiol.* **2021**, *12*, 602724.
149. Lucaβen, K.; Müller, C.; Wille, J.; Xanthopoulou, K.; Hackel, M.; Seifert, H.; Higgins, P.G. Prevalence of RND efflux pump regulator variants associated with tigecycline resistance in carbapenem-resistant *Acinetobacter baumannii* from a worldwide survey. *J. Antimicrob. Chemother.* **2021**, *24*, dkab079.
150. Evans, B.A.; Kumar, A.; Castillo-Ramírez, S. Genomic Basis of Antibiotic Resistance and Virulence in *Acinetobacter*. *Front. Microbiol.* **2021**, *12*. [CrossRef]
151. Pagdepanichkit, S.; Tribuddharat, C.; Chuanchuen, R. Distribution and expression of the Ade multidrug efflux systems in *Acinetobacter baumannii* clinical isolates. *Can. J. Microbiol.* **2016**, *62*, 794–801. [CrossRef]
152. Lôme, V.; Brunel, J.M.; Pagès, J.M.; Bolla, J.M. Multiparametric Profiling for Identification of Chemosensitizers against Gram-Negative Bacteria. *Front. Microbiol.* **2018**, *9*, 204. [CrossRef]
153. Scoffone, V.C.; Trespidi, G.; Barbieri, G.; Irudal, S.; Perrin, E.; Buroni, S. Role of RND Efflux Pumps in Drug Resistance of Cystic Fibrosis Pathogens. *Antibiotics* **2021**, *10*, 863. [CrossRef]
154. Hall, J.M.; Middleton, R.F.; Westmacott, D. The fractional inhibitory concentration (FIC) index as a measure of synergy. *J. Antimicrob. Chemother.* **1983**, *11*, 427–433. [CrossRef]
155. Kobylka, J.; Kuth, M.S.; Müller, R.T.; Geertsma, E.R.; Pos, K.M. AcrB: A mean, keen, drug efflux machine. *Ann. N.Y. Acad. Sci.* **2020**, *1459*, 38–68. [CrossRef]

156. Thakur, V.; Uniyal, A.; Tiwari, V. A comprehensive review on pharmacology of efflux pumps and their inhibitors in antibiotic resistance. *Eur. J. Pharmacol.* **2021**, *903*, 174151. [CrossRef]
157. Gilardi, A.; Bhamidimarri, S.P.; Brönstrup, M.; Bilitewski, U.; Marreddy, R.K.R.; Pos, K.M.; Benier, L.; Gribbon, P.; Winterhalter, M.; Windshügel, B. Biophysical characterization of *E. coli* TolC interaction with the known blocker hexa amminecobalt. *Biochim. Biophys. Acta Gen. Subj.* **2017**, *1861 Pt 11 A*, 2702–2709. [CrossRef]
158. Sharma, A.; Gupta, V.K.; Pathania, R. Efflux pump inhibitors for bacterial pathogens: From bench to bedside. *Indian, J. Med. Res.* **2019**, *149*, 129–145.
159. Laws, M.; Shaaban, A.; Rahman, K.M. Antibiotic resistance breakers: Current approaches and future directions. *FEMS Microbiol Rev.* **2019**, *43*, 490–516. [CrossRef]

MDPI
St. Alban-Anlage 66
4052 Basel
Switzerland
Tel. +41 61 683 77 34
Fax +41 61 302 89 18
www.mdpi.com

Antibiotics Editorial Office
E-mail: antibiotics@mdpi.com
www.mdpi.com/journal/antibiotics

www.ingramcontent.com/pod-product-compliance
Lightning Source LLC
LaVergne TN
LVHW070927170726
843515LV00005B/892